W0254193

PROCEEDINGS OF THE 30TH INTERNATIONAL GEOLOGICAL CONGRESS
VOLUME 26

COMPARATIVE PLANETOLOGY, GEOLOGICAL EDUCATION, HISTORY OF GEOLOGY

Proceedings of the 30th International Geological Congress

Volume 1 : Origin and History of the Earth
Volume 2 : Geosciences and Human Survival, Environment, and Natural Hazards
Volume 3 : Global Changes and Future Environment
Volume 4 : Structure of the Lithosphere and Deep Processes
Volume 5 : Contemporary Lithospheric Motion / Seismic Geology
Volume 6 : Global Tectonic Zones / Supercontinent Formation and Disposal
Volume 7 : Orogenic Belts / Geological Mapping
Volume 8 : Basin Analysis / Global Sedimentary Geology / Sedimentology
Volume 9 : Energy and Mineral Resources for the 21st Century / Geology of Mineral Deposits / Mineral Economics
Volume 10 : New Technology for Geosciences
Volume 11 : Stratigraphy
Volume 12 : Paleontology and Historical Geology
Volume 13 : Marine Geology and Palaeoceanography
Volume 14 : Structural Geology and Geomechanics
Volume 15 : Igneous Petrology
Volume 16 : Mineralogy
Volume 17 : Precambrian Geology and Metamorphic Petrology
Volume 18.A : Geology of Fossil Fuels - Oil and Gas
Volume 18.B : Geology of Fossil Fuels - Coal
Volume 19 : Geochemistry
Volume 20 : Geophysics
Volume 21 : Quaternary Geology
Volume 22 : Hydrogeology
Volume 23 : Engineering Geology
Volume 24 : Environmental Geology
Volume 25 : Mathematical Geology and Geoinformatics
Volume 26 : Comparative Planetology / Geological Education / History of Geosciences

PROCEEDINGS OF THE

30TH INTERNATIONAL GEOLOGICAL CONGRESS

BEIJING, CHINA, 4 - 14 AUGUST 1996

VOLUME 26

COMPARATIVE PLANETOLOGY, GEOLOGICAL EDUCATION, HISTORY OF GEOLOGY

EDITORS:

WANG HONGZHEN

CHINA UNIVERSITY OF GEOSCIENCES, BEIJING, CHINA

D.F. BRANAGAN

DEPARTMENT OF GEOLOGY AND GEOPHYSICS, UNIVERSITY OF SYDNEY, AUSTRALIA

OUYANG ZIYUAN

NATURE AND SOCIETY COORDINATION BUREAU, BEIJING, CHINA

WANG XUNLIAN

CHINA UNIVERSITY OF GEOSCIENCES, BEIJING, CHINA

UTRECHT, THE NETHERLANDS, 1997

VSP BV
P.O. Box 346
3700 AH Zeist
The Netherlands

First published in 1997

ISBN 90-6764-254-1

Printed in The Netherlands by Ridderprint bv, Ridderkerk.

CONTENTS

GEOLOGICAL EDUCATION

HISTORY OF GEOLOGY
Persons and institutes

HISTORY OF GEOLOGY:
Geoscience disciplines

FOREWORD

The Secretariat of the Thirtieth Geological Congress and VSP International Science Publishers of the Netherlands agreed to publish a series of Proceedings of the 30th IGC. It was decided to combine the Symposia 20 — Comparative Planetology, 21 — Geological Education, organized jointly by the COGEOED (Commission for Geoscience Education and Training) and 22 — History of Geology, organized jointly by INHIGEO (International Commission on the History of Geological Sciences), IUGS, and ICOGS (International Consortium of Geological Surveys) in one volume, the Volume 26. This volume accordingly contains these three symposia in their order, amounting to a total number of 26 articles. In the first two symposia the articles are arranged in the alphabetic order of author names respectively. In the third symposium, the History of Geology, the articles are separated into two parts, (a) Persons and Institutes, and (b) Geoscience Disciplines, the alphabetic order of author names being applied in each of the two parts. It is a pity that there are only three papers in geological education, especially as people are now beginning to appreciate the importance of geological education.

It may be specially noted that a very brief commemoration of the Semicentennial of the Passing away of Professor Amadeus W. Grabau was held in association with Session 22-2 during the Congress. We have on that account put the papers related to Grabau together in this volume.

We take the opportunity to express our deep thanks to all the contributors for their efforts to make the publication of this volume possible. We are also grateful for the cooperation of the VSP Publishers and for the help of the Secretariat of the 30th IGC in various ways.

The Editors
March 5, 1997

COMPARATIVE PLANETOLOGY

Proc. 30th Intern. Geol. Congr., Vol. 26, pp. 1-29
Wang *et al.* (Eds)

Influence of Continental Drift on the Tidal Evolution of the Earth-Moon System

MASANAO ABE   HITOSHI MIZUTANI
Institute of Space and Astronautical Science, Yoshinodai 3-1-1, Sagamihara, Kanagawa 229, Japan
MASATSUGU OOE
National Astronomical Observatory, Division of Earth Rotation, Hoshigaoka, 2-21, Mizusawa, Iwate 023, Japan

Abstract

Ocean tide was numerically calculated for various ocean continent configurations. The calculated tidal torque response for semidiurnal ocean tide is found to be bigger, when the continents are lined on the equator than when the continents are lined along the meridian. Since the ocean continent configuration significantly affects the tidal torque, variation of the Earth's spin rotation rate in the past should be associated with the continental drift. The present calculation suggests that the long and short term variation of l.o.d. (length of day) up to 600 Ma *b.p.* estimated from palaeontological studies is explained by temporal variation of the ocean continent configuration. Moreover it is highly probable from our result that the continents before Cambrian age were lined more or less on the meridian or of other configurations that gives smaller tidal torque than the Cambrian configuration.

Keywords: Earth-Moon System, Tidal Evolution, Continental Drift

INTRODUCTION

Many planetary scientists are interested in how the Earth-Moon system has evolved dynamically with time, because such venerable problem may provide the key to solve the Moon's origin and Earth's surface environment in their early history. On the view of the celestial mechanics, Kaula [1] and MacDonald [2] derived the equations for dynamical evolution due to the planet-satellite system. Using these equations, Goldreich [3] and Gerstenkorn [4] estimated the process of the dynamical evolution of the Earth-Moon system. Because of lack of quantitative knowledge on the tidal dissipation, however, they couldn't discuss about the time scale of the evolution.

On the other hand, from the geodesic observation and archaeo-astronomy it is known that the Earth's spin rotation and lunar orbital revolution are slowing down as time [5]. Assuming the constant phase lag of the Earth's tidal bulge to the Moon throughout the history, it is derived that the Earth-Moon distance was within a few Earth radii about 1.5 billion years before present [6]. Because there is no geological evidence for such a strong tide and tidal dissipation as expected in the short Earth-Moon distance at 1.5 billion years *b.p.*, the phase lag in the past is supposed to be much smaller than present value.

Tidal evolution occurs due to tidal torque between Earth and Moon or Sun. The Moon is a main tide generating object for the Earth and the Sun is the second. Moreover the source of tidal torque on the Earth is separated into the two parts. One is in the ocean and the other in the solid Earth. At present the tidal torque due to the ocean tide is about ten times larger than that due to the solid Earth tide [7]. The tidal torque on the ocean is sensitive to frequency of the tidal force. Webb [6] considered the frequency dependence of tidal dissipation solving numerically for a hemispherical ocean tide. His study led to the conclusion that the Earth-Moon distance become the smallest between 3.9 and 5.3 billion years before present. Ooe *et al.* [8] also emphasized the importance of the frequency dependence of tidal dissipation, calculating the ocean tide for realistic models of ocean-continent configurations. Ooe *et al.* modeled two ocean-continent configurations, the one corresponds to the present configuration and the other corresponds to the Permian configuration. They showed that the tidal responses of two ocean-continent configurations are significantly different. Krohn and Suendermann [9], although they didn't investigate the frequency dependency, calculated the ocean tide for six different ocean-continent configurations corresponding to those at six different ages (present, 0.2, 70, 240, 450 million years *b.p.*). They suggested that the tidal dissipation rate at the Permian (240 million years *b.p.*) was less than half of that at the past 450 million years, assuming the temporal variation of the Earth-Moon distance. Thus it seems clear from previous studies that the tidal dissipation in the past was different from the present one. But it is not clear that either the different geometry or the different frequency-dependent tidal response of the past ocean-continent configuration is a primary cause for it. Since the tidal dissipation depends on the frequency (period) of the tidal forcing which in turn depends on the Earth-Moon distance, it is necessary to combine the calculations of tides and dynamics of the Earth-Moon system for our better understanding the tidal evolution. In the present study, we calculate the frequency-dependent tidal (torque) response for three realistic and four simplified ocean-continent configurations. Comparison of the results of the realistic models with simplified models gives us better insight on how the ocean-continent configuration affects the tidal response. The result of tidal responses are then used in the secular perturbation equations of the Earth-Moon system. The combined results will lead us to how the Earth-Moon system evolved dynamically with time in the past.

DYNAMICAL EVOLUTION

Method of calculating the dynamical evolution

In our study, the procedures of calculating the tidal evolution are divided into two parts. First one is calculating the dynamical evolution of the Earth-Moon system. Second one is calculating the phase lag of the Earth. In this chapter, we describe about the first one.

We calculate the dynamical evolution referring to Goldreich [3]. We use the following set of the secular perturbation equations in the Earth-Moon system.

The temporal variation of the lunar semi-major axis $a_{\otimes\oplus}$ ($\otimes$and $\oplus$ denote the Moon and the Earth respectively) is

$$\frac{da_{\otimes\oplus}}{dt} = \frac{2a_{\otimes\oplus}}{h} T_{3(\otimes\rightarrow\oplus\rightarrow\oplus)} . \tag{1}$$

The temporal variation of the lunar mean motion $n_{\otimes\oplus}$ is

$$\frac{dn_{\otimes\oplus}}{dt} = -\frac{3}{2}\frac{n_{\otimes\oplus}}{a_{\otimes\oplus}}\frac{da_{\otimes\oplus}}{dt} . \tag{2}$$

The temporal variation of the angular velocity of the Earth's spin rotation $\theta_{\oplus}$ is

$$\frac{d\theta_{\oplus}}{dt} = \frac{1}{I_{\oplus} + 3C_{1\oplus}\theta_{\oplus}^{2}}[\sin i_{\otimes\oplus}(-T_{2(\otimes\rightarrow\oplus\rightarrow\otimes)} - T_{2(\ominus\rightarrow\oplus\rightarrow\otimes)})$$
$$+\cos i_{\otimes\oplus}(-T_{3(\otimes\rightarrow\oplus\rightarrow\otimes)})$$
$$+\sin i_{\oplus\ominus}(-T_{2(\otimes\rightarrow\oplus\rightarrow\ominus)} - T_{2(\ominus\rightarrow\oplus\rightarrow\ominus)})$$
$$+\cos i_{\oplus\ominus}(-T_{3(\ominus\rightarrow\oplus\rightarrow\ominus)})]. \qquad (3)$$

Here $i_{\otimes\ominus}$ is obliquity of the Earth's equator to the ecliptic, $i_{\otimes\oplus}$ is inclination of the lunar orbit to the Earth's equator. $I_{\oplus}$ is the inertia of the Earth in the absent of the rotational deformation and $C_{1\oplus}$ is the coefficient of variation of the inertia due to the change of the rotation rate, which is described by Munk and MacDonald [2] as follows:

$$C_{1\oplus} = \frac{2k_{s\oplus}R_{\oplus}^{5}}{9G}, \qquad (4)$$

where $k_{s\oplus}$ is Earth's secular Love number (we use $k_{s\oplus}$ =0.96), $R_{\oplus}$ is Earth's radius, G is gravity constant.

$T_{2(*\rightarrow*\rightarrow*)}$, $T_{3(*\rightarrow*\rightarrow*)}$ are the components of the tidal torque. At a set of marks(⊕, ⊗ and ⊙ represent the Earth, the Moon and the Sun respectively) in the parentheses of the subscript, the first mark means a tide-raising body, the second mark means a tide-raised body (and also a perturbing body), the third one means a body perturbed by the tidal torque.

Angular momentum is exchanged between the second body and the third body. Each component of the torque is decomposed as shown Fig. 1. When the tidal force is raise by the Moon, that is the torque is given by $T_{*(\otimes\rightarrow*\rightarrow*)}$, the basic plane is to be the lunar orbital plane. When the tidal force is raised by the Sun, that is the torque is given by $T_{*(\ominus\rightarrow*\rightarrow*)}$, the basic plane is to be ecliptic plane. T_1 offsets calculating in long time more than the precession period, then we ignore T_1.

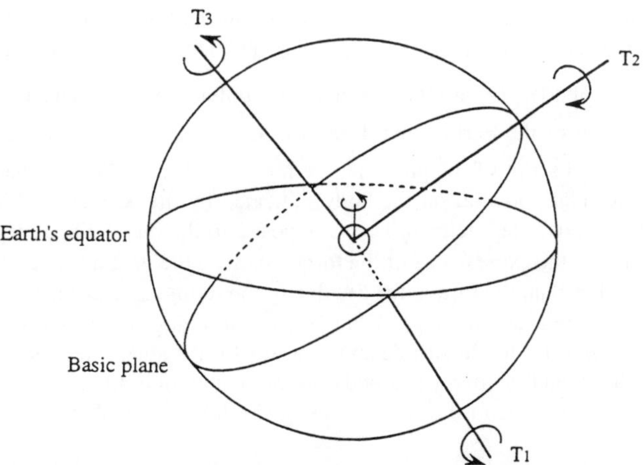

Figure 1. Three unit vector of the composition of the tidal torque.
T_1 is a component along the direction of the accending node of the basic plane to the Earth's equator. T_2 is a component vertical to T_1 on the basic plane. T_3 is perpendicular to the basic plane.

$T_{2(*\rightarrow*\rightarrow*)}$ and $T_{3(*\rightarrow*\rightarrow*)}$ are presented as follows:

$$T_{2(\otimes\to\oplus\to\otimes)} = \sum_{l=2}^{\infty}\sum_{m=0}^{l}\sum_{p=0}^{l}\sum_{-\infty}^{\infty} \frac{G m_{\otimes}^{2} k_{l\oplus}}{a_{\oplus\otimes}} \left(\frac{R_{\oplus}}{a_{\otimes\oplus}}\right)^{2l+1} (2-\delta_{0m}) \frac{(l-m)!}{(l+m)!} [F_{lmp}(i_{\otimes\oplus}) G_{lpq}(e_{\otimes\oplus})]^{2} \times \frac{m-(l-2p)\cos i_{\otimes\oplus}}{\sin i_{\otimes\oplus}} \sin \bar{E}_{lmpq(\oplus\to\otimes)}, \quad (5)$$

$T_{2(\Theta\to\oplus\to\Theta)}$ is obtained by exchanging $\otimes$ to Θ at (5).

$$T_{2(\Theta\to\oplus\to\otimes)} = \sum_{l=2}^{\infty}\sum_{m=0}^{l} \frac{G m_{\Theta} m_{\otimes} k_{l\oplus}}{a_{\oplus\Theta}} \left(\frac{R_{\oplus}}{a_{\oplus\Theta}}\right)^{l} \left(\frac{R_{\oplus}}{a_{\otimes\oplus}}\right)^{l+1} (2-\delta_{0m}) \frac{(l-m)!}{(l+m)!} F_{lm\frac{1}{2}}(i_{\oplus\Theta}) F_{lm\frac{1}{2}}(i_{\otimes\oplus}) \times G_{l\frac{1}{2}0}(e_{\oplus\Theta}) G_{l\frac{1}{2}0}(e_{\otimes\oplus}) \frac{m}{\sin i_{\otimes\oplus}} \sin[m(\Omega_{\oplus\Theta} - \Omega_{\otimes\oplus}) + \bar{E}_{lm\frac{1}{2}0(\oplus\to\Theta)}], \quad (6)$$

$T_{2(\otimes\to\oplus\to\Theta)}$ is obtained by exchanging $\otimes$ to Θ at (6).

$$T_{3(\otimes\to\oplus\to\otimes)} = \sum_{l=2}^{\infty}\sum_{m=0}^{l}\sum_{p=0}^{l}\sum_{-\infty}^{\infty} \frac{G m_{\otimes}^{2} k_{l\oplus}}{a_{\otimes\oplus}} \left(\frac{R_{\oplus}}{a_{\otimes\oplus}}\right)^{2l+1} (2-\delta_{0m}) \frac{(l-m)!}{(l+m)!} [F_{lmp}(i_{\otimes\oplus}) G_{lpq}(e_{\otimes\oplus})]^{2} \times (l-2p+q) \sin \bar{E}_{lmpq(\oplus\to\otimes)}, \quad (7)$$

$T_{3(\Theta\to\oplus\to\Theta)}$ is obtained by exchanging $\otimes$ to Θ at (7).

Here $m_{\otimes}$ and m_{Θ} are masses of the Moon and the Sun, $k_{l\oplus}$ is Earth's Love number of degree l, $a_{\oplus\Theta}$ is Earth-Sun distance, $e_{\otimes\oplus}$ and $e_{\oplus\Theta}$ are the eccentricities of the lunar orbit and the Earth's orbit, $\Omega_{\otimes\oplus}$ and $\Omega_{\oplus\Theta}$ are the ascending nodes of the lunar and solar orbits (with respect to the Earth's equatorial). $F_{lmp}(i)$ and $G_{lpq}(e)$ are the trigonometric polynomials described by Kaula [10]. $\bar{E}_{lmpq(*\leftrightarrow *)}$ is a phase lag of the Earth to be discussed at the very next section. A set of marks in the parentheses of the subscript of the phase lag defines as follows. The first mark indicates the body that generates the tidal potential on the body indicated by the second one. You should take care that the sign of the phase lag defined here is opposite to the one defined by Goldreich [3]. In the equations (5) ~ (7), the expressions of the torque include the parameters l, m, p, q. The torque for $l \geq 3$ is much smaller than the torque for l=2 due to the terms $R_{\oplus}/a_{\otimes\oplus}$ or $R_{\oplus}/a_{\oplus\Theta}$. If the inclination and the eccentricity are small, the $F_{lmp}(i)$ and $G_{lpq}(e)$ for large p, q become small. Therefore we use seven sets of l, m, p, q for the lunar tide and five sets for the solar tide as shown in Table 1. When the tide raising body and the perturbed body are not equal, only the torque for K_1 (principal daily declinational tide) and K_2 (semi-diurnal declinational tide) are available.

In precession period, inclinations and ascending nodes change periodically. Referring to Goldreich [3], we take into account of the variations of the inclinations $i_{\oplus\Theta}$, $i_{\otimes\oplus}$ and the ascending nodes $\Omega_{\otimes\oplus}$, $\Omega_{\oplus\Theta}$. In the calculation of the precession, we assume other variables are constant. As inclinations and ascending nodes change in the precession period, the amount of the tidal torque varies periodically. We calculate the average of the tidal torque in the precession period and put these values in the secular perturbation equations.

In Goldreich's paper, he assumes that lunar orbit is circular, but we consider the eccentricity. However, because eccentricity is very small, we use the same equations as Goldreich's equations in the calculation of the precession. For the longer time scale than precession period, the temporal variation of the eccentricity of the lunar orbit $e_{\otimes\oplus}$ is

$$\frac{de_{\otimes\oplus}}{dt} = \sum_{l=2}^{\infty}\sum_{m=0}^{l}\sum_{p=0}^{l}\sum_{-\infty}^{\infty} \frac{Gm_{\otimes}}{\sqrt{G(m_{\oplus}+m_{\otimes})a_{\otimes\oplus}}}\left(\frac{R_{\oplus}}{a_{\otimes\oplus}}\right)^{2l+1}(2-\delta_{0m})\frac{(l-m)!}{(l+m)!}$$
$$\times \frac{\sqrt{1-e_{\otimes\oplus}^{2}}}{a_{\otimes\oplus}e_{\otimes\oplus}}[\sqrt{1-e_{\otimes\oplus}^{2}}(l-2p+q)-(l-2p)][G_{lpq}(e_{\otimes\oplus})]^{2}$$
$$\times([F_{lmp}(i_{\otimes\oplus})]^{2}k_{l\oplus}\sin\overline{E}_{lmpq(\oplus\leftrightarrow\otimes)} + N_{l}[F_{lmp}(0)]^{2}k_{l\otimes}\sin\overline{E}_{lmpq(\otimes\to\oplus)}), \quad (8)$$

where $N_l=(m_{\oplus}/m_{\otimes})^2(R_{\otimes}/R_{\oplus})^{2l+1}$, $m_{\oplus}$is mass of the Earth, $k_{l\otimes}$ is lunar Love number of degree l, $\overline{E}_{lmpq(\otimes\to\oplus)}$ is the phase lag of the lunar tidal bulge to the Earth. In this equation we also put the mean value of $F_{lmp}(i_{\oplus\ominus})$ during precession period. We assume that the eccentricity of the Earth's orbit $e_{\oplus\ominus}$ is always constant and that the lunar spin axis is normal to the lunar orbit plane for simplicity. We use $k_{2\oplus}=0.3044$ and $k_{2\otimes}=0.027$.

Darwin nomenclature	l	m	p	q	Tide raising body	r_{lmpq}
M_2	2	2	0	0	Moon	0
N_2	2	2	0	1	Moon	0
K_2	2	2	1	0	Moon	0
L_2	2	2	0	-1	Moon	2
$2N_2$	2	2	0	2	Moon	0
K_1	2	1	1	0	Moon	1
O_1	2	1	0	0	Moon	-1
Q_1	2	1	0	1	Moon	-1
S_2	2	2	0	0	Sun	0
K_2	2	2	1	0	Sun	0
T_2	2	2	0	1	Sun	0
K_1	2	1	1	0	Sun	1
P_1	2	1	0	0	Sun	-1

Table 1. Using set of l, m, p, q for the tide and value of r_{lmpq}

Phase lag of the Earth

As mentioned in the very above section, $\overline{E}_{lmpq(*\to*)}$ is a phase lag of the Earth or the Moon. This is an index of the energy dissipation rate and the exchanging rate of angular momentum. This value changes with time. An assumption with a constant $\overline{E}_{lmpq(*\to*)}$ would make the discussion of the tidal evolution easy. In this case, however, the speed of evolution reveals to be so rapid that the Earth-Moon distance becomes a few earth-radii 1.5 billion year's ago. There is no evidence on the Earth's surface at that period indicating such an effect as expected from the close approach of the Moon. Hence, the assumption of the constant $\overline{E}_{lmpq(*\to*)}$ must not be appropriate. We need to consider the time variation of $\overline{E}_{lmpq(*\to*)}$.

$\overline{E}_{lmpq(\oplus\to *)}$ is divided by two components, which are the phase lag of the solid Earth, $\overline{E}_{solid,lmpq(\oplus\to *)}$, and the equivalent phase lag of the ocean, $\overline{E}_{ocean,lmpq(\oplus\to *)}$.

We assume the phase lag of the solid Earth is determined by the quality factor $Q_\oplus$ as follows:

$$\sin E_{solid,lmpq(\oplus\to *)} = \frac{1}{Q_\oplus}. \qquad (9)$$

Here the quality factor $Q_\oplus$ is for the global Earth and changes with frequency of the tidal force. Referring to Ooe's study [11], $Q_\oplus$ is determined as follows:

$$Q_\oplus^{-1} = 8.92 \times 10^{-4} \sigma^{-0.189}, \qquad (10)$$

where σ is the angular velocity in the unit of rad/sec. This equation corresponds to the present value for the semidiurnal tide, $Q_\oplus$ of 210.

Similarly, the phase lag of the Moon is represented by

$$\sin \overline{E}_{lmpq(\otimes\to\oplus)} = \frac{1}{Q_\otimes}. \qquad (11)$$

In this case, we assume $Q_\otimes$ to be constant 100 [12].

On the other hand, the equivalent phase lag of the ocean is obtained by numerical calculation of the ocean tide.

From the global ocean tide model, we obtain the amplitude D_{lmpq} and the phase lag ε_{lmpq} at first. How to obtain these values will be mentioned in the next part. The relation between the equivalent phase lag $\overline{E}_{ocean,lmpq(\oplus\to *)}$ and the phase lag obtained from the ocean tide model is derived by Lambeck [7] as follows:

$$\begin{aligned}\sin \overline{E}_{ocean,lmpq(\oplus\to *)} &= \frac{3m_\oplus}{m_\otimes}\frac{\rho_w}{\overline{\rho}}\frac{1+k'_{l\oplus}}{k_{l\oplus}}\frac{1}{2l+1}\left(\frac{a_{\otimes\oplus}}{R_\oplus}\right)^{l+1}\frac{1}{R_\oplus} \\ &\times \frac{(l+m)!}{(l-m)!(2-\delta_{0m})}\frac{1}{F_{lmp}(i_{\otimes\oplus})G_{lpq}(e_{\otimes\oplus})} \\ &\times D_{lmpq}\begin{bmatrix}\cos\\ \sin\end{bmatrix}_{l-m=odd}^{l-m=even}\{\pi(\frac{1}{2}r_{lmpq}+m)-\varepsilon_{lmpq}\},\end{aligned} \qquad (12)$$

where ρ_w and $\overline{\rho}$ are the mean density of the sea water and solid Earth, $k'_{l\oplus}$ is the l th order load Love number of the Earth. r_{lmpq} is an integer, which relates the argument used in the development of the orbital elements for tidal expansion, as given in Table 1. Equation(12) is for the case where the tide generating body is the Moon. If the case is for the Sun, we must change the variables corresponding to on the Sun.

Using above equation we can transform the result of the ocean tide calculation to the form to be used in the formulas of tidal torque (*eq*.(5) ~ (7)).

The phase lag of the Earth is represented as a sum of the phase lag of the solid Earth and equivalent phase lag of the ocean as follows:

$$\sin \overline{E}_{lmpq(\oplus\to *)} = \sin \overline{E}_{solid,lmpq(\oplus\to *)} + \sin \overline{E}_{ocean,lmpq(\oplus\to *)}. \qquad (13)$$

Both phase lag of the solid Earth and the equivalent phase lag of the ocean are small enough, so that the above equation is rewritten as:

$$\overline{E}_{lmpq(\oplus\to *)} = \overline{E}_{solid,lmpq(\oplus\to *)} + \overline{E}_{oean,lmpq(\oplus\to *)} . \quad (14)$$

In the equation(12), when we substitute the value of Table 1 to r_{lmpq}, we find the trigonometrical function always became a sine function. In the formula of equivalent phase lag, only D_{lmpq} sin ε_{lmpq} is available. So we call D_{lmpq} sin ε_{lmpq} " tidal torque response " hereafter.

OCEAN TIDE

Method of calculating the ocean tide

For the calculation of the dynamical evolution, we have to model the variation of the tidal torque response due to change of the angular velocity of the tidal potential. Since we can't get the tidal angular velocity exactly until the estimation of the dynamical evolution is made, it is helpful to get the frequency dependency of the " tidal torque response " for a rather wide frequency region at first. The " tidal torque response " is obtained by calculating the ocean tide numerically.

The ocean tide is calculated using the differential equations of motion and continuation as follows:

$$\frac{\partial V_\theta}{\partial t} = \frac{gH}{R_\oplus}\frac{\partial}{\partial\theta}(\alpha\eta - \beta\zeta) - B_\theta - E_\theta + 2\Omega V_\phi \cos\theta , \quad (15)$$

$$\frac{\partial V_\phi}{\partial t} = \frac{gH}{R_\oplus \sin\theta}\frac{\partial}{\partial\phi}(\alpha\eta - \beta\zeta) - B_\phi - E_\phi - 2\Omega V_\theta \cos\theta \quad (16)$$

and

$$\frac{\partial\zeta}{\partial t} = -\frac{1}{R_\oplus \sin\theta}\left\{\frac{\partial}{\partial\theta}(V_\theta \sin\theta) + \frac{\partial V_\phi}{\partial\phi}\right\}. \quad (17)$$

These equations are written in rotational spherical polar coordinates (r, θ, ϕ); r is the polar radius, θ is colatitude, ϕ is the east longitude. (Oceanographers usually use another spherical polar coordinate (λ, φ, r); λ is the east longitude, $\varphi(=1/2-\theta)$ is the north latitude, r is the polar radius.)

Furthermore t is the time, g is the gravity acceleration, H is the depth of the ocean, $R_\oplus$ is the Earth's radius, η is the amount of " primary " astronomical potential U divided by the gravity acceleration g; we call η " equilibrium tidal height ", and ζ is observed tidal height. Hereafter we call U " tidal potential " for simplicity. Ω is the angular velocity of the Earth's spin.

V_θ and V_ϕ express the current velocities integrated over the depth as follows:

$$V_\theta = \int_{z^b}^{z^s} \upsilon_\theta dz , \quad (18)$$

$$V_\phi = \int_{z^b}^{z^s} \upsilon_\phi dz , \quad (19)$$

where υ_θ and υ_ϕ are the southward and eastward current velocity, respectively, z is a new depth variable defined as $z=r-R_\oplus$, and z^s and z^b are the positions of the sea surface and bottom, respectively.

B_θ and B_ϕ stand for the bottom frictions, which depend on velocity and its square as follows:

$$B_\theta = \frac{B_1}{H} V_\theta + \frac{B_2}{H^2} V_\theta \sqrt{V_\theta^2 + V_\phi^2}. \tag{20}$$

B_ϕ is given by the corresponding replacements ($\theta \to \phi, \phi \to \theta$). Schwiderski [13] assumed B_1=b_1 sin θ to consider the latitude dependence of the bottom friction. In this study on $B_2 \neq 0$ we assume B_1=b_1, B_2=b_2.

E_θ and E_ϕ represent stresses that are induced by the eddy viscosity of the ocean. The complete expressions of E_θ, E_ϕ are shown in Schwiderski [13] paper. The eddy viscosity depends on the mean lateral cross section area of the flow. Therefore it is given by

$$\nu = \frac{a}{2} L H (1 + \sin\theta). \tag{21}$$

Where L is the mesh size along the equator, θ the colatitude and H the ocean depth.

The coefficients b_1, b_2, and a are determined in simulation studies of the present ocean tide. Adopted values are listed in Table 2. The result using the value of Type 2 much fits to present ocean tide than that of Type 1.

	a [1/s]	b_1 [m/s]	b_2
Type 1	0.0010	0.06941	0
Type 2	0.0003	0.01628	0.0015

Table 2. Adopted value of a, b_1, b_2

α is defined by $\alpha = 1 + k'_{l\oplus}$, where $k'_{l\oplus}$ are the l th order Earth's load Love number mentioned in the second section of the above part. β is the coefficient for yielding of the ocean bottom. In this study, we set α=0.69 and β=0.90.

As shown in the above equations, velocities are integrated over the depth of the ocean. So in our study the ocean tide model is given in a two-dimensional system.

Here, if we assume B_θ, B_ϕ and E_θ E_ϕ are zero and α and β are unity, these equations are reduced to the well-known classical `Laplace tidal equations'.

More detailed derivation of these equations is mentioned in the paper of Schwiderski(1980), who, however, assumed the bottom friction is to be linear. Schwiderski used observed tidal constants at the coast and some places of the offshore as the boundary condition. But in our calculation we simply assume no flows across coasts and free-slip along the shorelines, because of ignorance of the observational data in the palaeotide.

By the way, the equilibrium tidal height η is given by

$$\eta = \frac{U}{g} = \sum_{l=2}^{\infty} \sum_{m=0}^{l} \sum_{p=0}^{l} \sum_{q=-\infty}^{\infty} \frac{U_{lmpq}}{g}. \tag{22}$$

When the tide generating body is Moon,

$$U = \frac{Gm_{\otimes}}{a_{\otimes\oplus}}\left(\frac{r}{a_{\otimes\oplus}}\right)^{l}(2-\delta_{0m})\frac{(l-m)!}{(l+m)!}P_{lm}(\cos\theta)F_{lmp}(i_{\otimes\oplus})G_{lpq}(e_{\otimes\oplus})$$
$$\times\begin{bmatrix}\cos\\ \sin\end{bmatrix}_{l-m,odd}^{l-m,even}\{(l-2p0\omega_{\otimes\oplus}+(l-2p+q)M_{\otimes\oplus}+m(\Omega_{\otimes\oplus}-\theta^{\Psi})-m\phi\} \quad (23)$$

where $P_{lm}(\cos\theta)$ is the Legendre polynomials, $M_{\otimes\oplus}$ is the mean anomaly of the Moon, θ^{Ψ} is the sidereal angle of the Earth.

As the shown above, the tidal potential is represented by the sum of the species having different frequency. So the equilibrium tidal height is also represented by the sum of the species having different frequency as follows:

$$\eta = \sum_{l=2}^{\infty}\sum_{m=0}^{l}\sum_{p=0}^{l}\sum_{q=-\infty}^{\infty}\eta_{lmpq}. \quad (24)$$

At l=2, m=2, semidiurnal tide,

$$\eta_{22pq} = K_{22pq}\sin^{2}\theta\cos(\sigma_{22pq}t+\chi_{22pq}+2\phi). \quad (25)$$

At l=2, m=1, diurnal tide,

$$\eta_{21pq} = K_{21pq}\sin 2\theta\cos(\sigma_{21pq}t+\chi_{21pq}+\phi). \quad (26)$$

At l=2, m=0, long period tide,

$$\eta_{20pq} = K_{20pq}(3\sin^{2}\theta-2)\cos(\sigma_{20pq}t+\chi_{20pq}). \quad (27)$$

Tide	l	m	p	q	Origin	$\overline{K}_{lmpq}$, [m]	$\overline{\chi}_{lmpq}$, [deg]
M_2	2	2	0	0	Moon	0.242334	$2h_0-2s_0$
N_2	2	2	0	1	Moon	0.046398	$2h_0-3s_0+p_0$
K_2	2	2	1	0	Moon	0.020969	$2h_0$
L_2	2	2	0	-1	Moon	0.006850	$2h_0-s_0-p_0+180$
$2N_2$	2	2	0	2	Moon	0.00614	
K_1	2	1	1	0	Moon	0.096631	h_0+90
O_1	2	1	0	0	Moon	0.100514	h_0-2s_0-90
Q_1	2	1	0	1	Moon	0.019256	$h_0-3s_0+p_0-90$
S_2	2	2	0	0	Sun	0.112841	0
K_2	2	2	1	0	Sun	0.009735	$2h_0$
T_2	2	2	0	1	Sun	0.00676	
K_1	2	1	1	0	Sun	0.044850	h_0+90
P_1	2	1	0	0	Sun	0.046843	h_0-90

Table 3. Constants for major tidal constituents (partial tides).

In this study we consider diurnal and semidiurnal tide. We call K_{lmpq} and χ_{lmpq} "the amplitude of the equilibrium ocean tide" and "the astronomical argument of partial equilibrium tide". The present values of K_{lmpq}, χ_{lmpq} that are denoted by $\overline{K}_{lmpq}$ and χ_{lmpq} are listed in Table 3.

Here h_0, s_0 and p_0 are the mean longitudes of the Sun and Moon and the lunar perigee at Greenwich midnight.

At $l \geqslant 3$, K_{lmpq} is so small that we neglect this case.

σ_{lmpq} is the frequency of tidal potential which is given in the case of lunar tide as

$$\sigma_{lmpq} = m\left\{\dot{\theta}_{\oplus} - \frac{\partial\Omega_{\otimes\oplus}}{\partial t}\right\} - (l-2p)\frac{\partial\omega_{\otimes\oplus}}{\partial t} - (l-2p+q)n_{\otimes\oplus}, \tag{28}$$

where $\dot{\theta}_{\oplus}$ is the speed of the Earth's rotation that is equal to Ω defined in this chapter, and $\omega_{\otimes\oplus}$ is the argument of perigee of lunar orbit.

In the case of solar tide,

$$\sigma_{lmpq} = m\dot{\theta}_{\oplus} - (l-2p+q)n_{\otimes\oplus}. \tag{29}$$

Where $n_{\oplus\Theta}$ is mean motion of the Earth. In this case we neglect the temporal variations of the argument of perihelion $\omega\oplus\Theta$ and the ascending node $\Omega\oplus\Theta$.

$\partial\Omega_{\otimes\oplus} / \partial t$ and $\partial\omega_{\otimes\oplus} / \partial t$ are obtained by calculating three effects; Solar attraction, Earth's oblateness and tidal effect (See [14] and [10]). As present values of $\partial\omega_{\otimes\oplus} / \partial t$ and $\partial\Omega_{\otimes\oplus} / \partial t$ are much smaller than $n_{\otimes\oplus}$ or $\dot{\theta}_{\oplus}$, temporal variation of $\omega_{\otimes\oplus}$ and $\Omega_{\otimes\oplus}$ hardly affect the estimation of dynamical evolution in a few billion years.

Ocean-continent configuration model

In order to calculate the ocean tide, we need ocean-continent configuration models. These models are represented by giving the bathymetry of the ocean at each mesh point. In continental region, the bathymetry is defined zero. In this study mesh size is set to 4° × 4° . If we used more fine mesh, we could produce more detailed pattern of the ocean tide. This mesh size, however, is enough fine to discuss tidal evolution, because what we want to know is global pattern to estimate two degrees of spherical harmonics of " tidal torque response " .

In this study, as mentioned later, we use the palaeo-ocean-continent configuration. For the palaeo-ocean-continent configuration we can't construct so fine model. We conclude also the 4°× 4° meshes model is the best. We don't have many information on the bathymetry for the palaeo-oceans. So we divide the ocean into two parts, one is the continental shelf and the other the deep ocean, and we assume that the percentages of them to Earth's surface are equal to the present oceans. In this study we construct three real configuration models and four simple configuration models as Fig. 2 ~ Fig. 8. For the continental shelves of the simple configuration models, we make them around the continents. In standard case (Fig. 2 ~ Fig. 7) we assume that the bathymetry of the continental shelf is 200 m and deep ocean is 4,200 m depth. The land is assigned to 0 m depth. In each standard model the volume of water is kept constant.

Tidal torque response

Using above equations and ocean-continent configuration model we can obtain the value of the tidal torque response at each frequency and at each ocean-continent configuration for each partial tide. Here the " frequency " means σ_{lmpq} mentioned in the first section of this part. The procedure of how to obtain the tidal torque response or equivalent phase lag is mentioned in this section.

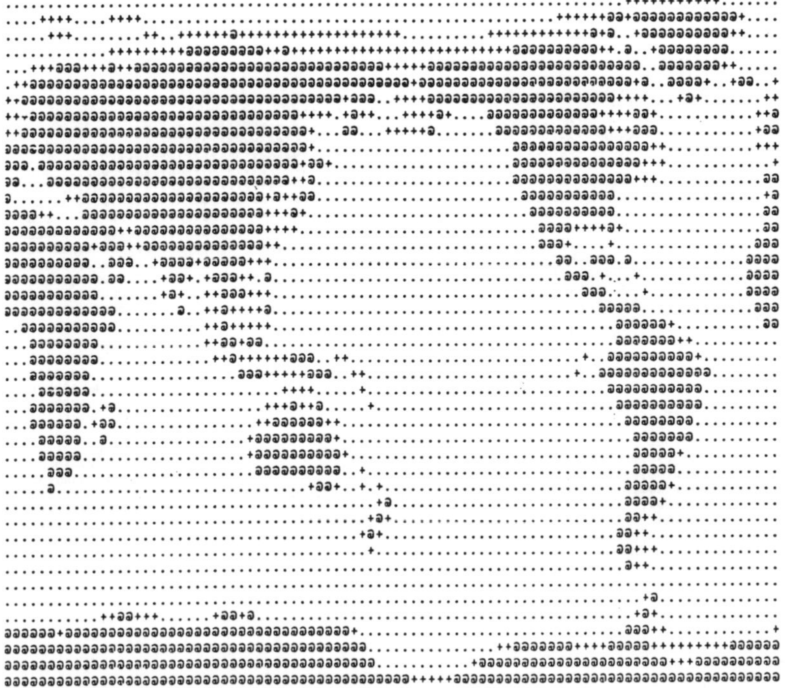

Figure 2. Present model corresponding to the present ocean-continent configuration. ∂ : contingent, +: continental shelf (200 *m* deep), · :deep ocean (4,200 *m* deep).

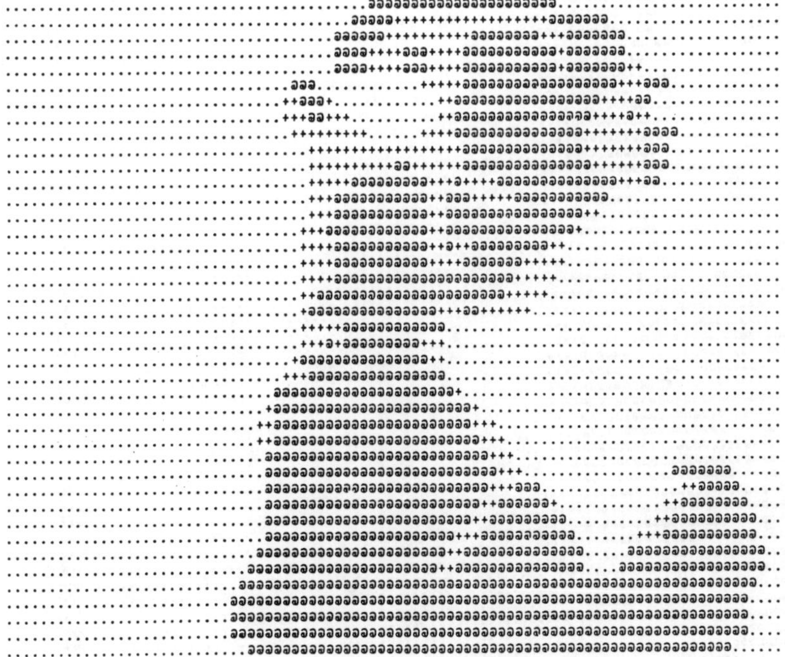

Figure 3. Permian model corresponding to the ocean-continent configuration at Permian age (250 Ma *b.p.*). ∂ : continent, +: continental shelf (200 *m* deep), · :deep ocean (4,200 *m* deep).

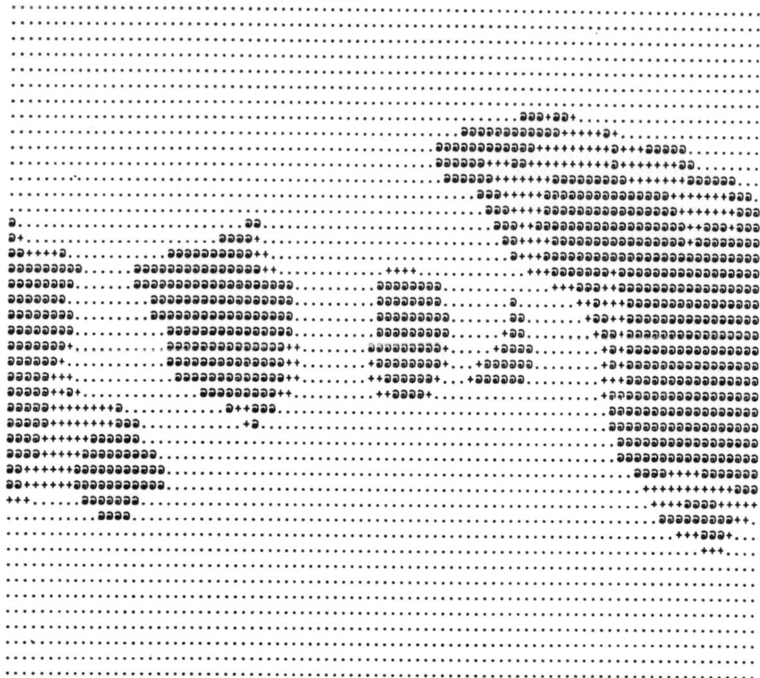

Figure 4. Cambrian model corresponding to the ocean-continent configuration at Cambrian age (500 Ma *b.p.*). ∂ : continent, +: continental shelf (200 *m* deep), · : deep ocean (4,200 *m* deep).

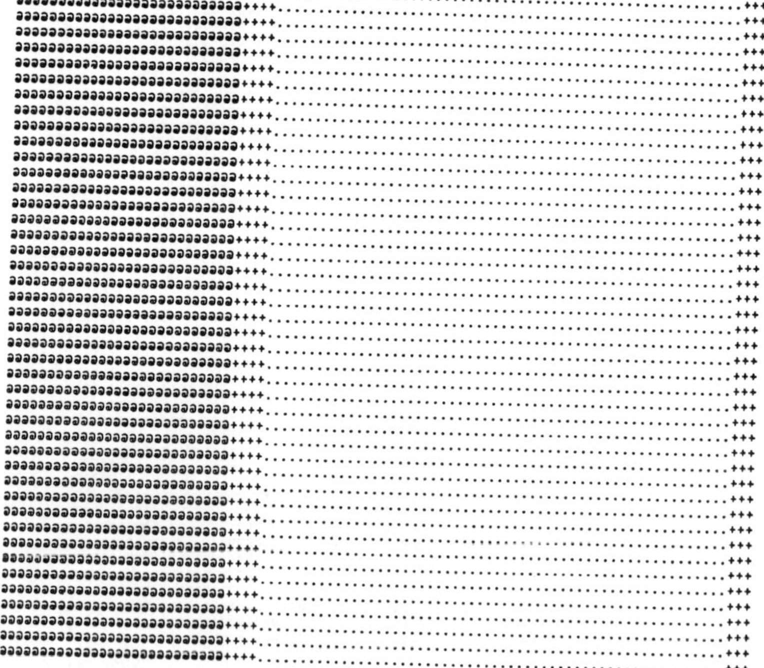

Figure 5. Meridian model. ∂ : continent, +: continental shelf (200 *m* deep), · : deep ocean (4,200 *m* deep).

Figure 6. Equator model. ∂ : continent, +: continental shelf (200 m deep), · : deep ocean (4,200 m deep).

Figure 7. Polar model. ∂ : continent, +: continental shelf (200 *m* deep), · : deep ocean (4,200 *m* deep).

Figure 8. All-Ocean model. All the meshes are set as deep ocean.

The " equilibrium tidal height " η_{lmpq} is given for each tidal frequency of σ_{lmpq} setting the index *lmpq* as the mentioned above. Integration of the discrete tidal equations is started by setting V_θ and V_ϕ to zero as initial values at all the meshes. Solution of tidal height distribution converges after about ten hundred times iterations of integration. During this period the tide generating body rotates around the Earth about a few ten times. The obtained tidal height is expressed as a function of time and position as follows:

$$\zeta_{lmpq}(\theta,\phi,T) = \zeta^0_{lmpq}(\theta,\phi)\cos[\sigma_{lmpq}T - \psi_{lmpq}(\theta,\phi)], \tag{30}$$

where ζ^0_{lmpq} is the local amplitude of the tidal height, $\psi_{lmpq}(\theta,\phi)$ is the local phase lag of the tidal height and T is time. Fig. 9 shows the result of the Present model for M_2 semidiurnal tide using present tidal potential.

σ_{lmpq} is equal to the frequency of given tidal force. This tidal height is decomposed into ((s, t) components of) spherical harmonics as follows:

$$\zeta_{lmpq}(\theta,\phi,T) = \sum_{s=1}^{\infty}\sum_{t=0}^{s}\sum_{+}^{-} D^{\pm}_{lmpq.st}\cos(\sigma_{lmpq}T \pm t\phi - \varepsilon^{\pm}_{lmpq.st})P_{st}(\cos\theta). \tag{31}$$

A summation $\sum_{+}^{-} D^{\pm}\cos(a \pm b - \varepsilon^{\pm})$ stands for

$$D^{+}\cos(a+b-\varepsilon^{+}) + D^{-}\cos(a-b-\varepsilon^{-}).$$

For the tidal evolution problem, only the case of $s = l$, $t = m$, and $\pm = +$ is important.

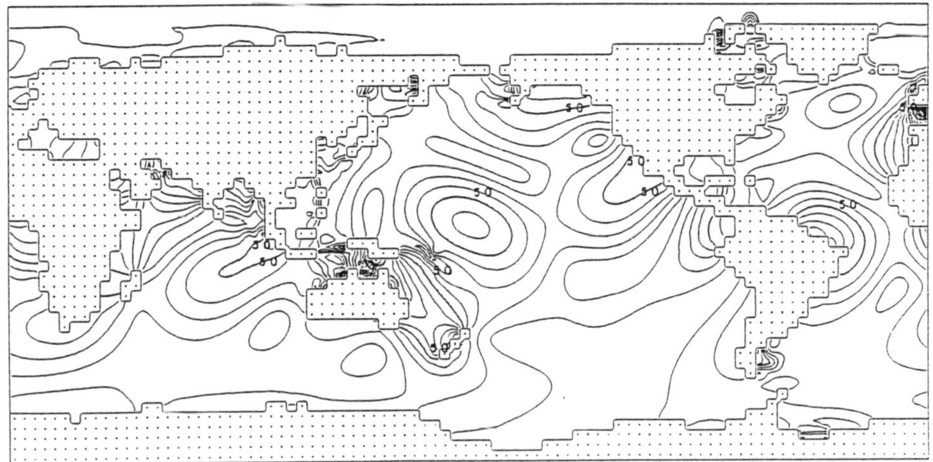

Figure 9. The distribution of the local amplitude for present-day M_2 ocean tide. The co-amplitude lines are drawn with ten centimeter intervals.

In this equation (31) $D^{+}_{lmpq,lm}$ and $\varepsilon^{+}_{lmpq,lm}$ are the amplitude and phase lag of the global ocean, which we want to obtain. D_{lmpq} and ε_{lmpq} defined in Section 2.2 presented by $D^{+}_{lmpq,lm}$ and $\varepsilon^{+}_{lmpq,lm}$ where the tidal torque response is given by $D_{lmpq} \sin \varepsilon_{lmpq}$.

In the Table 4 obtained tidal torque response for M2 tide for Type 1 or Type 2 friction model mentioned at Section 3.1 are listed. In this table we can compare our result with other studies. Lambeck [15] said Schwiderski's model most fits to the present ocean tide. Our results show nearly equal values to the tidal model of Schwiderski's. Type 2's result is more close to Schwiderski's than Type 1's.

	D_{2200} [cm]	ε_{2200}[deg]	$D_{2200} \sin \varepsilon_{2200}$ [cm]
Accad-Pekeris(1978)	3.10	125.7	2.52
Parke-Hendershott(1980)	3.88	133.2	2.47
Schwiderski(1980)	2.95	139.4	1.92
Zahel(1977)	4.80	130.4	3.65
Cartwright and Ray(1991)	3.45	131.6	2.58
Matsumoto *et al.*(1995)	3.25	130.2	2.49
This study Type 1	2.45	119.4	2.13
This study Type 2	2.76	134.9	1.95

Table 4. Amplitude and phase lag of M_2 ocean tide and tidal torque response

In recent decade the global sea surface topography data obtained by some artificial satellite are available. Cartwright and Ray [16] derived the tidal torque response from the Geosat altimetry data, and Matsumoto *et al.* [17] derived from TOPEX/POSEIDON data. In these two results the tidal torque response is slightly larger than the result of Schwiderski. These differences are considered to be related to luck of off shore data in Schwiderski's study. If the data obtained by the artificial satellites are more reliable, we may need to change the frictional coefficients to fit to the tidal torque response.

In our study, the above procedure is carried out for various frequencies on various species of partial tide. The amplitude of the equilibrium ocean tide K_{lmpq} and angular velocity of the Earth's spin Ω is unknown unless we calculate the dynamical evolution. Then we calculate the ocean tide using constant K_{lmpq} that is K_{lmpq} =0.1m.

Using this value of K_{lmpq} the obtained D_{lmpq} in the ocean tide calculation is different from real D_{lmpq}. We denote here the obtained amplitude of the global ocean at K_{lmpq} =0.1m " $D_{lmpq,0.1}$ " , and we call $K_{lmpq,0.1} \sin \varepsilon_{lmpq}$ " normalized tidal torque response " . At the same frequency of the tidal potential for the same ocean continent configuration D_{lmpq} is proportional to K_{lmpq}. That is,

$$D_{lmpq} = D_{lmpq,0.1} \frac{K_{lmpq}}{0.1m} \tag{32}$$

As mentioned in the first section of this part, the amplitude of equilibrium ocean tide K_{lmpq} is proportional to the amount of the tidal potential U_{lmpq}. If the tide generating body is Moon, it is

$$K_{lmpq} \propto \frac{F_{lmp}(i_{\otimes\oplus}) G_{lpq}(e_{\otimes\oplus})}{a_{\otimes\oplus}^{l+1}} . \tag{33}$$

Then K_{lmpq} is expressed with the present amplitude of the equilibrium ocean tide $\overline{K}_{lmpq}$ as

$$K_{lmpq} \propto \overline{K}_{lmpq} \left(\frac{\overline{a}_{\otimes\oplus}}{a_{\otimes\oplus}}\right)^{l+1} \frac{F_{lmp}(i_{\otimes\oplus}) G_{lpq}(e_{\otimes\oplus})}{F_{lmp}(\overline{i}_{\otimes\oplus}) G_{lpq}(\overline{e}_{\otimes\oplus})} . \tag{34}$$

The real D_{lmpq} is given by

$$D_{lmpq} \propto D_{lmpq,0.1} \frac{\overline{K}_{lmpq}}{0.1m} \left(\frac{\overline{a}_{\otimes\oplus}}{a_{\otimes\oplus}}\right)^{l+1} \frac{F_{lmp}(i_{\otimes\oplus}) G_{lpq}(e_{\otimes\oplus})}{F_{lmp}(\overline{i}_{\otimes\oplus}) G_{lpq}(\overline{e}_{\otimes\oplus})} . \tag{35}$$

Moreover if we substitute these equation to the *eq*.(12) in the second section of the second part, we get,

$$\begin{aligned} \sin \overline{E}_{ocean,lmpq} = & \frac{3m_{\oplus}}{m_{\otimes}} \frac{\rho_w}{\rho} \frac{1+k_l'}{k_l} \frac{1}{2l+1} \left(\frac{\overline{a}_{\otimes\oplus}}{R_{\oplus}}\right)^{l+1} \frac{1}{R_{\oplus}} \frac{(l+m)!}{(l-m)!(2-\delta_{0m})} \frac{1}{F_{lmp}(\overline{i}_{\otimes\oplus}) G_{lpq}(\overline{e}_{\otimes\oplus})} \\ & \times D_{lmpq,0.1} \frac{\overline{K}_{lmpq}}{0.1m} \begin{bmatrix} \cos \\ \sin \end{bmatrix}_{l-m=odd}^{l-m=even} \{\pi(\frac{1}{2} r_{lmpq} + m) - \varepsilon_{lmpq}\} , \end{aligned} \tag{36}$$

In these procedures, we can get the value of the equivalent phase lag of the ocean from the normalized tidal torque response without knowing the result of the dynamical evolution.

If tide raising body is the Sun, we must exchange the variables for the Sun in *eq*.(33) through *eq*.(36). The lunar mean motion, which corresponds to any Earth spin velocity, was determined using the conservation equation of angular momentum in the Earth-Moon system as follows:

$$C_{\oplus} \Omega \cos i_{\oplus\odot} + \frac{m_{\oplus} m_{\otimes}}{m_{\oplus} + m_{\otimes}} \{G(m_{\oplus} + m_{\otimes})\}^{\frac{2}{3}} n_{\otimes\oplus}^{-\frac{1}{3}} \cos i_{\otimes\odot} = C_0 = \mathrm{cost.}, \tag{37}$$

therefore,

$$n_{\otimes} = -\frac{\frac{G^2(m_{\oplus}m_{\otimes})^3}{m_{\oplus}+m_{\otimes}}\cos^3 i_{\otimes\ominus}}{(C_0 - C_{\oplus}\Omega\cos i_{\oplus\ominus})^3}, \tag{38}$$

where $i_{\otimes\ominus}$ is inclination of the lunar orbit with respect to the ecliptic.

Assuming $C_{\oplus}$, cos $i_{\oplus\ominus}$, and cos $i_{\otimes\ominus}$ are constant, we can get $n_{\otimes\oplus}$ without any calculation of the dynamical evolution. Moreover using *eq.*(28) and neglecting motion of ascending node and perigee,

$$\sigma_{lmpq} = m\Omega - ln_{\otimes\oplus}. \tag{39}$$

As shown in the above equation, we distinguish only the diurnal tide from the semidiurnal tide. For example, we don't distinguish O_1 from K_1 on calculating the dynamical evolution.

Simple configuration models

In this section, we use the simple ocean-continent configuration models to discuss the variation of the tidal torque due to the change of the ocean-continent configurations and the angular velocity of the tidal potential.

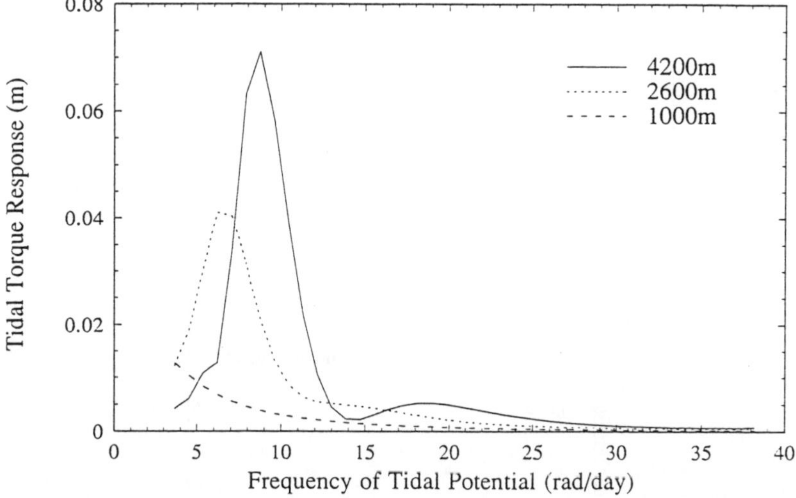

Figure 10. The response curve of All-ocean model: for M_2 ocean tide $(l, m, p, q)=(2, 2, 0, 0)$ $K_{2200}=0.1$m, using Type 1. Solid line is the response curve for the 4200m model of All-ocean model. Broken line is for 2600m model and dashed line is for 1000m model. The horizontal axis is the frequency of the tidal potential σ_{lmpq} and vertical axis is the normalized tidal torque response $D_{lmpq, 0.1}\sin\varepsilon_{lmpq}$.

At first we used the All-ocean models (Fig. 8) with various sea depths. Sea depth was set at 4200m, 2600m and 1000m. 4200m is the same depth as the deep ocean of the Present model (Fig. 2). The sea water volume of the model with 2600m is same as that for the Present model. We set the sea depth of 1000m in order that the difference between second one and third one is same that between first one and second one. Fig. 10 is the result of the ocean tide for various frequencies of tidal potential using three models mentioned above. This is a result for M_2 semidiurnal ocean tide, that is $(l, m, p, q)=(2, 2, 0, 0)$, with $K_{2200}=0.1m$ for the dissipation coefficients of Type 1. Three curves are the responses of each model. Each curve shows a peak at the resonance frequency of the ocean. The resonance frequency is a function of the determined ocean configuration and sea water depth.

For the All-ocean model, the resonance frequency is solved analytically. From Longuet-Higgins's study [18], the resonance frequency for semidiurnal tide is given by

$$\sigma \cong 3.4 \frac{\sqrt{gH}}{R_{\oplus}}, \tag{40}$$

where g is gravity acceleration H is the sea water depth and $R_{\oplus}$ is Earth radius. From above equation we get the values 9.3, 7.3 and 4.6 rad/day for the three models. From our calculations the resonance frequencies are 8.8, 7.1 and 3.7 rad/day, respectively. These values are agree with the analytical values. The peak height seems to be proportional to the volume of the sea water.

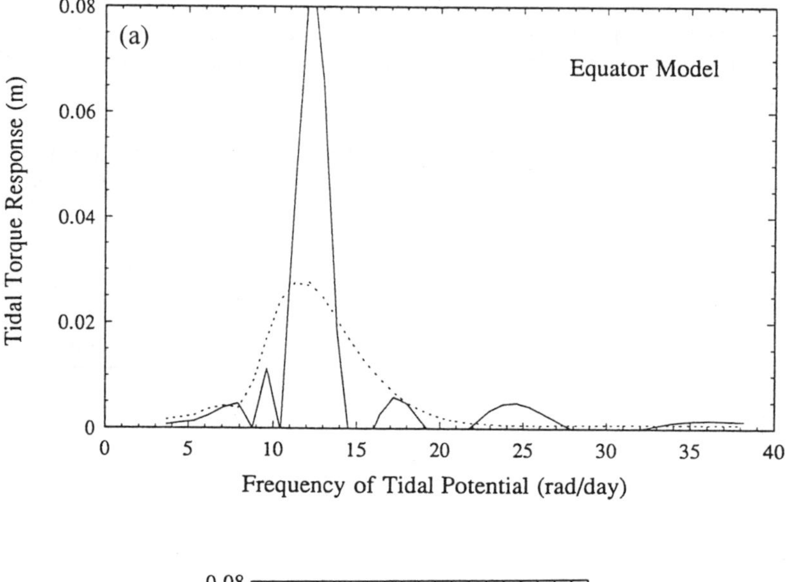

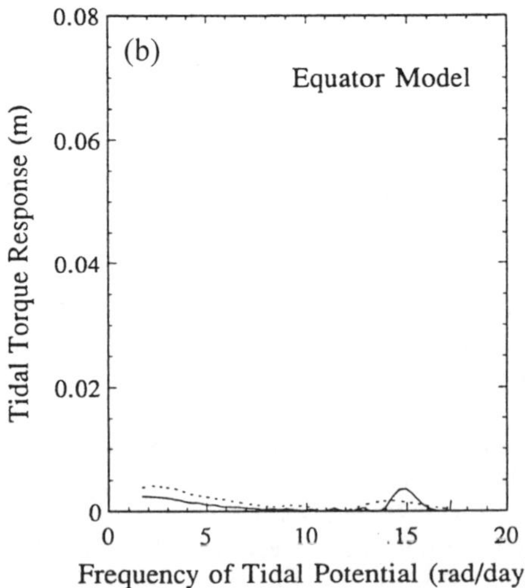

Figure 11. The variation of normalized tidal torque response for the Meridian model. X-axis is the angular velocity of the tidal potential. Y-axis is (a) $D_{2200,0,1}\sin\varepsilon_{2200}$ of the M_2 semidiurnal ocean tide, (b) $D_{2100,0,1}\sin\varepsilon_{2100}$ of the O_1 diurnal ocean tide. The angular velocity in present-day is shown by an arrow. The dotted line is the result using Type 1. The solid line is the result using Type 2.

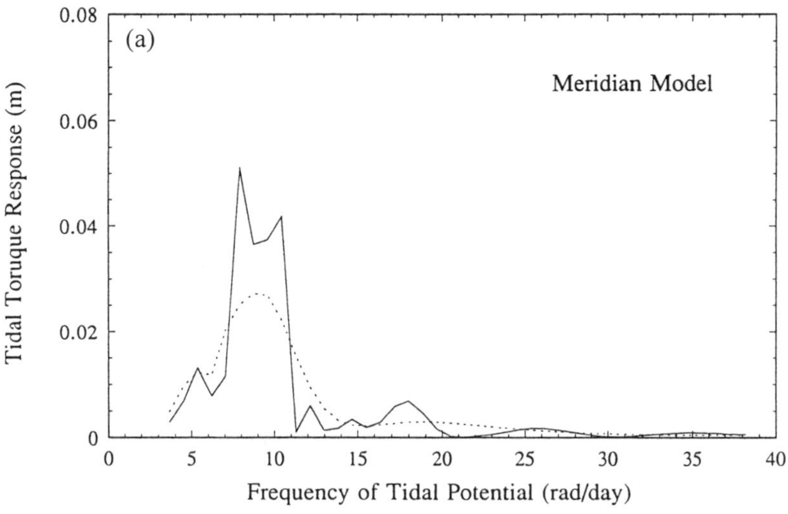

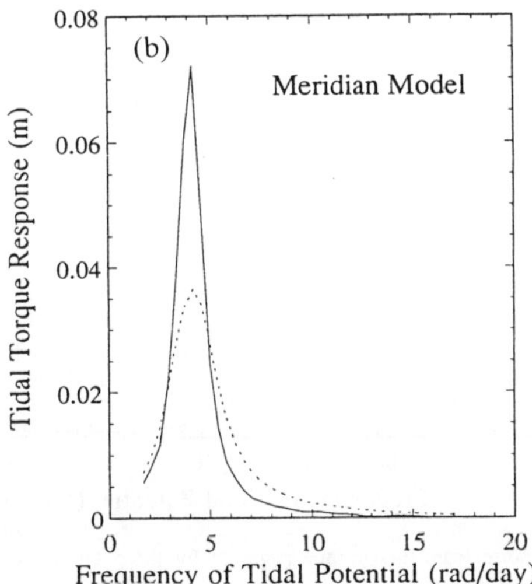

Figure 12. The variation of normalized tidal torque response for the Equator model. X-axis is the angular velocity of the tidal potential. Y-axis is (a) $D_{2200,0,1}\sin\varepsilon_{2200}$ of the M_2 semidiurnal ocean tide, (b) $D_{2100,0,1}\sin\varepsilon_{2100}$ of the O_1 diurnal ocean tide. The angular velocity in present-day is shown by an arrow. The dotted line is the result using Type 1. The solid line is the result using Type 2.

Next we used the hypothetical configuration model shown in Fig. 5 ~ Fig. 7. One is the Meridian model (Fig. 5) in which continent of 108 degree width in longitude lies along the meridian with continental shelves of 12 degree width at both sides. Fig. 6 is the Equator model in which continent of 32 ~ 36 degree width in latitude lies along the equator with continental shelves of 4 ~ 8 degree width at both sides. The reason why the coastline of the Equator model is not straight is to coincide the area of continental shelf and deep sea with the area of the present sea. Fig. 7 is the Polar model in which continents gather around the north pole. Although in the Polar model we don't set continental shelf around the continent, the volume and depth of the deep ocean equal to that of the Present model.

Fig. 11(a) and Fig. 11(b) are the results for the ocean tide with various frequencies of the tidal potential using Meridian model. Fig. 11(a) is the calculated result for M_2 ocean tide, that is $(l, m, p, q)=(2, 2, 0, 0)$, with $K_{2200}=0.1m$. Fig. 11(b) is calculated for O_1 ocean tide, that is $(l, m, p, q)=2, 1, 0, 0)$, with $K_{2100}=0.1m$. Two curves in these figures are the response curves using the set of Type 1 and Type 2 in Table 2 for the coefficients of the dissipation. As shown the figures the peak positions are same on both sets of the coefficients. For Type 1 the peaks are reduced in amplitude and become broader than that for Type 2, because the effect of the friction is larger. The peak position of the resonance curve for Meridian model is almost same as that for 4200m model of All-ocean model using the set of Type 1.

As the case of the Meridian model, Fig. 12(a) and Fig. 12(b) are the results of the ocean tide using the Equator model. In this case also the peak positions are equal for Type 1 and Type 2. These positions are at faster frequency of the tidal potential than those for 4200m All-ocean model and Meridian model.

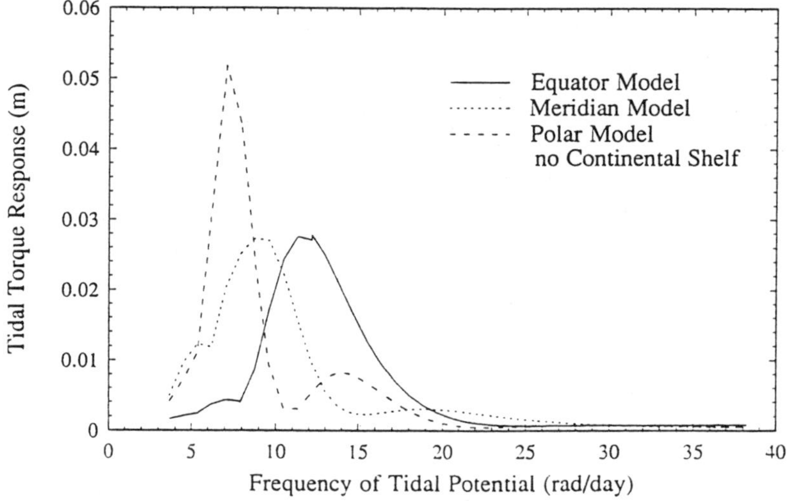

Figure 13. The variation of normalized tidal torque response for three simple configuration model. X-axis is the angular velocity of the tidal potential. Y-axis is $D_{2200,0,1}\sin \varepsilon_{2200}$ of the M_2 semidiurnal ocean tide using Type 1. The solid line is the result for the Equator model. The dotted line is for the Meridian model. The broken line is for the Polar model.

In the case using the Polar model, we calculate only the semidiurnal tide using Type 1. In Fig. 13 the result is shown with the results for the Meridian model and the Equator model. The peak position of the resonance curve for the Polar model is at slower frequency of the tidal potential than that for the Meridian model.

The reason why the peak positions are difference for these models seems to have a relation to the ocean width of North-South direction. It should be noticed that the peak position is at faster as the ocean width in the direction of north-south is narrower. It is difficult to discuss the resonance frequency for these configuration models more concretely, so we need further discussion.

Real configuration models

In this section, we show the results of the ocean tide using the present and the past real ocean-continent configuration models. Used configuration models are shown in Fig. 2 ~ Fig. 4. Fig. 3 follows the model given by Suendermann and Brosche [19], which is designed from the map by Dietz and Holden [20]. This configuration is named " Permian model " . Fig. 4 is based on the map by Scotese *et al.* [21]. This configuration is named " Cambrian model " . In each model the volume of water is kept constant and equal to that of the Present model.

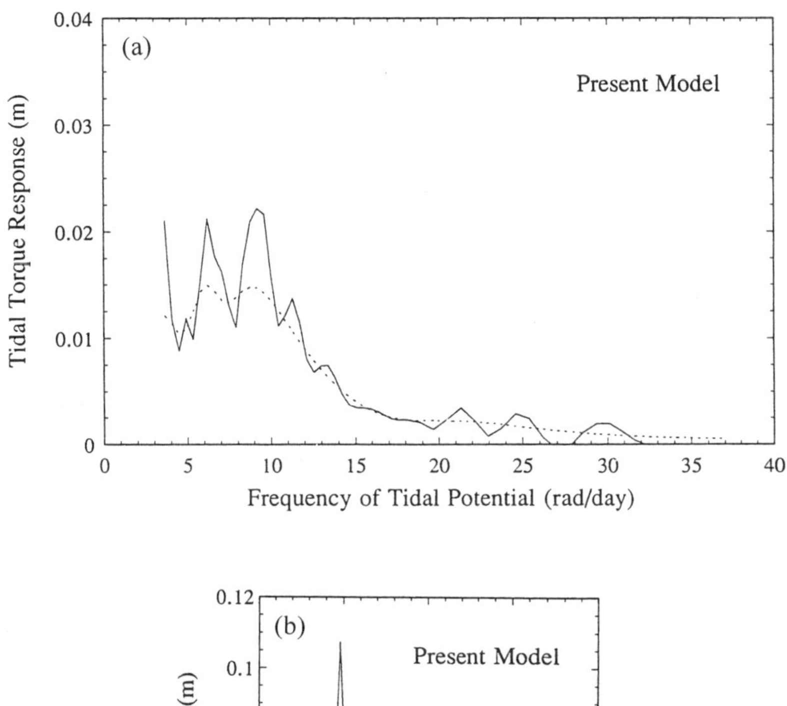

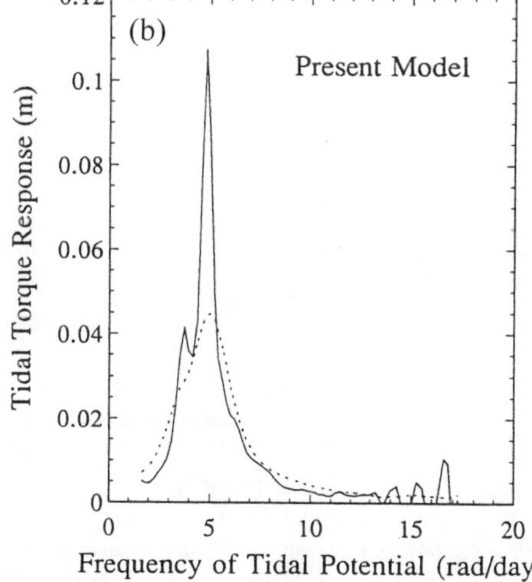

Figure 14. The variation of normalized tidal torque response for the Present model. X-axis is the angular velocity of the tidal potential. Y-axis is (a) $D_{2200.0.1}\sin\varepsilon_{2200}$ of the M_2 semidiurnal ocean tide, (b) $D_{2100.0.1}\sin\varepsilon_{2100}$ of the O_1 diurnal ocean tide. The angular velocity in present-day is shown by an arrow. The dotted line is the result using Type 1. The solid line is the result using Type 2.

It should be noticed that the continent on Permian almost lies along the meridian as the Meridian model. On the other hand the continents on Cambrian lie along the equator as the Equator model. Fig. 14(a) and Fig. 14(b) are the obtained response curves of M2 semidiurnal and O_1 diurnal tide for the Present model with K_{lmpq}=0.1m in our simulation studies. Fig. 15(a) and Fig. 15(b) are the obtained response curves for the Permian model. Fig. 16(a) and Fig. 16(b) are for the Cambrian model as well.

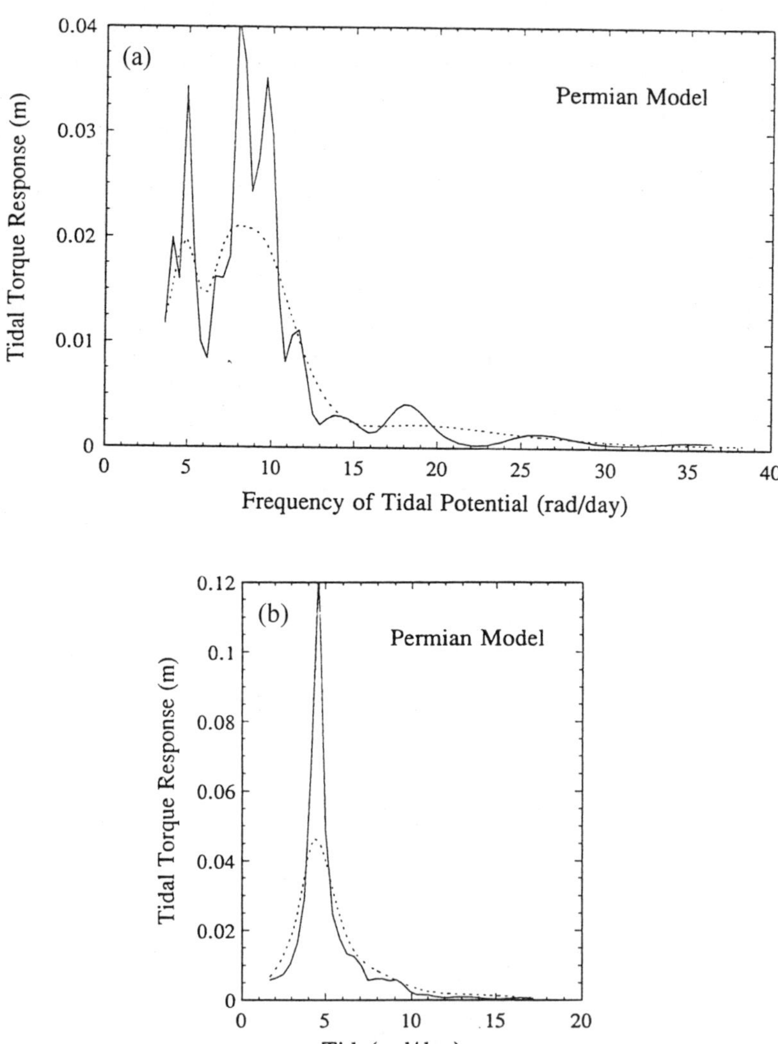

Figure 15. The variation of normalized tidal torque response for the Permian model. X-axis is the angular velocity of the tidal potential. Y-axis is (a) $D_{2200,0,1}\sin\varepsilon_{2200}$ of the M_2 semidiurnal ocean tide, (b) $D_{2100,0,1}\sin\varepsilon_{2100}$ of the O_1 diurnal ocean tide. The angular velocity in present-day is shown by an arrow. The dotted line is the result using Type 1. The solid line is the result using Type 2.

It should be emphasized here that the peak position of resonance curve for the Permian model is similar to that for the Meridian model and that that for the Cambrian model is similar to that of the Equator model. It is found that if the global pattern of the ocean-continent configuration is the same, though the detail pattern of the sea shore lines is difference, the variation of the tidal torque has the same frequency response. As sea shore lines are more complicated, the peak height of the resonance curve is slightly reduced and the peak width becomes broader.

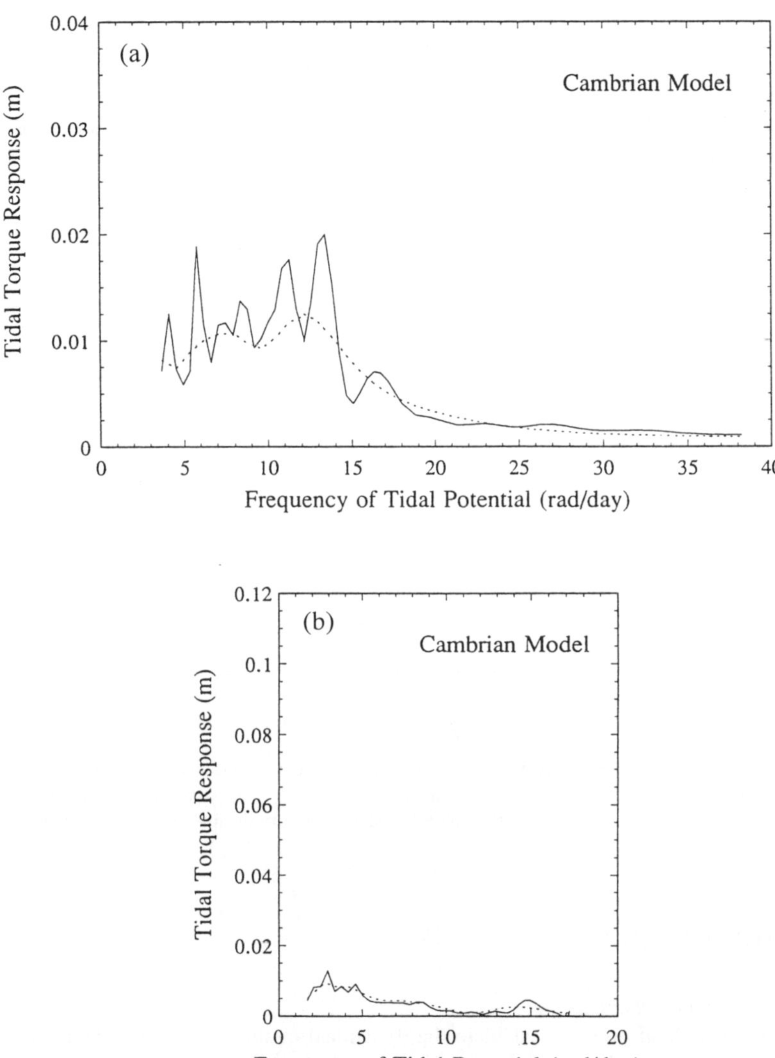

Figure 16. The variation of normalized tidal torque response for the Cambrian model X-axis is the angular velocity of the tidal potential. Y-axis is (a) $D_{2200,0,1}\sin\varepsilon_{2200}$ of the M_2 semidiurnal ocean tide, (b) $D_{2100,0,1}\sin\varepsilon_{2100}$ of the O_1 diurnal ocean tide. The angular velocity in present-day is shown by an arrow. The dotted line is the result using Type 1. The solid line is the result using Type 2.

The peak position for the Equator or the Cambrian model is at the larger angular velocity than the present angular velocity of M_2 tide that is pointed in the figures by arrows. On the contrary, the peak position for the Polar, the Meridian, of the Permian model is at the smaller angular velocity than the present angular velocity of M_2 tide. Because the angular velocity of the tidal potential becomes larger as tracing back toward the past, the ocean having the similar configuration to the Equator model or the Cambrian model should experience the resonance state in the past. It should be noticed that the ocean having the similar configuration to these models will experience the resonance state in the future.

We calculated the ocean tide with the Present model removing continental shelf. Fig. 17 shows the resonance curve of the semidiurnal tide with friction constants of Type 1. In this figure it is noticed that the resonance curves of the tidal torque are almost the same whether the configuration model has the continental shelf or not. Although the continental shelf affects the tidal height pattern around the sea shore line, it does not have an affect on the tidal torque response. The tidal torque response seems to be determined by the global pattern of the ocean-continent configuration.

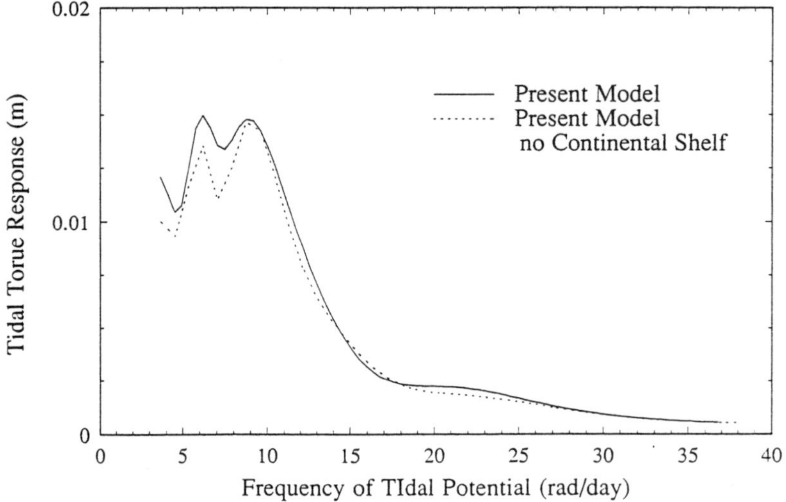

Figure 17. The comparison of normalized tidal torque response for between Present model and no continental self model. X-axis is the angular velocity of the tidal potential. Y-axis is $D_{2200,0,1}\sin\varepsilon_{2200}$ of the M_2 semidiurnal ocean tide using Type 1. The solid line is the result for the Present model having continental shelf. The dotted line is for the Present model without continental shelf.

VARIATION OF L.O.D

calculating the dynamical evolution

As mentioned in p. 2, in order to calculate the dynamical evolution, it is necessary to know the time variation of the (normalized) tidal torque response. From the discussion in the very above part, if the corresponding angular velocity of the tidal potential is given, we can obtain the tidal torque response at three ages (Present, Meridian, and Cambrian) directly. In order to know the tidal torque response at other age, we need to calculate the ocean tide for the configuration at that age. Because of the difficulty of calculating the ocean tide for all configurations in the past, we interpolate the results of the ocean tide for three models in order to obtain the tidal torque response at the optional age.

The procedure of the interpolation is mentioned as follows. If the age when we want to know the tidal torque response is $T \times 10^8$ year before present, we assume that the tidal torque response at that age is

$$\begin{pmatrix}\text{Tidal torque response} \\ \text{at } T \times 10^8 \text{ year b.p.}\end{pmatrix} = \frac{(2.5 - T)A + TB}{2.5} \qquad 0 \leq T \leq 2.5 \tag{41}$$

$$= \frac{(5 - T)B + (T - 2.5)C}{5 - 2.5} \qquad 2.5 \leq T \leq 5 \tag{42}$$

$$= \frac{(7.5 - T)C + (T - 5)B}{7.5 - 5} \qquad 5 \leq T \leq 7.5 \tag{43}$$

where *A, B, C* are the value of the tidal torque response for three real configuration models, that is Present model, Permian model, Cambrian model respectively, at the corresponding angular velocity of the tidal potential at that time, that is $T \times 10^8$ year before present. The reason, why the tidal torque response is set same in equation (43) as that of Permian model at 7.5×10^8 year before present, is baseless so much. According to the papers by Hartnady [22] and Hoffman [23] there was supercontinent lying along the meridian at 7 ~ 7.5×10^8 years age. So we may consider that the configuration at 7.5×10^8 year *b.p.* is similar to the Permian model. This is discussed later.

our result

In this study we carry out the calculation of the dynamical evolution from present of past. In Fig. 18 we show the result of the variation of the Earth's l.o.d. by 600 million years before present. The solid curve is our result. Many symbols shown in this figure mean the data inferred from palaeontological studies summarized by Williams [24]. This figure shows that the l.o.d. variation calculated here fits the palaeontological data very well. First it is noticed that the whole changing rate of the l.o.d. (3 hour/5×10^8 year) in our result is same as the palaeontological data. Secondly a rapid change of the l.o.d. at 400 million years before present and a moderate change at 150 million years before present are consistent with the palaeontological data. The change of the slope of the curve is also due to the change of ocean-continent configuration.

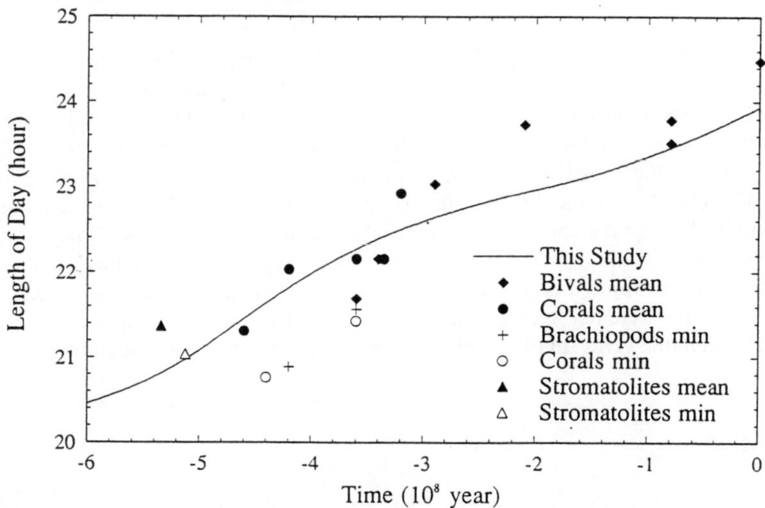

Figure 18. The calculated result of the variation of l.o.d. from present to 600 Ma *b.p.* Observational data of l.o.d. from palaeontological studies summarized by Williams [24] are also shown by various symbols.

In this calculation, we can obtain the tidal torque response at three ages, that is Present, Permian, Cambrian. We find that the amount of the tidal torque at the Permian is about half of that at the present. The amount at the Cambrian is found to be five times as large as that at the Permian. This result agrees with the result of Krohn *et al*. Our result seems to be a bit larger than Krohn's one.

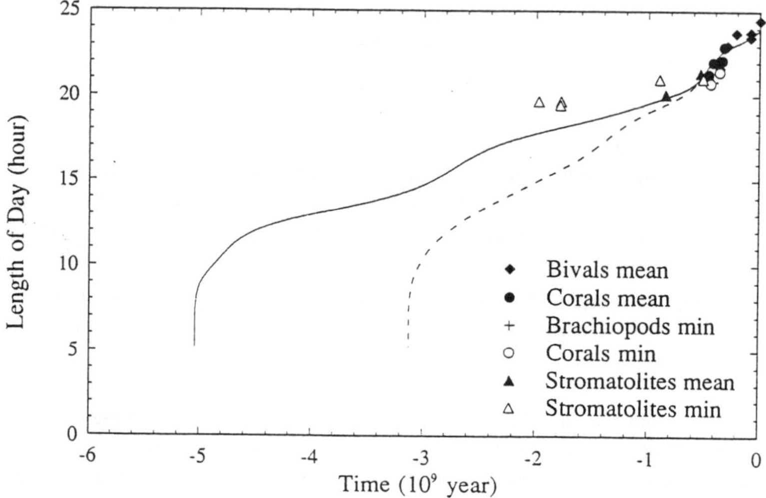

Figure 19. The calculated result of the variation of l.o.d. The solid line is the result of Type A. The broken line is the result of Type B. Observational data of l.o.d. from palaeontological studies summarized by Williams [24] are also shown by various symbols.

In this study we carry out the calculation backward moreover. Because of no confidence about the ocean-continent configuration, we consider the two extremity cases. One is that the continent is same as the Permian model all the time before 750 million years ago. Another one is that the continent is same as the Cambrian model all the time before 750 million years ago. We name the former Type A, and the latter Type B. Type A is the example of that the tidal torque was comparatively small in the past, Type B is the example of the comparatively large tidal torque. How to calculate the tidal torque response from present to more than 750 million years before present is similar to the equation (41) ~ (43) in the Section 4.1. However on the case of Type B we exchange B for C in the equation (43). The calculated result is shown in Fig. 19. The solid line is the result of Type A and the broken line is the result of Type B. Many symbols are the palaeontological data similar to Fig. 18. In Fig. 19 it should be noticed that Type A is closer the stromatolites data at 2 billion years before present than Type B. This suggests that it is highly probable that the tidal torque was comparatively small at and before 750 million years ago. In other words, the continents were not lined on the equator before Cambrian age but on the meridian or of other configuration that gives smaller tidal torque than the Cambrian configuration.

DISCUSSION

As mentioned in p. 24, we can obtain the variation of the l.o.d. in the past. Our calculated results of variation of the l.o.d. fit well to the palaeontological data up to about 500 million years before present. As shown in Fig. 19, however, because of the uncertainty of the ocean-continent configuration before Cambrian age, the time when the Earth-Moon distance was very short is uncertain by about two billion years. In order to calculate the dynamical evolution more accurately, it is necessary to know how the ocean-continent configuration has varied. It is not necessary, however, to know the configuration in a small time scale as Krohn *et al.* showed, but it is surely enough to consider the configuration and the location only when the continents formed a supercontinent or they were dispersing and changing their positions. It is highly possible that the gathering and breaking up of continents have occurred in the Wilson cycle. We consider that the suitable time interval that we need to construct the ocean-continent configuration model is about half of the Wilson cycle. If the Wilson cycle is a few hundred million years, the time interval that is inferred to be suitable to construct the ocean-continent configuration model is 2.5 hundred million years. By means of interpolation we can obtain the tidal torque at other ages.

Recently Maruyama [25] raised a concept of "Plume tectonics". In his book he argues that there was a supercontinent five times since 1.9 billion years *b.p.* to present. He also suggests that in the Archean there were numerous island arcs on the "micro-plates", which grew up into small continents between 2.5 billion years *b.p.* and 1.9 billion years *b.p.* According to Scotese [26] there were three supercontinents in ~ 1000 Ma, ~ 600 Ma, and ~ 300 Ma *b.p.* Moreover, he suggests that fragments of continents crossed South pole approximately 750 Ma *b.p.* In our study, we did not construct the ocean-continent configuration model after Maruyama [25] and Scotese [26]. If we adopt their configuration histories, we can calculate the plausible dynamical evolution.

In the glacial period, the sea water volume decreased and seashore line retreated. In our calculation of the dynamical evolution we did not consider the effect of the glacial period. However, from our results, we can estimate this effect. Because of the retreat of the seashore line, the continental shelf disappears. As the existence of the continental shelf does not affect the global tidal torque so much as shown in Fig. 17, the change of the sea water volume would have a dominant effect on the tidal torque as shown in the Fig. 10. As mentioned on p. 17, if the sea water volume decreases, the tidal torque seems to decrease in proportion to the sea water volume. If the sea level falls down by 200 m, that is the sea water volume decreases by a few percentages, the tidal torque is expected to decrease by a few percentages. Therefore the effect of the glacial period would be to decrease the rate of the dynamical evolution.

When the Earth-Moon distance was short in the past, the angular velocity of the tidal potential was very large. As the angular velocity of the tidal potential becomes larger, the normalized tidal torque response of the ocean tide becomes smaller as shown previously. If the angular velocity becomes more than 20 rad/day, the differences due to the ocean-configuration become small rapidly as shown in Fig. 13. Our calculation of the dynamical evolution shows that the date when the angular velocity becomes 20 rad/day is about 3.5 billion years before present. Therefore, in order to calculate the dynamical evolution up to the time when the Earth-Moon system had been initiated, we consider that it is enough to obtain the information of the ocean-continent configuration every half of a Wilson cycle up to at least 3.5 billion years before present.

Using the method discussed on p. 3, we can obtain not only l.o.d. but also the Earth-Moon distance or the lunar orbital elements and so on. In Fig. 20 we show the calculated results of the variation of Earth-Moon distance in two cases (Type A2, Type A2). Both cases assume that continental configurations are same as Type A from 750 Ma *b.p.* to the present and same as All-Ocean model before 3 billion years ago. The difference of two cases is the continental configuration at 2.5 billion years *b.p.* In Type Al (solid line) it is assumed to be similar to the Permian model. In Type A2 (broken line) it is similar to the Cambrian model. As shown Fig. 20, if the continental configuration is assumed, we can calculate the evolution of Earth-Moon distance. In these cases, due to uncertainty of the continental configurations, the time when the Earth-Moon distance was very short is still uncertain by about one billion years. We want to show the other data, for example the inclination of the lunar orbit and obliquity of the Earth, if we calculate the dynamical evolution more accurately in the near future.

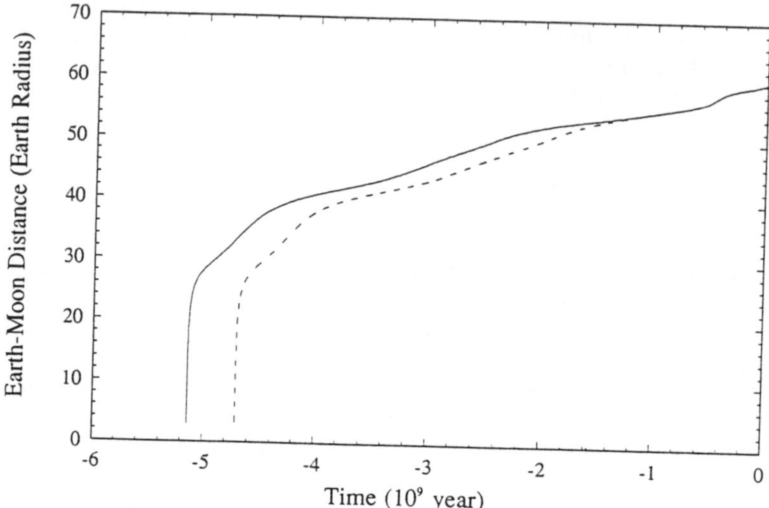

Figure 20. The calculated result of the variation of Earth-Moon distance. The solid line is the result of Type A1. The broken line is the result of Type 2.

Acknowledgements

This research is supported by Fellowships of the Japan Society for the Promotion of Science for Japanese Junior Scientists and a Grant in Aid for Scientific Research (No. 03640253) of the Ministry of Education Science and Culture of Japan. The authors appreciate Dr. Y. Tamura and Mr. K. Matsumoto for helpful discussion.

REFERENCES

1. W. M. Kaula. Tidal Dissipation by Solid Friction and the Resulting Orbital Evolution, *Rev. Geophys.* **2**, 661-685 (1964).
2. G. J. F. MacDonald. Tidal Friction, *Rev. Geophys.* **2**, 467-541 (1964).
3. P. Goldreich. History of the Lunar Orbit, *Rev. Geophys. Space Phys.* **4**, 411-439 (1966).
4. H. Gerstenkorn. On the Controversy over the Effect of Tidal Friction upon the History of the Earth-Moon System, *Icarus* **7**, 160-167 (1967).
5. F. R. Stephenson and L. V. Morrison. Long-term changes in the rotation of the Earth: 700 B.C. to A. D. 1980, *Phil. Trans. R. Soc. London Ser. A* **313**, 47-70 (1984).
6. D. J. Webb. Tides and the evolution of the Earth-Moon system. Geophys, *J. R. Astr. Soc.* **70**, 261-271 (1982).
7. K. Lambeck. Tidal dissipation in the oceans: astronomical, geophysical and oceanographic consequences, *Philos. Trans. R. Soc. London Ser. A* **287**, 554-594 (1977).
8. M. Ooe, H. Sasaki and H. Kinoshita. Effect of the tidal dissipation on the Moon's orbit and the

Earth's rotation, *AGU Monograph* **59**, 51-57 (1991).
9. J. Krohn and J. Suendermann. Palaeotides Before the Permian. In: *Tidal Friction and Earth's Rotation* **2**. P. Brosche and J. Suendermann (Eds). pp. 190-209 (1982).
10. W. M. Kaula. *Theory of Satellite of Geodesy.* Blaisdell Publ. Company (1966).
11. M. Ooe. Tidal Deformation of the Earth's and Evolution of the Moon's Orbit, *Proc. of NAO Symp. on the Earth and Planetary Interiors*, 107-115 (1989) (in Japanese).
12. P. Goldreich and S. Soter. Q in the Solar System, *Icarus* **5**, 375-389 (1966).
13. E. W. Schwiderski. Ocean Tides, Part I: Global Ocean Tidal Equations, *Marine Geodesy* **3**, 161-217 (1980).
14. W. M. Smart. *Celestial Mechanics.* Longmans (1953).
15. K. Lambeck. *Geophysical Geodesy.* Oxford Sci. Publ. (1988).
16. D. E. Cartwright and R. D. Ray. Energetics of Global Ocean Tides From Geosat Altimetry, *J. Geophys. Res.* **96**, 16897-16912 (1991).
17. K. Matsumoto, M. Ooe, T. Sato and J. Segawa. Ocean tide model obtained from TOPEX/POSEIDON altimetry data, *J. Geophys. Res.* (1995) (in press).
18. M. S. Longuet-Higgins. The eigenfunctions of Laplace's tidal equations over a sphere, *Phil. Trans. Roy. Soc. London Ser. A* **262**, 511-607 (1968).
19. J. Suendermann and P. Brosche. Numerical Computation of Tidal Friction for Present and Ancient Oceans. In: *Tidal Friction and the Earth's Rotation.* P. Brosche and J. S ü ndermann (Eds). pp. 125-144. Springer-Verlag (1978).
20. R. S. Dietz and J. C. Holden. Reconstruction of Pangea: Breakup and dispersion of continents, Permian to present, *J. Geophys. Res.* **75**, 4939-4956 (1970).
21. C. R. Scotese, R. K. Bambach, C. Barton, R. Van Der Voo and A. M. Ziegler. Palaeozoic base maps, *J. Geol.* **87**, 217-277 (1979).
22. C. J. H. Hartnady. About turn for supercontinents, *Nature* **352**, 476-478 (1991).
23. P. F. Hoffman. Did the Breakout of Laurentia Turn Gondwanaland Inside-Out? *Science* **252**, 1409-1412 (1991).
24. G. E. Williams. Tidal rhythmites: geochronometers for the ancient Earth-Moon system, *Episodes* **12**, 162-171 (1989).
25. S. Maruyama. *What has the Earth done in 4.5 billion years*? Iwanami (1993) (in Japanese).
26. C. R. Scotese. Late Precambrian and Palaeozoic Palaeogeography. *1993 SEPM meeting Abstracts with Program* 44-45 (1993).

Proc. 30th Intern. Geol. Congr., Vol. 26, pp. 31-54
Wang *et al.* (Eds)

Multiple Impacts at the KT Boundary and the Death of the Dinosaurs

SANKAR CHATTERJEE
Museum of Texas Tech University, Lubbock, Texas 79409-3191, U.S.A.

Abstract

The Cretaceous/Tertiary (KT) mass extinction has been correlated with both asteroid impact and Deccan volcanism. The physical evidence for large asteroid impact(s) at the KT boundary is overwhelming and the Chicxulub structure in Yucatán Peninsula, Mexico, is now considered as the most likely site of the crater. A second KT impact scaróthe Shiva Crateró has been identified recently from subsurface data at the India-Seychelles plate assembly. Its size and shape are revealed by geophysical and structural anomalies, as well as by oil wells drilled inside the structure. This buried oblong crater is 600 km long, 450 km wide, and 12 km deep and may represent the largest impact structure of Phanerozoic age. The KT boundary age of the crater is inferred from its Deccan lava floor, Palaeocene age of the overlying sediments, isotope dating (~65 Ma) of presumed melt rocks, and the Carlsberg rifting event (chron 29R) that split the crater into two-halves. The crater shows the morphology of a complex impact structure and basin, with a distinct central uplift in the form of a series of peaks, an annular trough, and a slumped rim. The oblong shape of the crater and the asymmetric distribution of fluid ejecta indicate oblique impact in a SW-NE trajectory. It is estimated that a 40-km diameter meteorite crashed on the western continental shelf of India around 65 Ma, excavating the Shiva Crater, shattering the lithosphere, and triggering the India-Seychelles rifting. The crater becomes narrow in the form of a teardrop to the NE or downrange where the fluid ejecta was emplaced radially outside the crater rim by the impact shock. A third impact site may lie on the Pacific plate as revealed from the discovery of a tiny fragment of the KT bolide and cosmic spinel from DSDP site 576. The evidence for multiple impacts is also supported by different nature and sorting of ejecta components, with a Shiva impact giving iridium anomaly and glassy spherule of the lower member of the KT doublet, a Chicxulub impact giving shocked quartz of the upper KT member, and a Pacific impact giving nickel spinel fall out and the bolide chip. Although both impacts and Deccan volcanism may have contributed to the KT biotic crisis, impacts appear to be the main extinction cause.

Keywords: dinosaurs, KT extinctions, multiple impacts, Chicxulub Crater, Shiva Crater, India-Seychelles plate, Pacific plate, Deccan volcanism

INTRODUCTION

The Cretaceous/Tertiary (KT) mass extinction brought a close to the Mesozoic and led to the transition from the age of the dinosaurs to the age of mammals. The sudden extinction of dinosaurs at the KT boundary has puzzled both scientists and public for more than a century. Along with dinosaurs, pterosaurs, plesiosaurs, mosasaurs as well as several families of birds and marsupial mammals, and hundreds of other plants were also suddenly wiped out at this time. Two-thirds of all marine animal species including calcareous planktons, ammonites and rudists also vanished, collapsing the oceanic food chain. The biosphere of the Earth was devastated.

What triggered this Cretaceous crisis? The cause remains controversial, but recently two competing views have been proposed to explain this KT extinction: meteorite impact hypothesis and volcanic hypothesis. The impact view [1-8] postulates that the environments were lethally altered or destroyed by the collision of a large asteroid leading to biotic crisis. The volcanic view [9, 10, 11, 12] argues that the pollution in the atmosphere and oceans by the massive outpourings of Deccan flood basalt in India had devastating effect on ecology.

THE IMPACT MODEL

In 1980, the Alvarez group [2] proposed the asteroid impact theory to explain the sudden demise of dinosaurs at the KT boundary. They discovered an abnormally high concentration of iridium (about 30 times more than the surrounding rocks) at the KT boundary level of Gubbio, Italy. Soon a comparable iridium anomaly was found globally at different KT boundary sections [13]. Since iridium is a very rare element in the earth's crust, but fairly abundant in chondritic meteorites, the Alvarez team proposed that the iridium spike at the KT boundary is cosmic in origin, settled out of a global dust cloud triggered by the impact of an asteroid 10 km in diameter. They proposed that this giant asteroid crashed into the Earth with a velocity of 90,000 km/hour and produced a crater with a diameter of 200 km. This impact lofted so much debris from the target rock into Earth's atmosphere as to decrease solar insolation and create a "nuclear winter" that caused much of the life on Earth to perish [1]. A blackout of sun would kill both terrestrial and marine plants and destory the food chain. Other killing mechanisms of impact include atmospheric pollution, partial destruction of ozone layer, acid rain and global fire [14, 15, 16, 17]. The massive impact, proposed by the Alvarez group, would generate a 100-million megaton blast of energyó1000 times more powerful than the explosion of all nuclear arsenal of the world [1, 18]. The direct impact, and the subsequent environmental damage, would have a devastating effect on terrestrial lifeóespecially the larger land animals like dinosaurs.

The impact theory was bolstered by additional evidence such as shocked quartz [19, 20], stishovite [21], micro-diamonds [22], impact glasses [23], osmimium isotope ratios [24], Ni-rich spinels [25], rhodium [26], carbon soots [17], tsunami deposits [27], and extraterrestrial amino acids [28] in the KT boundary layer at different sites. Among all this cumulative evidence, the shocked quartz is a distinctive signature of impact event as it can form at a force more than 10 gigapascal (GPa) that travels through quartz-bearing grains of the target rock to porduce microscopic shock lamellae [29].

The most conclusive evidence for the impact event left by the KT bolide comes from the discovery of the Chicxulub Crater at the Yucatán Peninsula of Mexico. A second point of collision appears to be the Shiva Crater at the India-Seychelles rift margin. Both craters preserve hallmarks of a complex crater: bowl-shaped basin, raised rim, central peak, and annular trough. With the discovery of these two craters, the physical evidence for bolide impact at the KT boundary is overwhelming.

KT IMPACT CRATERS

The Chicxulub Crater

The most promising candidate for the KT impact event is the Chicxulub Crater on the northern margin of the Yucatán Peninsula, Mexico [30, 31, 32]. It is a circular structure about 180 km in diameter, buried under 1,100 m of carbonate strata, extending out under the Gulf of Mexico, and defined by magnetic and gravity anomalies. Presence of shocked quartz, impact melt, brecciation and iridium anomaly within the crater itself is compatible with an impact origin for the Chicxulub structure [33, 34]. If Chicxulub is indeed a very large KT impact scar, it should be surrounded by extensive impact ejecta and tsunami deposits. This is indeed the case. Distribution of proximal ejecta components and tsunami deposits at the KT sections in Haiti, Mexico, Texas, Alabama, the Caribbean and adjacent areas strengthen the point of collision at Chicxulub [27, 35, 36, 37, 38].

The Shiva Crater

Recently, a second KT impact structure, the Shiva Crater, has been identified at the India-Seychelles plate margin, almost antipodal to the Chicxulub structure [39]. Today, the Seychelles microcontinent is separated from the western coast of India by 2,800 km because of continental rifting along the Carlsberg Ridge (Fig. 1), but at the KT boundary time, they were joined together. Part of the crater was first identified by Hartnady [40] who suggested that the Amirante Basin, south of the Seychelles Island, may be a possible KT impact site. The basin has a subcircular shape of about 300 km in diameter, bounded on the northeast by the Seychelles Bank and partially ringed on the southwest by the structure of the Amirante Arc. Sediments from the adjacent Amirante Passage have yielded Late Maastrichtian foraminifera *Abathomphalus mayorensis* [41], while basalt samples dredged from the Amirante Arc look like Deccan Trap and have a similar radiometric age [42]. Both palaeontologic and radiometric age indicate that the arc was formed near the KT boundary. However, its arcuate structure is enigmatic. It does not appear to be a recent or ancient trench, as it lacks volcanic activity, seismicity, and any significant accretionary sedimentary prism on its "landward" side [41]. Thus the interpretation of the Amirante Arc as a crater rim is a distinct possibility. The proposed impact also may explain the puzzling jump of the Carlsberg Ridge at the KT boundary during the rifting of India and the Seychelles. Hartnady [40] noticed that the Carlsberg Ridge between the Seychelles and Madagascar jumped more than 500 km to the northeast to lie between India and the Seychelles and initiate rifting between these two landmasses. He could not find any evidence for plate reorganization in the Atlantic or Pacific Oceans during this time. He attributed this major plate tectonic adjustment to the enormous force of a large meteorite. As additional evidence, he pointed to massive tsunami deposits in the KT boundary section of Somalia and Kenya, which may be linked to this impact event.

Alt *et al.* [43] concurred with Hartanady [40] that the western rim of the crater survives in the Amirante Arc, but the eastern rim lies along the west coast of India, hidden by the overlying Deccan Traps. They speculated that the impact was forceful enough to create not only the enormous crater approximately 600 km in diameter, but also to cause pressure-release melting in the asthenosphere. Basalt then filled the crater basin to form an immense lava lake, the terrestrial equivalent of a lunar mare.

Chatterjee [44, 45] identified the eastern rim of the crater along the Panvel Flexure, near the Bombay coast (Fig. 1). The Panvel Flexure is an arcuate segment of the crater about 120 km long on the Deccan Traps, and it is difficult to explain in terms of conventional tectonics. It is marked by a line of hot springs, dikes, deep crustal faults, and seismicity [46, 47]. Since the Indian shield is usually asiesmic, the seismicity along the flexure is unusual, indicating tectonic instability. The geothermal gradient is abnormally high along this flexure (36-78°C km^{-1}) with evidence of thinned lithosphere (31-39 km), suggesting melting conditions at shallow depths [48]. The Panvel Flexure exercises tectonic control on the attitude of the Deccan lavas. To the east of the flexure, the basaltic flows are horizontal; to the west of the flexure, the basaltic flows dip west to west-southwest at 50° to 60° toward the coast. The abrupt change of dip along the flexure axis may indicate the slope of the eastern crater wall, which is now concealed by Deccan lavas. Seismic data indicate that the basement topography below the Deccan lava west of the flexure has a crater-like depression [46]. The reconstruction of the India-Seychelles fit at the KT boundary time reveals the nature of this impact scar (Fig. 2). Completing the oval by combining the Amirante Arc and the Panvel Flexure, the extent of the crater can be extrapolated. It is a giant oval crater, 600 km long and 450 km wide, showing the morphology of a complex impact scar. Chatterjee [45] named this impact structure the Shiva Crater, after the Hindu god of destruction and renewal (Fig. 1).

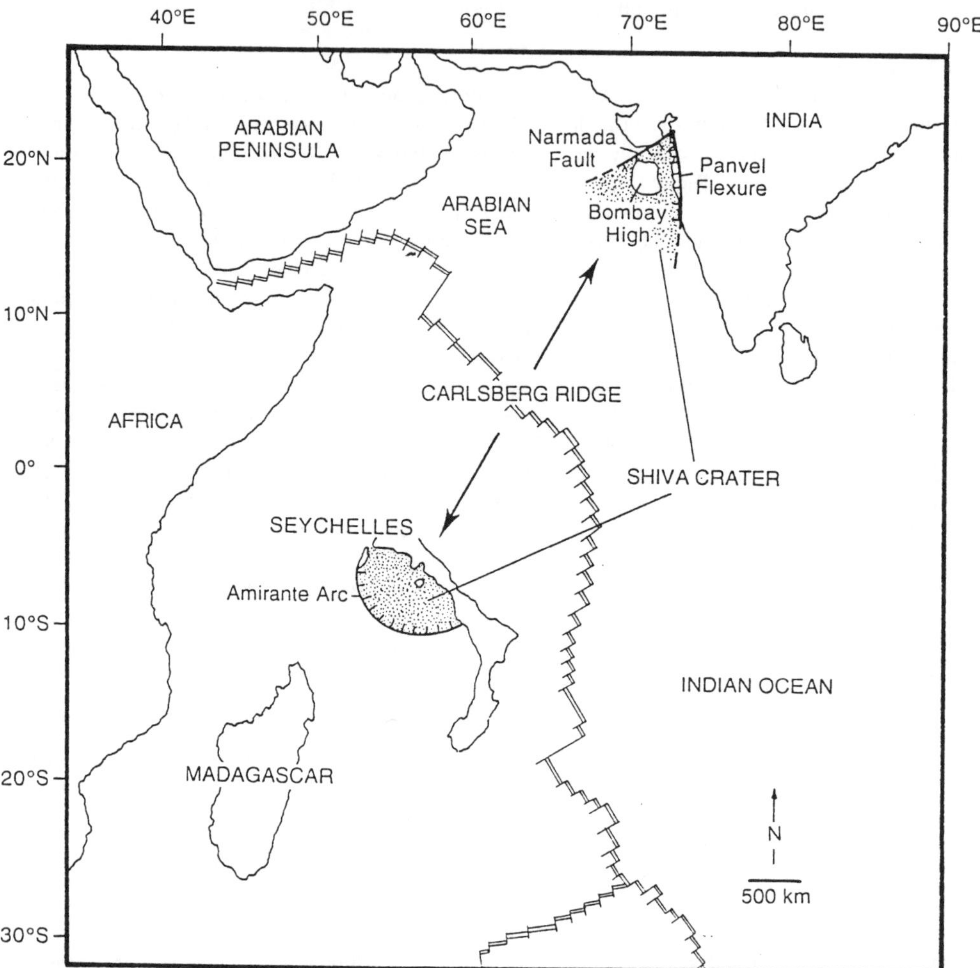

Figure 1. Sketch map showing present day location of the split Shiva Crater in reference to India and Seychelles on either side of the Carlsberg Ridge. Today, part of the Shiva Crater is attached to the southern part of the Seychelles, the other-half to the western part of India. The crater was joined 65 million years ago when the Seychelles was part of India before the spreading of the Carlsberg Ridge.

Since the Shiva Crater was spilt by the Carlsberg Ridge, and each half is now buried under a thick pile of lava flows, and because the structure is largely submarine, geophysical exploration and drilling data are essential to understanding its morphology and structure. Morever, overlying lava flows and thick sediments obstruct a direct examination of various impact signatures such as shock metamorphic effects, breccia and impact melt that are generally associated with complex craters. Today, one part of the crater is attached to the western coast of India, the other to the Seychelles (Fig. 1), but of course both parts were joined at the KT boundary (Fig. 2).

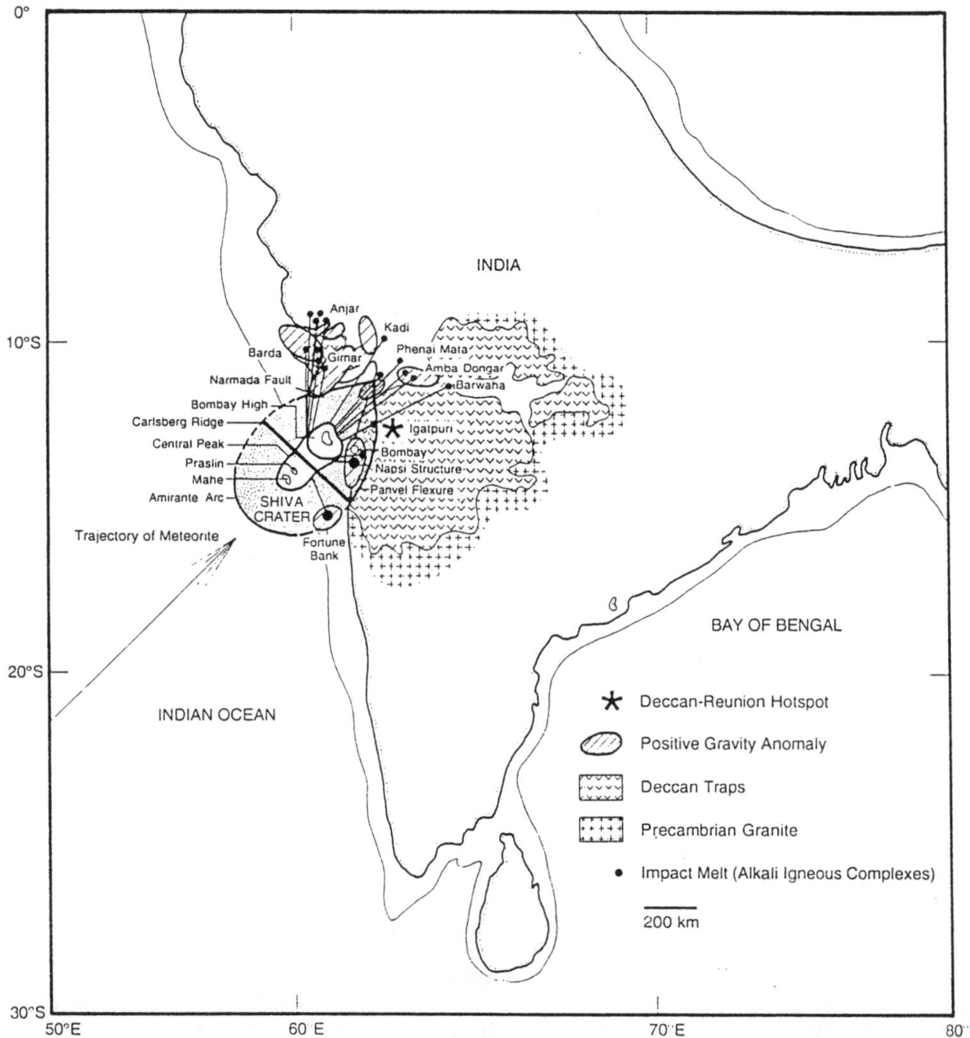

Figure 2. Sketch map showing location of the Shiva Crater at the India-Seychelles plate assembly during the KT boundary; arrow indicates trajectory of meteorite; radial, asymmetric distributions of alkaline igneous complexes (impact melt rocks) downrange of the Shiva Crater are shown by closed circles [64]; areas of positive gravity anomaly [55] coincide with the ejecta melt distribution; asterisk indicates location of the Deccan-Reunion hotspot.

The most critical evidence regarding this impact structure comes from recent oil exploration in the Bombay High offshore basin, which represents the eastern-half of the crater. Bombay High is a giant offshore oilfield (~120,000^2 km) located 160 km west of Bombay in the Arabian Sea at a depth of about 75 m (Fig. 3). The structure is 60 km long and 20 km wide, trending WNW-ESE with a faulted eastern flank. The stratigraphy of this enigmatic structure is known from extensive drilling and seismic data by the Oil and Natural Gas Commission [49, 50, 51]. The Bombay offshore basin shows part of the crater rim (Panvel Flexure and Narmada Fault), annular trough (Surat Basin, Dahanu Depression,and Panna Depression) and the central uplift (Bombay High) (Fig. 3A). The morphology of the western-half of the Shiva Crater around the Seychelles microcontinent is also known from oil exploration data [52, 53, 54]. Here the crater rim is represented by the

Amirante Arc, annular trough by the Amirante basin, and the central peaks by the Mahe and Praslin granitic cores. When the Shiva structure is restored to KT boundary time (Fig. 4A), it shows the structure and morphology of a giant, complex crater, oval in outline with (1) a collapsed outer ring, 600 km long and 450 km wide; (2) an annular trough, presumably filled with KT melt rocks; and (3) a distinct central core in the form of linear uplifted peaks of older Precambrian granite.

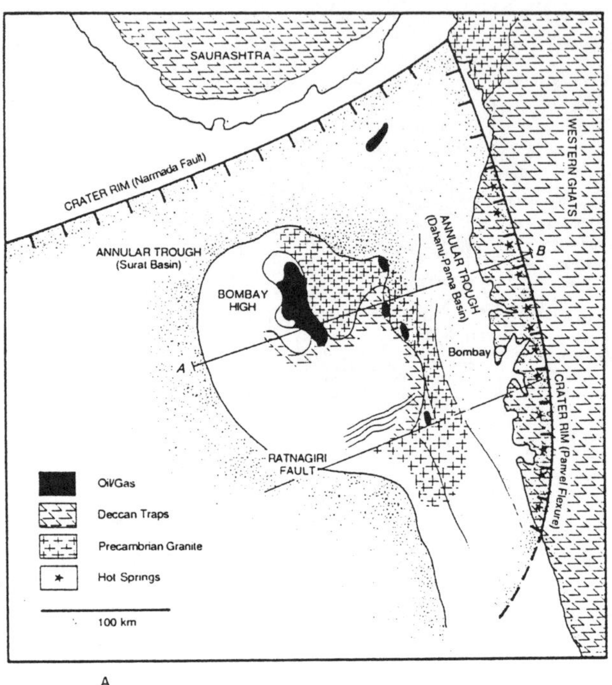

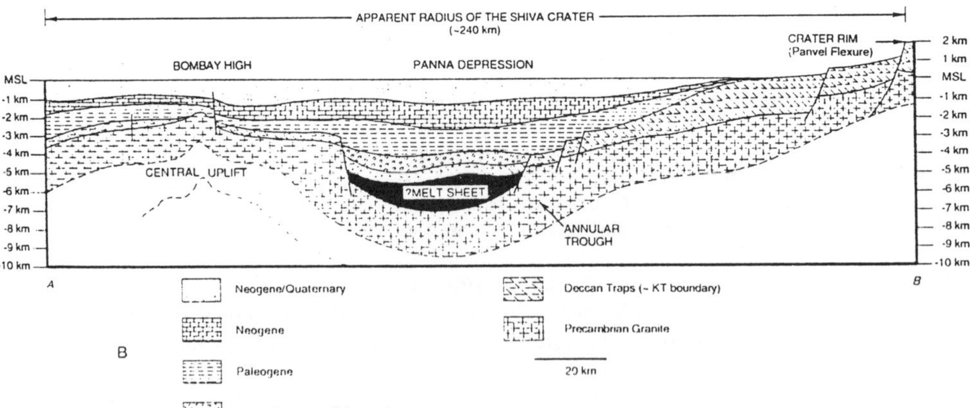

Figure 3. A, Location and tectonic framework of Bombay High in relation to India representing the eastern-half of the Shiva Crater; B, seismic cross-section along the line AB in Fig. A to show the central uplift (Bombay High), annular trough (Panna Depression), and the crater rim (Panvel Flexure) (simplified from [51]).

ANATOMY AND AGE OF THE SHIVA CRATER

The Shiva Crater is defined by a collapsed, faulted outer ring, which is partially preserved in the form of the Amirante Arc, Panvel Flexure, and the Narmada Fault (Figs. 1 and 3). The outer rim is surrounded by a gravity high, especially in both the Panvel Flexure and Narmada Fault areas [55], and may be linked to the distribution impact melt. Part of the outer ring along the northwest and southeast edges was obliterated by spreading of the Carlsberg Ridge. The outer ring is followed by the annular trough which was largely filled with ponded Deccan lava (and possibly by suevite and impact melt). Part of the trough is preserved in the Surat Basin, Dahanu Depression, and Panna Depression around the Bombay High, and the Amirante Basin, southwest of the Seychelles (Fig. 4). The central peaks, all composed of older Precambrian granitic cores, are represented by the Bombay High on the western coast of India, and the Praslin and Mahe Islands of the Seychelles. Jansa [56] pointed out that the central uplift in oceanic impact sites is generally cylindrical in shape, as is the Shiva Crater. The lateral continuity of the central peaks was disrupted by the Carlsberg Rift and may indicate a collapse structure. The central uplift associated with the 600-km Shiva Crater is estimated to be 150 km wide, about one-fourth of the crater's final diameter, as is expected in a complex crater [57]. Figure 4B is a geologic cross-section of the 600 km diameter Shiva structure along the longitudinal axis.

The depth of the crater can be estimated from the thickness of the sedimentary basins around the annular trough as well as from the height of the central peaks. The sedimentary fill around the trough of Bombay High consists of shallow marine Tertiary sediments exceeding 5,000 m, overlying the Deccan basalt floor [4]. Such thick sediments indicate that the crater basin is more than 5 km deep above the Deccan floor. The thickness of the Deccan lava, and the presence of impact melt within the crater are unknown, but Meyerhoff and Kamen-Kaye [54] have described a well log on the Saya de Malha Bank which penetrated 832 m of basalt overlain by 2400 m of upper Palaeocene to Quaternary sediments. Based on their seismic work, they suggested that approximately 2 km of Deccan basalt overlie the granitic basement some 80 km southwest of the Mahe Peak. If to this we now add the 2 km thick lava pile in the Western Ghats section near the Panvel Flexure, the depth from rim to actual floor may exceed 10 km (Fig. 4B).

Although the age of the Shiva Crater is not precisely known, combined evidence from various components of the structure, such as the formation of the rim, biostratigraphy of the annular trough, age of the melt sheets, timing of the central uplift, association of Deccan volcanics, and the Carlberg rifting event suggests that the crater formed at the KT boundary. As discussed earlier, the preserved rim of the Shiva Crater, such as the Amirante Arc [43, 58], the Panvel Flexure [45, 59], and the Narmada Fault [55] evolved at KT boundary time. Seismic stratigraphy has identified the basement rock as reflection-free or chaotic Precambrian granite in the form a central uplift with a thin veneer of Deccan lava at the base [49, 50, 51]. The oldest sediment overlying the Deccan Trap or crystalline basement is the Panna Formation of Palaeocene age (Fig. 3B). The upper boundary of the Panna Formation coincides with H-4 seismic horizon. The Panna Formation is composed of poorly sorted, angular sandstone, claystone and trap fragments at the botton, followed by shale and coal sequence. This unit is relatively thin on the uplift, but attains a large thickness (75 m) on the flank. Seismic data indicate that the formation may be as thick as 500 m in the annular trough region, but wells have not penetrated the bottom layer. Although the formation is mostly unfossiliferous, it has yielded *Globorotalia pseudomenardii* from the middle of the sequence corresponding to P4 planktic foraminiferal zone of Late Palaeocene, indicating a short palaeontologic hiatus after the KT event. Since the impact took place in a shallow-marine setting, the hiatus may be linked to erosion by a megatsunami generated by the impact. The KT boundary section lies farther down at the bottom of the sequence. Jansa [56] suggested that in oceanic impacts most of the fall-out breccia is reworked back into the crater cavity. If so, the lowest unit of the

Panna Formation, if recovered in future, should be investigated for an iridium anomaly and shock metamorphic minerals (Fig. 3B). The crater floor is composed of younger Deccan flows from the adjoining Western Ghat section that took place around ~65 Ma [9, 60]. The Deccan floor at the annular trough and the overlying Panna Formation narrow down its age to close to the KT boundary. Recent isotopic dating of the Shiva melt rock, as discussed in the following section, has yielded an $^{40}Ar/^{39}Ar$ date of 65 Ma [61]. Finally, the Carlsberg rifting within the crater basin was formed during the magnetic chron 29R [62].

Although it is generally believed that the Bombay High was formed during the breakup of India-Seychelles at the KT boundary [49], no tectonic mechanism has been offered to explain this spectacular uplift (~ 10 km high) of the Precambrian granite at the passive margin of the Bombay shelf (Fig. 4B). We propose that this structural high represents one of the central peaks of the Shiva impact that originally underlay the transient crater. The other two central peaks, Praslin and Mahe Islands in the Seychelles (Fig. 4A), also rebounded upward at the same time as they contain similar Precambrian granites and younger KT intrusive rocks [53]. These central peaks are composed of deformed and fractured rocks that have been stratigraphically uplifted distances comparable to the crater depth [57]. The fractured nature of the central peak of the Bombay High can be seen in Figure 3B. The crystalline rocks beneath the Shiva Crater are shattered, as in the Ries Crater of Germany, as inferred from the low seismic velocity beneath the H4 horizon [51]. Similarly, Baker [63] reported a "megablock zone" ñ a chaotic assemblage of gigantic blocks of granite, each up to 13 m high, often marked with surface fluting at the Mahe uplift in the Seychelles. Evidence from drill holes, geochronology, and seismicity suggest that the magmatism at the Seychelles Bank occurred at the KT boundary, producing both Deccan lavas and alkaline igneous complexes [53].

The Shiva impact must have produced enormous volumes of impact melt, breccias and shocked materials between the central structure and the rim. Yet, these impact signatures are difficult to interpret because of mobilization and mixing of thick lava flows from Carlsberg Ridge, subsequent burial by the Deccan lava, as well as lack of drilling and seismic data below the Deccan floor. However, there is some indirect evidence that indicates a lavalike impact melt was emplaced radially within and outside the crater.

One of the most intriguing features following the Deccan flood basalt volcanism is the occurrence of several post-tholeiitic alkali igneous complexes of nepheline-carbonatite affinities, along the radii of the Shiva Crater (Figs. 2, 5). They are manifested in plug-like bodies and minor intrusions in the western and northwestern part of the Deccan volcanic province and are limited in space and volume compared to the vast expanse of tholeiitic lavas [64, 65]. They are clearly defined by zones of gravity highs [55]. Extrusive rocks are relatively rare except in the Kutch rift zone, indicative of fissure eruption. Devey and Stephens [53] described contemporary alkaline intrusives in several islands in the Seychelles microcontinent, especially in Mahe, North Island and Silhouette Island within the Deccan lavas. Geochemically they are very similar to the Murud alkaline dikes of Bombay. Recent $^{40}Ar/^{39}Ar$ dating of these alkaline complexes indicates 65 Ma, precisely coinciding with the KT boundary [53, 61, 66]. The spatial and temporal correlations of these alkaline igneous complexes with the Shiva Crater are intriguing, making a causal relationship likely. As discussed elswhere, the alkaline igneous rocks around the Shiva Crater [39] represent melt ejecta produced by impact-induced mixing and melting of the target rocks, and were emplaced radially on downrange side of the trajectory (Fig. 5A). The close isotopic and age relationship of the impact melt with the younger Deccan volcanics such as the Ambenali Formation [53, 66] indicates remelting of these rocks. We believe three groups of rocks, younger Deccan volcanics, platform carbonates and evaporites, and Precambrian basement were involved in the genesis of the melt rocks of the Shiva Crater.

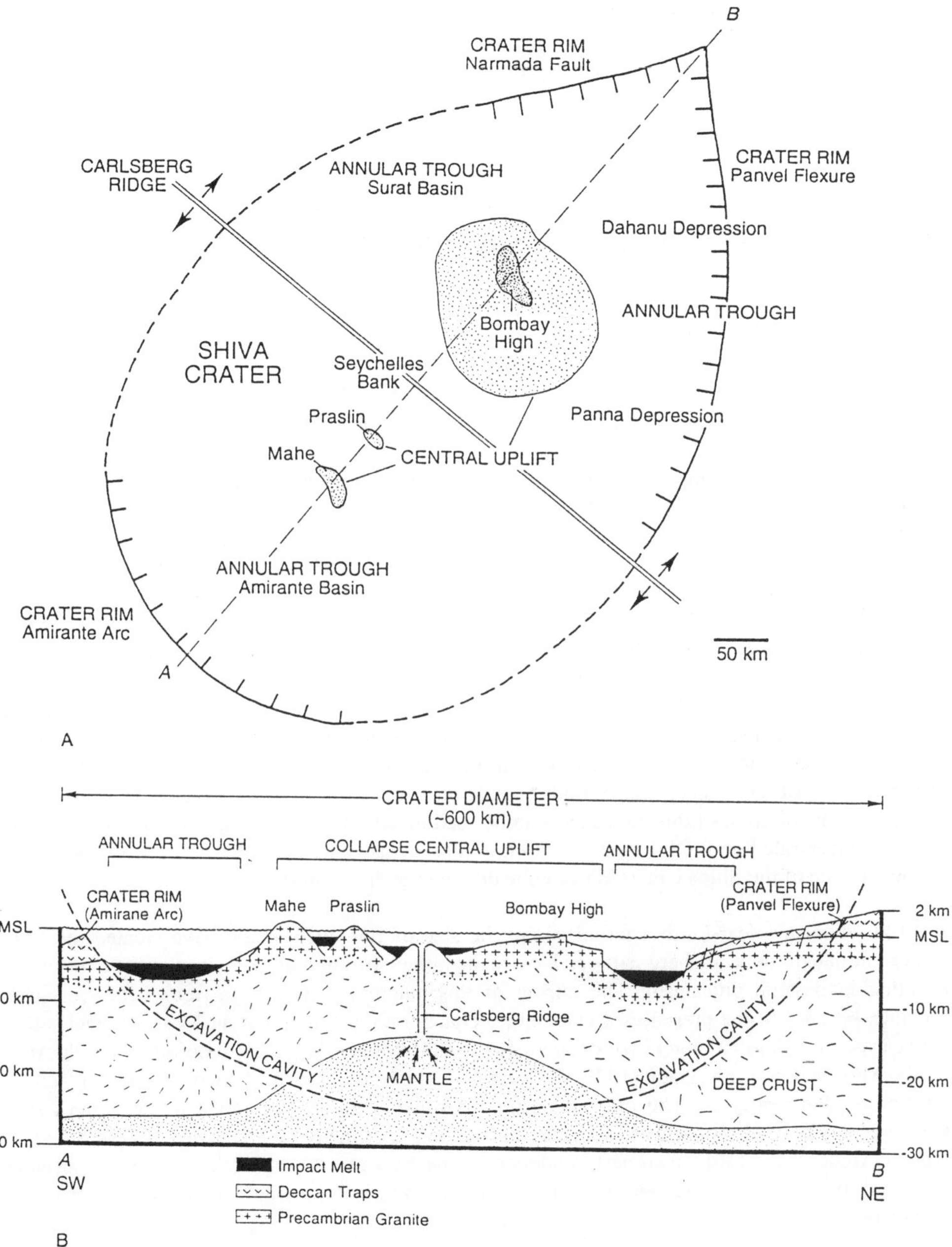

Figure 4. Morphology of the Shiva Crater; A, plan view of the Shiva Crater showing a central uplift (Bombay High, Praslin and Mahe granitic core); an annular trough (Surat Basin, Dahanu Depression, Panna Depression, and Amirante Basin); and a slumped outer rim (Narmada Fault, Panvel Flexure and Amirante Arc). The oblong crater is about 600 km long, 450 km wide, and more than 12 km deep; it is bisected by the Carlsberg Ridge B, schematic cross-section of the Shiva Crater along the line AB in Fig. A; the post-impact Deccan lava flows are removed to show the morphology of the crater and possible sites of impact melt sheets; (seismic profile data from [46, 51]).

The radial distribution of melt sheets in the down-range direction of the Shiva Crater is intriguing (Fig. 5A). They might have developed either during the passage of the shock waves when fluid ejecta was emplaced downrange, or during the collapse of the transient crater when radial fractures were formed. Melosh [57] discussed the mechanism by which radial fracture patterns develop during the collapse of a large, multi-ring crater. Radiating fractures from the central uplift to the rim are known from the Manicougan Crater of Canada [67] and the Wells Creek Structure of Tennessee [68].

SIZE AND TRAJECTORY OF THE SHIVA BOLIDE

It is estimated that out of 1,000 near-Earth asteroids (NEA) that have diameters greater than 1 km, three exceed the 10 km range [69]. These are 1866 Sisyphus (~11 km), 433 Eros (~20 km) and 1036 Ganymed (~40 km). These asteroids pose the greatest threat from space, with their potential for global environmental catastrophe and mass extinctions if they hit the Earth. Any of these large impacts would be lethal enought to wipe out the human civilization.

Although hypervelocity impacts normally create circular craters, impacts at a low angle (~15° from the horizontal) often generate elongate craters such as the Messier and Schiller Craters of the Moon [70], and the Rio Cuarto Crater in Argentina [71]. Craters formed by artificial oblique impact are generally oblong [72, 73]. The shape of an artificial crater formed by oblique impact at 15° is like a teardrop [74], where the pointed end indicates the downrange direction (Fig. 5B). In an oblique impact, the crater and its ejecta are bilaterally symmetrical about the plane of the trajectory, but the distribution of ejecta is concentrated asymmetrically on the downrange side. Surprisingly the shape of the Shiva Crater and the distribution of melt ejecta are almost identical to those of the artitificial crater (Fig. 5A). If so, the impact that produced the Shiva Crater was probably oblique along a SW-NE trajectory as evident from the direction of the longer diameter of the oblong crater; the tip of teardrop indicates that the downrange direction was NE. Howard and Wilshire [75] described flows of impact melt of large lunar craters both outside on crater rims and inside on crater walls, where asymmetric distribution of this melt sheet can be used to determine impact trajectory. The rim pools tend be concentrated on the inferred downrange side.The asymmetric concentrations of sheet melts on the NE side of the Shiva Crater indicate the downrange direction (Fig. 5A).

A 40 km diameter near-Earth asteroid, about the size of Ganymed, could have created the Shiva Crater, initiated the Carlsberg rift, and excavated the crustal materials into mantle reservoirs resulting in basaltic volcanism. The impact of this magnitude may be a rare event, but seems possible, because of the presence of even larger craters on the Moon and on Mars. Currently there are at least three known craters which are close to 200 km across [67]: Vredefort in South Africa (2,000 Ma), Sudbury in Canada (1849 Ma), and Chicxulub (65 Ma) in Mexico. The Shiva Crater is not unique, but nearly so. The only older similar structure in its size range is the eastern shore (Nastapoka Arc) of Hudson Bay in Canada: this 600 km wide depression is believed to be an impact scar of Archean age [76]. If properly understood, the Shiva Crater is the largest impact structure produced in the Phanerozoic, and it is consistent with the environmental havoc wreaked at the KT boundary.

Could a large bolide impact create ridge boundary and initiate ocean-floor spreading? Asteroids strike the Earth at an average speed of 25 km/sec and transfer considerable kinetic energy to the target rocks [77]. The pressures exerted on the meteorite and target rock can exceed 100 GPa; temperatures can reach several thousand degrees Celsius; and impacting energy would generate 100-million megaton blast [29]. Such a hypervelocity impacting body penetrates the target rocks to two or three times its radius [78]. An asteroid of 40 km diameter would produce cratering and associated tectonic rebound-collapse effects sufficient to shatter the 80-km thick lithosphere that could form plate boundaries and continental rifts. This may be the case for the origin of the Carlsberg Ridge and the rifting of the India -Seychelles microplates at the point of collision of the KT impact. This concept opens up a new field of research to investigate whether plate tectonics may be influenced by impacts of large bodies.

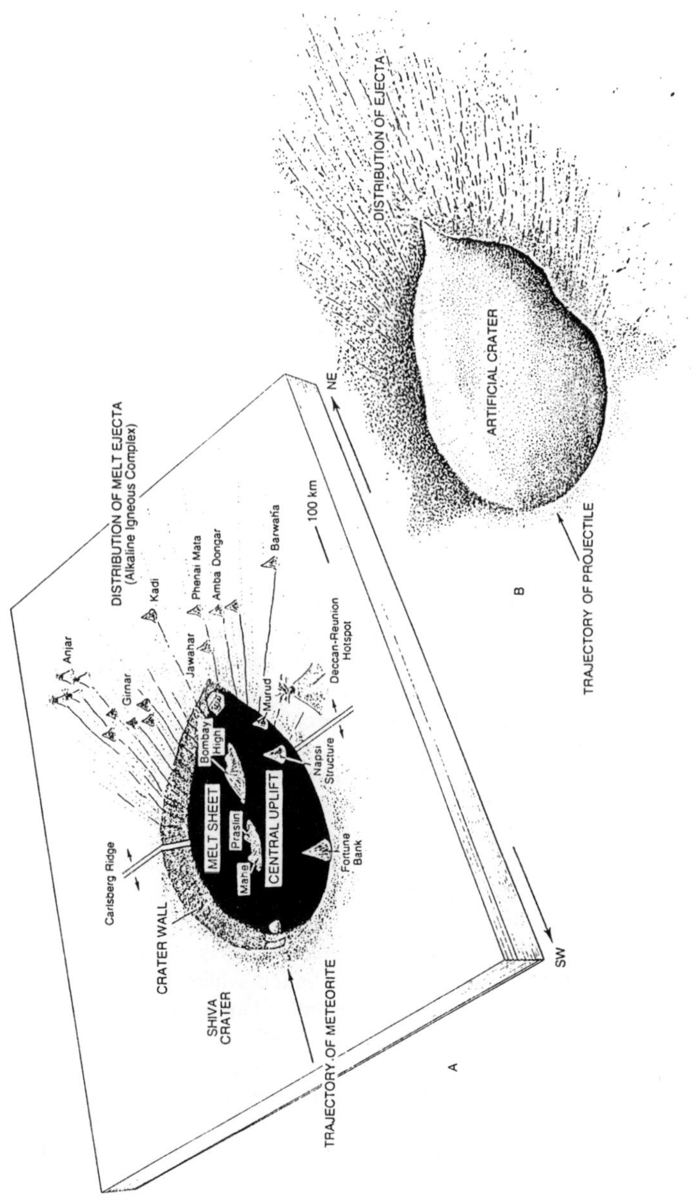

Figure 5. A, Schematic three-dimensional view of the Shiva Crater showing the radial distribution of the impact melt rocks in the form of volcanic plugs; younger Deccan volcanics are removed to show the morphology of the crater; oblong, teardrop shape of the crater and the asymmetric distribution of the melt rocks consistent with an oblique impact event along the NE downrange direction [39]; B, artificial crater produced by low-angle (~15°) oblique impact in the laboratory mimics the shape and ejecta distribution of the Shiva Crater (simplified from [74]).

A MULTIPLE-IMPACT MODEL

If both the Chicxulub and the Shiva Craters are real and were formed at KT boundary time, is there any genetic link between them? Hartnady [58] suggested that if a low-angle, oblique primary impact occured in the Southern Hemisphere near India, then bolide ricochet may have resulted in secondary impacts a few minutes later in the Northern Hemisphere. However, since the trajectory of the Shiva Crater is from SW to NE, Hartnady's model fails to explain the origin of both the Shiva Crater and the Chicxulub Crater by oblique impact ricochet process. However, the recent crash of 21 fragments of comet Shoemaker-Levy 9 on (S-L 9) Jupiter inspired scientists searching for similar multiple impacts on other bodies in the solar system. Crater chains were soon discovered on Jupiter's satellites and on the Moon [79]. The cometary fragments of S-L 9 did not collide simultaneously on Jupiter but spread in time over 5 days. If the original KT meteorite broke into several fragments, as in the case of Shoemaker-Levy, and the larger one formed the Shiva Crater, the second impact, almost after 12 hours (or odd multiple thereof), could create the Chicxulub as the Earth rotated anticlockwise around its axis. If the collisions of two fragments were spread over 12 hours, two antipodal craters could be formed by meteorite fragments around a great circle. If so, one can predict that additional KT impact scars, if discovered in future, should lie on this great circle joining the Shiva and the Chicxulub structures. This great circle has been named the **Alvarez Impact Belt** in honor of Luis and Walter Alvarez for their pioneering work on cratering process [39] (Fig. 6).

Various authors have predicted a third KT impact site on the Pacific plate which surprsingly lay along the Alvarez Impact Belt. Frank Kyte (pers. comm.) made a dramatic discovery of a tiny fragment (~ 3 mm) of the KT bolide in a drill core from DSDP site 576 in the western North Pacific. The bolide chip held micrometer-size metallic grains that are upto 87% nickel and rich in iridium. From the geochemical signature of the chip, Kyte speculates that the KT projectile is probably an asteroid, not a comet, that slammed into Earth at a shallow angle. This is the first direct evidence regarding the carrier for iridium. The chip may have broken off of the asteroid before it crashed on the Pacific plate. At 65 Ma, the Pacific impact site was located midway between the Chicxulub and the Shiva Craters right on the Alvarez Impact Belt (Fig. 6). Kyte *et al.* [80] also identified additional five KT boundary sites on the Pacific Plate around 576, characterized by an iridium anomaly, spherules and shocked quartz; they all cluster along this DSDP site 576. The point of collision of the KT projectile on the Pacific plate is further supported by other evidence. Robin *et al.* [25] concluded from spinel compositions at the KT boundary sites of the Pacific plate that multiple impacts might have occurred from a single disrupted bolide, where the largest objects would have impacted in the Pacific and Indian Ocean. Their spinel distributions and proposed impact site at the Pacific fit nicely with the Alvarez Impact Belt. Finally, all the known localities with an iridium anomaly and shocked quartz related to the KT impact event [6] cluster symmetrically on either side of the Alvarez Impact Belt. Distribution of impact-related minerals and trace elements along this belt coincides nicely with biogeographic selectivity of extinction at low latitudes [81]. The high latitudes probably served as a refuge for many organisms, especially for plants. If so, the distribution of several impact sites, and variation of composition of impact-generated minerals along the Alvarez Impact Belt could be explained by multiple impact events on a rotating Earth, rather than by a single large impact around the equator. Such a latitudinal gradient, with most extinctions occurring in the tropics, is exactly what is to be expected by chain of three impact sites in India, Mexico and Pacific plate respectively. These sites were formed when a large parent body broke apart into a string of smaller asteroidal fragments in the inner solar system and crashed into the planet.

Figure 6. Possible links between the Shiva and Chicxulub Craters. Locations of the Chicxulub and the Shiva Craters along the "Alvarez Impact Belt" at KT boundary time; note, impact debris and extinction events are concentrated along this belt. Both craters might have originated when two fragments from a larger meteorite crashed on a rotating earth over the course of 12 hours. A third impact site may lie on the Pacific plate as evident from the tiny fragment of the bolide chip and meteoric spinel from DSDP site 576 [39].

Recently, Alvarez *et al.* [8] raised three paradoxes in KT boundary sediments which are apparently difficult to reconcile in a single impact hypothesis. First, in North America the KT boundary reveals the double layer of ejecta: the lower layer consists mainly of iridium spike and altered spherules, wheras the upper one contains shocked quartz. Second, shocked quartz is more abundant and coarser grained at longitudes west than east of the Chicxulub Crater. Various KT boundary sites in Europe, Africa, and Asia show little or no evidence of shocked quartz and may represent "forbidden zone" for shocked quartz distribution. Third, the proposed impact energy (> 100 GPa) released by collision of a 10-km diameter asteroid at Chicxulub Crater would produce impact melt, not the moderate-pressure shock lamellae. Alvarez *et al.* linked all these three anomalies to differential timing of emplacement of target rocks. However, the double layer ejecta at KT boundary sediments reflects two different sources of impact events: the lower iridium spike and impact spherules layer might have come from the earlier Shiva impact, whereas shocked quartz layer was emplaced during the Chicxulub event (Fig. 7). This travel time sorting at two different impact sites may explain stratigraphic superposition of the double layer ejecta in North America and lack of upper layer in the distant areas of Europe, Asia and Africa. The energy released at the Shiva impact site was hundreds of gigapascals because of collision of a larger bolide fragment that produced mainly impact melt emplaced around the crater vicinity, not the shocked quartz. From this impact site, spherules and iridium spike were emplaced distally and globally. This scenario may explain why I could not detect any shocked quartz grains at the KT boundary sections in India after repeated search, as well as a general lack of shocked quartz in adjacent landmasses in Asia, Europe and Africa because the impact pressure was too high in this site. On the other hand, the impact energy at the Chicxulub site was moderate to the tune of a few tens of gigapascals that produced mainly shocked quartz ejecta that emplaced on the western side of the crater as the Earth rotated anticlockwise. Occurrence of multiple iridium spikes at some KT boundary sections [11] may be attributed to different timing of arrival from separate sources of impact sites.

Figure 7. A multiple-impact model at the KT boundary to explain the origin of the Chicxulub and the Shiva craters. To produce two distinct craters, the asteroids broke apart well in advance of collision; the larger projectile (~40km) created the Shiva Crater, the smaller one (~10 km) produced the Chicxulub Crater. Another small fragment crashed on the Pacific plate as revealed from the local distribution of tiny bolide chip and the meteoric spinel. The double layer of ejecta at the KT boundary may indicate separate sources from two impact sites: the lower layer consisting of mainly of iridium anomaly and melt spherules may come from the Shiva impact, whereas the upper layer containing shocked quartz derived from the Chicxulub site. Patterns of asteroid trajectory and ejecta distribution are shown by arrows.

WHAT KILLED THE DINOSAURS?

At the end of the Cretaceous period, there was an unprecedented ecological crisis leading to extinctions of many groups of organisms, from the giant dinosaurs to microscopic plankton. Although both impacts and Deccan volcanism coincided with the KT boundary, their relative importance for global climatic instability and disruption of life is still hotly debated. Basically, there are three possible ways in which extinctions could have been brought about: Either, it was an abrupt cataclysm triggered by the impact. Or, it was a gradual extinction process induced by the prolonged Deccan volcanism that polluted the atmosphere and changed the world's climate. Finally, the combined effects of both impact and volcanism might have contributed to environmental crisis leading to mass extinction. The fossil record of victims and survivors across the KT boundary have been used by the proponents of both impact and volcanic models to champion their views.

Volcanic-Extinction Mechanism

Many palaeontologists are skeptical about the scenario of impact holocaust. They argue that the KT extinction was neither global, nor instantaneous, but occurred over an extended period of time, because different organisms disappear at different levels at or near the KT boundary. They look for extinction agents that could explain a gradual extinction event and favor terrestrial causes, such as prolonged Deccan volcanism and regression of sea level, as main contributing factors for the biotic crisis [9, 10, 12, 82, 83, 84].

However, the gradual extinction pattern seen among some organisms may be an artifact of preservation and declining sampling quality. Signor and Lipps [85] showed theoretically how a sudden catastrophic extinction would appear to have been gradual in the fossil record, if the record was not dense. The Signor-Lipps effect weakens the case for a gradual extinction. Furthermore, the instantaneous extinction in both continental and marine organisms is not a necessary corollary of the impact theory [4-7]. A large impact could trigger environmental crisis, and the latter could, in turn, cause extinctions spread over 10^4 to 10^5 years [86]. The most ecologically-sensitive organisms were probably the first victims, with progressively more tolerant groups succumbing in the later stages. The short term physical event may lead to long term biological events.

There are many examples of pre-boundary and post-boundary extinction events [83] which led many palaeontologists to advocate a close relationship between the biotic crisis and Deccan volcanism. There is no doubt that such a massive volcanic outburst over an extended period would have deleterious environmental consequences. Based on data known from the Laki eruption in 1783, the Deccan eruption must have pumped large volumes of SO_2, HCl, and ash into the atmosphere to cause global cooling, immense amounts of acid rain, reduction in alkalinity and pH of the surface ocean, and ozone layer depletion that might be harmful to both terrestrial and marine organisms [10]. Deccan volcanism must have produced large-scale aerosol clouds of sulfur and CO_2. The reduction in temperatures caused by sulfur aerosols in the stratosphere act in the opposite direction to the greenhouse gases such as carbon dioxide. Emissions of CO_2 from the Deccan volcanism would directly increase the atmospheric and oceanic CO_2, leading to sustained global warming.

Biotic Effects in India

Since India was ground zero for both impact and Deccan volcanism, this would be an ideal place to search for the evidence of crisis on local biota. The vast thickness of Deccan lava flows were not extruded all at once. In between the lava flows are fluvial or lacustrine deposits of Lameta beds that contain abundamt remains of plants, invertebrates, dinosaurs and their eggs. Damming of local drainage by lava flows had created new lakes in subaerial environments. Such lakeshores were centers for dinosaur communities and became the favorable nesting sites [87]. I could not detect any evidence of biotic crisis in these fossil assemblages during the episodic volcanic activity. On the other hand, fossil evidence indicates that dinosaurs thrived during the recurrent Deccan eruptions (Fig. 8). Moreover, no bones or eggs from this locality show any pathological abonormality. The extrusion of Deccan volcanism certainly affected the local flora and fauna by habitat destruction and pollution, but it had little direct effect on the major decimation of terrestrial organisms. Deccan volcanism consisted largely of nonexplosive tholeiitic eruptions similar to Kilauea and Reunion emissions, though on a grander scale. Recurrent eruptions do not disrupt much of the rich biota on the islands of Hawaii and Reunion. Similarly, Deccan volcanism cannot be the proximate cause of the mass extinction. Within the Lameta beds, magnificent fossil accumulations indicate that life was resilient in this hostile environment of Deccan volcanism. Although pollution from the Deccan volcanism might have had some adverse effect on vertebrate populations, Prasad *et al.* [88] could not detect any change in the composition or population of biota from infratraps to intertraps, during episodic volcanic activity. They concluded that the Deccan volcanism was not detrimental to life. There is another line of evidence which supports this theory. Midoceanic ridges spewing tholeiitic lava, toxic gases, and sulfides, support a rich oasis of life, including tubeworms, polychaetes, crustaceans and mollusks. In fact, submarine vents provide a refuge for many relict forms that might have escaped the KT extinction [89].

However, Deccan volcanism must have played a significant role in the initial breakdown of the marine food chain on a global scale. With the emissions of Deccan pollutants, the countdown of disruption of oceanic biota started. Deccan volcanism must have pumped several trillion tons of toxic gases into the upper atmosphere, which would result in global climatic perturbations disrupting the ecosystem. Within the trappean beds, especially in the Lameta beds, the greenhouse effect can be identified in the sediments, by the climate becoming more severe, with marked seasonal changes and aridity [87].

The effect of impact on local Indian biota is more clear-cut. The dinosaurs died out suddenly at the KT boundary of Anjar section, Gujrat, which is defined on the basis of the iridium anomaly and isotopic dating [90]. Here, the KT boundary section is a 1 cm thick, pinkish clay layer above a cross-bedded limestone unit which contains associated titanosaurid bones. The impact layer contains a significant enrichment of iridium (348 parts per trillion). No dinosaur bones occur above the impact layer. This is the first unequivocal evidence that the dinosaur extinction occurred precisely at the time of meteoritic impact [39, 90].

Impact-Extinction Mechanisms

One of the most important questions about the KT extinctions concerns its timing and duration. The extinction of six groups of marine microfossils ñ planktic and benthic forams, coccoliths, radiolarians, dinoflagellates, and diatoms ñ took place precisely at the impact horizon and are compatible with this scenario [5, 91, 92]. Brachiopods [93] and ammonites [94] also show a similar abrupt pattern of extinction at the KT boundary. Similarly, pterosaurs, plesiosaurs and mosasaurs in the marine realm became extinct suddenly at the end of the Maastrichtian [95]. There is no question that the oceanic food chain collapsed by environmental changes around the KT boundary. In terrestrial communities, the sharp floral event exactly at this boundary strongly suggests a causal link with the impact scenario [96]. Careful sampling indicates that dinosaurs in Montana died out suddenly at the KT boundary [97]. We see the similar evidence of abrupt disappearance of dinosaurs precisely at the iridium KT boundary layer in Anjar section of Gujrat [39]. Thus, the palaeontological record favors a sudden and simultaneous extinction pattern at the KT boundary among certain groups of organisms.

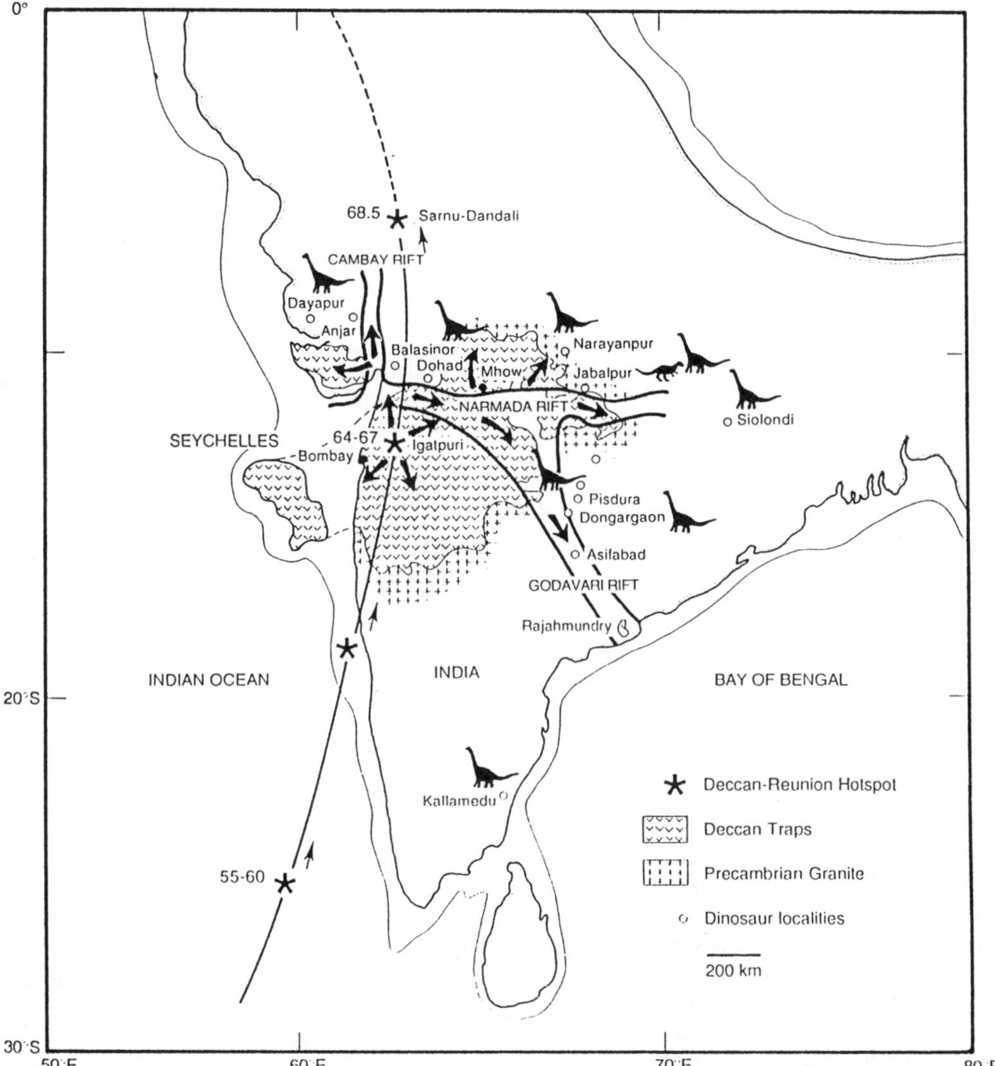

Figure 8. Sketch map showing the localities of Maastrichtian dinosaurs around the Deccan volcanic province before the impact event. Distribution and abundance of dinosaur fauna suggest that Deccan volcanism had little effect on their sudden disappearance at the KT boundary. Asterisks indicate the Deccan-Reunion hotspot track during the rapid northward drift of India.

If, as seems likely, one or more meteorite impacts did play a role in the mass extinction event, what mechanism was involved? How devastating was the collision and how did it affect life? Various models have been proposed in recent times to explain the killing mechanism. A massive impact would generate earthquakes of high magnitude, inject a large volume of dust into the stratosphere, darken the skies, halt photosynthesis, ignite global fires, initiate tsunamis, produce acid rains, decrease ocean surface alkalinity, and devastate the biosphere. Millions of organisms would die instantly from the direct effect of the impact ñ shock heating of the atmosphere by the expanding fireball [1, 3, 15]. Comet Shoemaker-Levy 9 provided proof that the KT impact ignited global fires [79]. As the ejecta fell back into the atmosphere after 20 minutes of each impact, they reentered with a release of titanic energy; the heat from this reentry was so intense it was easily detected from Earth. Similarly, when the KT ejecta reentered into Earth's atmosphere, they ignited terrestrial forests and plant covers. The wildfires consumed oxygen and poisoned the atmosphere with CO. This massive impact would generate a 100-million megaton blastó1000 times more powerful than the explosion of all nuclear arsenal of the world. The impact force would have vaporized not only the meteorite but also melted the target rock. This melt ejecta would spread in large waves, until it had formed a large crater. Huge tsunamis produced by the oceanic impact could destroy coastal habitats across the globe [18].

Available data suggest that the extinction of carbonate-secreting plankton was especially severe at the KT boundary [83, 85]. Acid trauma is likely a consequence of many biotic crisis in the oceans. Various impact mechanisms have been proposed to produce large volumes of HNO_3 and H_2SO_4 for acidification of surface marine waters. Shock heating of the atmosphere would cause nitrogen and oxygen to combine with steam to form nitric acid [16]. The sheer size of the projectile cannot by itself account for the devastation. The unique composition of the crash sites provided additional arsenal. Because the target rock at the Yucatán Peninsula (and western coast of India) has a thick-cover of $CaSO_4$, an impact there could have released as much as a trillion tons of SO_2 [14]. Release of devolatilized SO_2 would form sulfuric acid aerosols which would contribute to a rapid decline in global surface temperature and halt photosynthesis. The subsequent showers of acid rain must have been agents for decreased ocean alkalinity and destruction of life. All of these events had cascading effects through the terrestrial and marine ecosystems to collapse the food chain.

All the extinction mechanisms cited above are based on a 10 km-sized meteorite. If we consider the impact of a 40-km meteorite, the devastation would be much more traumatic on the biosphere. One obvious consequence of a larger oceanic impact would be partial vaporization of the photic zone, raising the atmosphere's temperature to hundreds of degrees, and destoying all life in the ocean surface. Thus, a large body impact may explain the marine regression by vaporization of water which is generally associated with the KT event. The impact-induced thermal radiation would ignite global wildfires on land and destroy ecological niches [15].

Combined Extinction mechanisms

Some of the consequences of an asteroid impact and of massive volcanism would be quite similar, such as pollution of the atmosphere, darkness resulting from dust (either ejecta or ash), suppression of photosynthesis, acid rain, global cooling, carbonate crisis in the ocean waters, environmental stress, devastation of ecosystems, and the collapse of the food chain. The scenario of extinction by food chain disruption is compatible with both impact and volcanic hypotheses [98]. It is generally believed that the biotic crisis was most severe and sometimes limited to tropical-subtropical regions while high latitude faunas and floras were little affected [81]. The geographic locations of tmultiple impact sites, Deccan volcanism, and the distribution of impact deposits along the Alvarez Impact Belt may explain this geographic selection of KT extinction (Fig. 6B). The sudden cooling by H_2SO_4 aerosol and warming by CO_2 would result in chaotic climatic perturbations which would be especially traumatic for global organisms. However massive volcanism would lack some of the titanic lethal features that are generally associated with a large body impact, such as gigantic shock effects leading to vapor plume, global fire, thermal pulse, evaporation of the photic zone and concomitant sea regression, chondritic metal toxicity, as well volatilization of target rocks. Impacts probably made more dramatic changes to KT environments globally and a more traumatic crisis to the ecosystem than volcanic emissions. Impacts also perturbed the environment so suddenly and catastrophically that most organisms failed to adapt to these changes in such a short time and perished. In contrast, prolonged volcanism allowed enough time for some organisms to adapt, for others to disappear gradually. Global fire must have decreased the amounts of atmospheric oxygen which would further exacerbate the hostile conditions already developing among terrestrial and marine life. On the other hand, selectivity and a step-wise extinction pattern can be better explained by the volcanic model. Deccan volcanism could have been an accomplice, not the main killer, for the massive destruction of life. It seems that impact was the proximate cause to the biotic crisis at the KT boundary, whereas long-term volcanism produced harmful changes that enhanced climatic stress and finished off the extinction process. Both impacts and Deccan volcanisms contributed heavily to the breakdown of stable ecological communities and disrupted the biosphere. The KT extinctions was a compound crisis, induced both by impact and volcanism.

A cause-and effect connection between impact and Deccan volcanism has been the subject of extensive discussion and speculation. Since the Shiva Crater was proximate to the Reunion hotspot and created the vast outpouring of the Deccan lava at the KT boundary, the near-coincidence of timing and space is striking (Fig. 5). Could Deccan volcanism be triggered by the Shiva impact [45]? Probably not. Deccan volcanism seems to have begun erupting at least 1 Ma before the impact event, making a causal link unlikely [90]. However, the impact might have shaken the Earth's mantle violently to enhance the spectacular Deccan volcanic outburst precisely at the time of the KT boundary [39].

Acknowledgements

I thank Dhiraj K. Rudra for continued help and collaboration in field works in India, Frank Kyte for sharing his unpublished work, and Moses Attrep, Jr. for iridium analysis. I thank Ralph E. Molnar and Fernando E. Novas for permission to reproduce some of the drawings. The research was supported by the National Geographic society, Dinosaur Society, Texas Tech University, and Indian Statistical Institute.

REFERENCES

1. L. W. Alvarez. Mass extinctions caused by large bolide impacts, *Phys. Today* **40:7**, 24-33 (1987).
2. L. W. Alvarez, W. Alvarez, F. Asaro and H. V. Michel. Extraterrestrial cause for the Cretaceous-Tertiary extinction. *Science* **208**, 1095-1108 (1980).
3. W. Alvarez. Toward a theory of impact crisis, *EOS, Trans. Amer. Geophys. Un.* **67:35**, 649-658 (1986).
4. W. Alvarez, L. W. Alvarez, F. Asaro and H. V. Michel. Current status of the impact theory for the terminal Cretaceous extinction, *Geol. Soc. Amer. Sp. Pap.* **190**, 305-315 (1982).
5. W. Alvarez, L. W. Alvarez, F. Asaro and H. V. Michel. The end of the Cretaceous: sharp boundary or gradual transition? *Science* **223**, 1183-1186 (1984).
6. W. Alvarez and F. Asaro. An extraterrestrial impact, *Scient. Amer.* **263:4**, 78-84 (1990).
7. W. Alvarez, F. Asaro, P. Claeys, J. M. Grajales-N., A. Montnari and J. Smit. Developments in the KT impact Theory since Snowbird II, *Lunar Planet. Contr.* **825**, 3-4 (1994).
8. W. Alvarez, P. Claeys and S. W. Kieffer. Emplacement of Cretaceous-Tertiary boundary shocked quartz from Chicxulub Crater, *Science* **269**, 930-935 (1995).
9. V. Courtillot. A volcanic eruption, *Scient. Amer.* **263:4**, 85-92 (1990).

10. A. Hallam. End-Cretaceous mass extinction event: Argument for terrestrial causation, *Science* **238**, 1237-1242 (1987).
11. C. B. Officer, A. Hallam, C. L. Drake, J. D. Devine and A. A. Meyerhoff. Late Cretaceous paroxysmal Cretaceous/Tertiary extinctions, *Nature* **326**, 143-149 (1987).
12. S. M. Stanley. *Extinction.* W. H. Freeman, San Francisco (1987).
13. C. J. Orth. Geochemistry of the bio-event horizons. In: *Mass Extinctions: Processes and Evidence*. S. K. Donovan (Ed). pp. 37-72. Columbia University Press, New York (1989).
14. S. D'Hondt, H. Sigurdsson, A. Hanson, S. Carey and M. Pilson. Sulfate volatilization, surface-water acidification and extinction at the KT boundary, *Lunar Planet. Inst. Contr.* **825**, 29-30 (1994).
15. H. J. Melosh, N. M. Schneider, K. J. Zahnle and D. Latham. Ignition of global wildfires at the Cretaceous/Tertiary boundary, *Nature* **343**, 251-254 (1990).
16. R. G. Prinn and B. F. Fegley, Jr. Bolide impacts, acid rain, and biospheric traumas at the Cretaceous-Tertiary boundary, *Earth Planet. Sci. Lett.* **83**, 1-15 (1987).
17. W. S. Wolbach, I. Gilmour, E. Anders, C. J. Orth and R. R. Book. Global fire at the Cretaceous-Tertiary boundary, *Nature* **324**, 665-669 (1988).
18. V. L. Sharpton and P. D. Ward (Eds). *Global catastrophes in earth history. Geol. Soc. Amer. Spec. Pap.* **247**, 1-631 (1990).
19. B. F. Bohor, E. E. Foord, P. J. Modreski, and D. M. Triplehorn. Minerologic evidence for an impact event at the Cretaceous-Tertiary boundary, *Science* **224**, 867-869 (1984).
20. M. R. Owen and M. H. Anders. Evidence from cathodluminiscence for non-volcanic origin of shocked quartz at the Cretaceous/Tertiary boundary, *Nature* **334**, 145-147 (1988).
21. J. F. McHone, R. A. Nieman, C. F. Lewis and A. M. Yates. Stishovite at the Cretaceous-Tertiary boundary, Raton, New Mexico, *Science* **243**, 1182-1184 (1989).
22. D. B. Carlisle and D. R. Braman. Nanometric-size diamonds in the Cretaceous/Tertiary boundary clay of Alberta, *Nature* **352**, 708-709 (1991).
23. G. Izett. The Cretaceous/Tertiary boundary interval, Raton Basin, Colorado and New Mexico, *Geol. Soc. Amer. Spec. Pap.* **249**, 1-100 (1990).
24. K. K. Turekian. Potential of $^{187}Os/^{186}Os$ as a cosmic versus terrestrial indicator in high iridium layers of sedimentary strata, *Geol. Soc. Amer. Spec. Pap.* **190**, 243-249 (1982).
25. E. Robin, J. Gayraud, L. Froget and R. Rocchia. On the origin of the regional variation in spinel compositions at the KT boundary, *Lunar Planet. Inst. Contr.* **825**, 96-97 (1994).
26. G. I. Bekov, V. S. Letokhov, V. N. Radaev, D. D. Badyukov and M. A. Nazarov. Rhodium distribution at the Cretaceous/Tertiary boundary analyzed by ultrasensitive laser photoionization, *Nature* **332**, 146-148 (1988).
27. J. Bourgeois, T. A. Hansen, P. L. Wilberg and E. G. Kauffman. A tsunami deposit at the Cretaceous-Tertiary boundary in Texas, *Science* **241**, 567-570 (1988).
28. D. B. Carlisle and D. R. Braman. Extraterrestrial amino acids at the Cretaceous-Tertiary boundary of Alberta, *Nature* **361**, 551-553 (1993).
29. R. A. F. Grieve. Impact cratering on the Earth, *Scient. Amer.* **261:4**, 66-73 (1990).
30. A. R. Hildebrand, G. T. Penfield, D. A. King, M. Pilkington, A. Z. Camaro, S. B. Jacobson and W. B. Boynton. Chicxulub Crater: a possible Cretaceous/Tertiary boundary impact crater on the Yucatan Peninsula, Mexico, *Geology* **19**, 867-871 (1991).
31. A. R. Hildebrand, M. Pilkington, M. Connors, C. Ortiz-Aleman and R. E. Chavez. Size and structure of the Chicxulub Crater revealed by horizontal gravity gradients and cenotes, *Nature* **376**, 415-417 (1995).
32. G. T. Penfield and Z. A. Camagro. Definition of a major igneous zone in the ventral Yucatán platform with aeromagnetics and gravity, *Geophys.* **47**, 448-449 (1982).
33. V. L. Sharpton, G. B. Darlymple, L. E. Martin, G. Ryder, B. C. Schuaraytz and J. Urrutia-Fucugachi. New links between the Chicxulub impact structure and the Cretaceous/Tertiary boundary, *Nature* **359**, 819-821 (1992).

34. C. C. Swisher, J. M. Grajales-Nishimura, S. V. Margolis, P. Claeys, W. Alvarez, P. Renne, E. Cedello-Pardo, F. J. M. R. Maurasse, G. H. Curtis, J. Smit and M. O. McWilliams. ^{40}Ar/^{39}Ar ages of 65.0 million years ago from Chicxulub crater melt rock and Cretaceous-Tertiary boundary tektites, *Science* **257**, 954-958 (1992).
35. A. R. Hildebrand and W. V. Boynton. Proximal Cretaceous-Tertiary boundary impact deposits in the Caribbean. *Science* **248**, 843-847 (1990).
36. G. Izett, G. B. Dalrymple and L. W. Snee. ^{40}Ar/^{39}Ar age of the Cretaceous-Tertiary boundary tektites from Haiti, *Science* **252**, 1539-1541 (1991).
37. J. Smit, A. Montari, N. H. M. Swinburne, A. P. Hildebrand, S. V. Margolis, P. Claeys, W. Lowrie and F. Asaro. Tektite-bearing, deepwater clastic unit at the Cretaceous-Tertiary boundary in northeastern Mexico, *Geology* **20**, 99-103 (1992).
38. J. Smit, T. B. Roep, W. Alvarez, S. Montanari and P. Claeys. Stratigraphy and sedimentology of KT clastic beds in the Moscow Landing (Alabama) outcrop: evidence for impact-related earthquakes and tsunamis, *Lunar Planet. Inst. Contr.* **825**, 119-120 (1994).
39. S. Chatterjee and D. K. Rudra. KT events in India: impact, rifting, volcanism and dinosaur extinction. In: *Gondwana Dinosaurs: Phylogeny and Palaeobiogeograpgy*. R. E. Molnar and F. E. Novas (Eds). pp. 103-146. *Mem. Queensland Mus.*, Australia (1996).
40. C. J. H. Hartnady. Amirante Basin, western Indian Ocean: possible-impact site at the Cretaceous-Tertiary extinction bolide? *Geology* **14**, 423-426 (1986).
41. D. A. Johnson, W. A. Berggren and J. E. Damuth. Cretaceous ocean floor in the Amirante Passage: tectonic and oceanographic implications. *Marine Geol.* **47**, 31-343 (1982).
42. R. L. Fisher, C. G. Engel and T. W. C. Hilde. Basalts dredged from the Amirante Ridge, western Indian Ocean, *Deep Sea Res.* **15**, 521-534 (1968).
43. D. Alt, J. M. Sears and D. W. Hyndman. Terrestrial maria: the origins of large basalt plateaus, hotspot tracks and spreading ridges, *Jour. Geol.* **96**, 647-662 (1988).
44. S. Chatterjee. A possible K-T impact site at the India-Seychelles boundary, *Abs. Lunar Planet. Sci. Conf.* **21**, 182-183 (1990).
45. S. Chatterjee. A kinematic model for the evolution of the Indian plate since the Late Jurassic. In: *New Concepts in Global Tectonics*. S. Chatterjee and N. Hotton (Eds). pp. 33-62. Texas Tech University Press, Lubbock (1992).
46. K. L. Kaila, P. R. K. Murty, V. K. Rao and G. E. Kharetchko. Crustal structure from deep-seismic surroundings along the Koyna II (Kelsi-Loni) profile in the Deccan Trap area, India, *Tectonophys.* **73**, 365-384 (1981).
47. K. B. Powar. Lineament fabric and dyke pattern in the western part of the Deccan volcanic province. In: *Deccan Volcanism*. K. V. Subbarao and R. N. Sukeshwala (Eds). pp. 45-57. *Geol. Soc. India Mem.* **3**, Bangalore (1981).
48. J. G. Negi, P. K. Agarwal, A. P. Singth and O. P. Pandey. Bombay gravity high and eruption of Deccan flood basalts (India) from a shallow secondary plume, *Tectonophys.* **206**, 341-350 (1992).
49. D. N. Basu, A. Banerjee and D. M. Tamhane. Facies distribution and petroleum geology of the Bombay offshore basin, India, *Jour. Petrol. Geol.* **5:1**, 51-75(1982).
50. L. L. Bhandari and S. K. Jain. Reservoir geology and its role in the development of the L-III reservoir, Bombay High Field, India, *Jour. Petrol. Geol.* **7:1**, 27-46 (1984).
51. R. P. Rao and S. N. Talukdar. Petroleum geology of Bombay High Field, India, *Amer. Assoc. Petrol. Geol. Mem.* **30**, 487-506 (1980).
52. M. Kamen-Kaye. Mesozoic columns below the Seychelles Bank, western Indian Ocean, *Jour. Petrol. Geol.* **8:3**, 323-329 (1985).
53. C. W. Devey and W. E. Stephens. Deccan-related magmatism west of the Seychelles-India rift. In: *Magmatism and the Causes of Continental Break-up*. B. C. Storey, T. A. Alabaster and R. J. Pankhurst (Eds). pp. 271-291. *Geol. Soc. Spec. Publ.* **68**, London (1992).

54. A. A. Meyerhoff and M. Kamen-Kaye. Petroleum prospects of Saya de Malha and Nazareth Banks, Indian Ocean, *Amer. Assoc. Petrol. Geol. Bull.* **65**, 1344-1347 (1981).
55. S. K. Biswas. Structure of the western continental margin of India and related igneous activity. In: *Deccan Flood Basalts*. K. V. Subbarao (Ed). pp. 371-393. *Geol. Soc. India Mem.* **10**, Bangalore (1988).
56. L. F. Jansa. Cometary impacts into ocean: their recognition and threshold constraints for biological extinctions, *Palaeogeo. Palaeoclimatol. Palaeoecol.* **104**, 271-286 (1993).
57. H. J. Melosh. *Impact Cratering: A Geologic Process*. Oxford University Press, New York (1989).
58. C. J. H. Hartnady. Oblique impact hypothesis relates Deccan volcanism to multiple northern hemisphere craters to Cretaceous-Tertiary boundary, *Abs. 28th Int. Geol. Cong.* **2**, 33-34 (1989).
59. J. B. Auden. Dykes of western India: a discussion of their relationship with the Deccan Traps, *Trans. Nat. Inst. Sci. India*. **3**, 123-157 (1949).
60. R. A. Duncan and D. G. Pyle. Rapid eruption of the Deccan flood basalts at the Cretaceous/Tertiary boundary. *Nature* **333**, 841-843 (1988).
61. A. R. Basu, P. R. Renne, D. K. Dasgupta, F. Teichmann and R. J. Poreda. Early and late alkali igneous pulses and high-^{3}He plume origin for the Deccan Flood Basalts, *Science* **261**, 902-906 (1993).
62. B. R. Naini and M. Talwani. Structural framework and evolutionary history of the continental margin of western India, *Amer. Assoc. Petrol. Geol. Mem.* **34**, 167-193 (1982).
63. B. H. Baker. The Precambrian of the Seychelles Archipellago. In: *Precambrian Research*. K. Rankama (Ed). pp. 122-132. Interscience, New York (1967).
64. M. K. Bose. Alkaline magmatism in the Deccan volcanic province, *Jour. Geol. Soc. India* **21**, 317-329 (1980).
65. A. De. Late Mesozoic-Lower Tertiary magma types of Kutch and Saurashtra. In: *Deccan Volcanism*. K. V. Subbarao and R. N. Sukeshwala (Eds). pp. 327-339. *Geol. Soc. India Mem.* **3**, Bangalore (1981).
66. K. Pande, T. R. Venkatesan, K. Gopalan, P. Krishnamurthy and J. D. Macdougall. ^{40}Ar/^{39}Ar ages of alkali basalts from Kutch, Deccan volcanic province, India. In: *Deccan Flood Basalts*. K. V. Subbarao (Ed). pp. 145-150. *Geol. Soc. India Mem.* **10**, Bangalore (1988).
67. R. A. F. Grieve, C. A. Wood, J. B. Garvia, G. McLaughlin and J. F. McHone, Jr. *Astronaut's Guide to Terrestrial Impact Craters*. Lunar and Planetary Institute, Houston (1988).
68. R. G. Stern. The Wells Creek structure, Tennessee. In: *Shock Metamorphism in Natural Minerals*. B. M. French and N. M. Short (Eds). pp. 323-337. Mono Book Corporation, Baltimore (1968).
69. G. W. Wetherill and G. M. Shoemaker. Collision of astronomically observable bodies with the Earth, *Geol. Soc. Amer. Spec. Pap.* **190**, 1-13 (1982.).
70. D. E. Wilhelms. The geologic history of the moon, *U.S.G.S. Prof. Pap.* **M1348**, 1-293 (1987).
71. P. H. Schultz and R. E. Lianza. Recent grazing impacts on the Earth recorded in the Rio Cuarto Crater field, Argentina, *Nature* **355**, 234-237 (1992).
72. D. E. Gault and J. A. Wedekind. Experimental studies of oblique impact, *Proc. 9th Lunar Planet. Sci. Conf.* 3843-3875, (1978).
73. H. H. Moore. Missile impact centers (White Sands Missile Range, New Mexico) and application to lunar research, *U.S.G.S. Prof. Pap.* **812B**, B1-B47 (1976).
74. P. H. Schultz and Gault, D. E. Prolonged global catastrophes from oblique impacts, *Geol. Soc. Amer. Spec. Pap.* **247**, 239- 261 (1990).
75. K. A. Howard and H. G. Wilshire. Flows of impact melt at lunar craters, *Jour. Res. U.S.G.S.* **3**, 237-251 (1975).
76. R. S. Dietz. Review of terrestrial evidence for major cosmic impacts (abs.), *74th Ann. Meet. Pacific Divi. AAAS, Progr. with Abs.* 40 (1993).

77. E. M. Shoemaker. Asteroid and comet bombardment of the Earth, *Ann. Rev. Earth Sci.* **11**, 461-494 (1983).
78. R. A. F. Grieve. Terrestrial impact structures, *Ann. Rev. Earth Planet. Sci.* **15**, 245-270 (1987).
79. D. K. Levy, E. M. Shoemaker and C. S. Shoemaker. Comet Shoemaker-Levy 9 meets Jupiter, *Scient. Amer.* **273:2**, 84-91 (1995).
80. F. T. Kyte, J. A. Bostwick and L. Zhou. The KT boundary on the Pacific plate, *Lunar Planet. Inst. Contr.* **825**, 64-65 (1994).
81. G. Keller. Global biotic effects of the KT boundary event: mass extinction restricted to low latitudes? *Lunar Planet. Inst. Contr.* **825**, 57-58 (1994).
82. W. A. Clemens. Pattern of extinction and survival of the terrestrial biota during the Cretaceous/Tertiary transition. *Geol. Soc. Amer. Spec. Pap.* **190**, 407-413 (1982).
83. G. Keller. Extended period of extinctions across the Cretaceous/Tertiary boundary in planktonic foraminifera of continental-shelf sections; implications for impact and volcanism theories, *Geol. Soc. Amer. Bull.* **101**, 1408-1419 (1989).
84. W. J. Zinsmeister, R. M. Feldmann, M. O. Woodburne and D. H. Elliot. Latest Cretaceous/earliest Tertiary transition on Seymour Island, Antarctica, *Jour. Paleont.* **63**, 731-738 (1989).
85. P. W. Signor and J. M. Lipps. Sampling bias, gradual extinction patterns, and catastrophes in the fossil record, *Geol. Soc. Amer. Spec. Pap.* **190**, 291-296 (1982).
86. K. J. Hsu, J. A. McKenzie and Q. X. He. Terminal Cretaceous environmental and evolutionary changes, *Geol. Soc. Amer. Spec. Pap.* **190**, 317-328 (1982).
87. A. Sahni, S. K. Tandon, A. S. Jolly, S. Bajpai, A. Sood and S. Srinivasan. Upper Cretaceous dinosaur eggs and nesting sites from the Deccan-volcano-sedimentary province of peninsular India. In: *Dinosaur Eggs and Babies*. K. Carpenter, K. F. Hirsch and J. R. Horner (Eds). pp. 204-226. Cambridge University Press, Cambridge (1994).
88. G. V. R. Prasad, J. J. Jaeger, A. Sahni, E. Gheerbrant and C. K. Khajuria. Eutherian mammals from the Upper Cretaceous (Maastrichtian) intertrappean beds of Naskal, Andhra Pradesh, India, *Jour. Vert. Paleont.* **14**, 260-277 (1994).
89. V. Tunnicliffe. The biology of hydrothermal vents: ecology and evolution, *Marine Biol. Ocean. Ann. Rev.* **29**, 319-407 (1991).
90. N. Bhandari, P. N. Shukla, Z. G. Ghevaria and S, M. Sundaram. KT boundary in Deccan intertrappeans: chemical anomalies and their implications, *Lunar Planet. Inst. Contr.* **825**, 10-11 (1994).
91. H. R. Thierstein. Terminal Cretaceous plankton extinctions: a critical assessment, *Geol. Soc. Amer. Spec. Pap.* **190**, 385-405 (1982).
92. J. Smit. Extinction and evolution of planktonic foraminifera at the Cretaceous/Tertiary Boundary, *Geol. Soc. Amer. Spec. Pap.* **190**, 329-352 (1982).
93. F. Surlyk and M. B. Johnson. End-Cretaceous brachiopod extinctions in the chalk of Denmark, *Science* **23**, 1174-1177 (1984).
94. P. D. Ward, W. J. Kennedy, K. G. MacLeod and J. F. Mount. Ammonite and inoceratid bivalve extinction patterns in Cretaceous/Tertiary boundary section of the Biscay region (southwestern France, northern Spain), *Geology* **19**, 1181-1184 (1991).
95. E. Buffetaut. Vertebrate extinctions and survival across the Cretaceous-Tertiary boundary, *Tectonophys.* **171**, 337-345 (1990).
96. R. M. Tschudy and B. D. Tschudy. Extinction and survival of plant life following the Cretaceous/Tertiary boundary event, western interior, North America, *Geology* **14**, 667-670 (1986).
97. P. M. Sheehan, D. E. Fastovsky, R. G. Hoffman, C. B. Berghaus and D. I. Gabriel. Sudden extinction of the dinosaurs: latest Cretaceous, Upper Great Plains, U. S. A., *Science* **254**, 835-839 (1991).

98. F. L. Sutherland. Volcanism around K/T boundary time and its role in an impact scenario for the K/T extinction events, *Earth Sci. Rev.* **36**, 1-26 (1994).

Proc. 30th Intern. Geol. Congr., Vol. 26, pp. 55-66
Wang *et al.* (Eds)

Hugoniot Data for Jilin Ordinary Chondrite and Nandan Iron Meteorite

DAI CHENGDA   WANG DAODE
Guangzhou Institute of Geochemistry, Chinese Academy of Sciences, Guangzhou 510640, P. R. China
FU SHIQIN   SHI SHANGCHUN   JIN XIAOGANG
National Laboratory of Shock Wave and Detonation Physics, Chinese Academy of Engineering Physics, Chengdu 610003, P. R. China.

Abstract

Hugoniot data for Jilin ordinary chondrite and Nandan iron meteorite are measured by the electric-contact pin technique under one-dimensional strain condition on the dynamic high pressure device equipped with a two-stage light gas gun. During the pressure range from 10 to 60GPa, the shock wave velocity D *versus* particle velocity u of Jilin ordinary chondrite can be expressed as $D = 3.895(\pm0.252) + 1.175(\pm0.114)u$ (0.5km/s<u<2.6km/s). During the pressure range from 60 to 204GPa, the shock wave velocity D *versus* particle velocity u of Nandan iron meteorite can be expressed as $D = 3.884(\pm0.006) + 1.840(\pm0.005)u - 0.12267(\pm0.00105)u^2$ (1.3km/s<u<3.2km/s), where the number in parentheses is one standard deviation uncertainty. Equation-of-states describing their P-V-E relations are chosen, and their values of parameters are evaluated. Demonstrated by the comparison between P-V curves of equation-of-states and experimental data points, the P-V relation of Jilin ordinary chondrite can be described by the universal form of equation-of-state, of which parameters are zero-pressure bulk modulus K_{0S} =52.80(±0.22)GPa, its pressure derivative K'_{0S} =3.70(±0.46). That of Nandan iron meteorite can be described by the three-term form of equation-of-state, of which parameters of matters are Q=41.23531GPa, q =12.27179. The equation-of-state defined by Hugoniot data provides strong empirical successes in describing the compression of meteorites to high pressure.

Keywords: Jilin ordinary chondrite, Nandan iron meteorite, Hugoniot data, equation-of-state, shock compression

INTRODUCTION

Hugoniot data for meteorites is one of the most fundamental physical properties when we study hypervelocity impact processes of meteorites. So far, however, Hugoniot data for only a few ordinary chondrites have been measured experimentally [1, 2]. These Hugoniot data are inconsistent, which apparently is an effect of the inhomogeneity of the chondrites used in these experiments [3]. But rocks with inhomogeneity can be approximate shock wave velocity *versus* particle velocity by linear relationship [4]. Therefore, we think that scattered Hugoniot data for chondrites are principally resulted from thinness of chondritic discs used in experiments. The thickness of chondritic disc, about 2mm, may be too thin to represent a bulk physical property of the specimen. So we try to design maximum thickness of chondritic disc constrained by lateral and presetting rarefaction of the shock wave propagation in chondritic disc.

In addition, based on the Hugoniot data for ordinary chondrites and iron meteorites, their equation-of-states fit for describing its *P-V-T* or *P-V-E* relation can be defined. Their equation-of-states perhaps play an important role in constraining the composition of the earth's interior. Some differentiated meteorites and some undifferentiated chondrites are predominated materials forming terrestrial planets [5]. Silicate constituents of the earth's mantle are similar to those of chondritic silicate to a certain degree. Equation-of-states of ordinary chondrites are useful to define compositions and thermodynamic properties of the earth's mantle by comparative methods. Furthermore, iron is considered to be the predominate component of the earth's core [6]. However, iron meteorites mainly composed of iron-nickel alloy are pieces of metal cores of asteroids. Therefore, in order to define principal constituents of the earth's core, it is much more reasonably to choose iron meteorite than pure iron as the analogue to the earth's core. So we measure Hugoniot data for Jilin ordinary chondrite and Nandan iron meteorite, and evaluate the values of parameters of their equation-of-states.

As we know, equation-of-states play important roles in thermodynamic calculations. However, not all equation-of-states of materials can be defined by theoretical methods. A set of well-developed experimental methods and techniques can be used to measure equation-of-states [7]. The popular way to determine equation-of-states is to measure the pressure-volume relation of the matter studied. For the reason that $P=P(V, T)$ or $P=P(V, E)$ equation-of-state defines the surface in the *P-V-T* or *P-V-E* dimension space, but *P-V* relation determined by experimental methods just represents pressure-volume curve of the equation-of-state projected on *P-V* plane. Only given the *P-V* compression curve will it be impossible to define the surface corresponding to its equation-of-state. To solve the difficulty, a general approach is to develop reasonable mathematical expressions fit for describing its *P-V-T* or *P-V-E* relation of the matter studied to high pressure. As for minerals, rocks and meteorites, several mathematical expressions of equation-of-states fit for these geological condensed matters, have been developed, including Bridgman [8], Murnaghan [9], Gruneisen [7], Slater [10], density polynomial [11], universal [12], finite strain [13] and three-term forms of equation-of-states [7], *etc.*

HUGONIOT DATA MEASUREMENT

Experimental Method

The meteoritical specimen used for Hugoniot data measurement were Jilin ordinary chondrite and Nandan iron meteorite. The samples were made into ϕ 5mm×2mm discs. The average initial densities of Jilin ordinary chondrite and Nandan iron meteorite are 3.469 g/cm^3 and 7.780 g/cm^3, respectively. Jilin ordinary chondrite is constituted mineralogically of olivine (29.52wt%), pyroxene (32.96wt%), plagioclase (9.1wt%), iron-nickel metal (18.66wt%) and troilite (6.21wt%) while Nandan iron meteorite are principally composed of Fe (92.5wt%), Ni (6.8wt%) and Co (0.47wt%) in chemical composition. The planar shock wave producer was a two-stage light gas gun (ϕ 23mm inner caliber) in National Laboratory of Shock Wave and Detonation Physics, Chinese Academy of Engineering Physics. In the experiment only shock wave velocity of meteorite D and impact velocity W (for calculation of particle velocity u) of flyer need to be measured. Shock wave velocity D was measured by the electric-contact pin technique [7]. Fig. 1 is the schematic diagram of shock wave velocity measurement for meteorite discs. The electric-pins used were copper-pins of 1mm in diameter. Little springs were fixed on pins for the purpose of close contact between centrical-pin and sample, outlying electric-pin and driver plate. 6μm-thick Mylar insulation films were respectively set on between centrical electric-pin and sample, outlying electric-pin and driver plate in order not to form a circuit when a voltage was loaded. When flyer impacted target, a shock wave front reached outlying electric-pin , and destroyed Mylar insulation film, and so the pulse time t_1 of the outlying electric-pin represents the time when shock wave reached the front surface of sample disc while that

t_2 of the centrical electric-pin represents the time when it reached the back surface of sample disc. Therefore, time interval Δt can be expressed as $\Delta t = t_2 - t_1$. If the sample thickness d was measured in advance, shock wave velocity $D = d/\Delta t$. Time intervals were determined by 32-channel CAMAC time interval recorder. A centrical electric-pin was fixed on the sample center while three outlying electric-pins distributed in a equilateral triangle around the sample disc were fixed on the driver plate, and so three time intervals between outlying electric-pins and centrical electric-pin can be recorded. Thus three shock wave velocities can be given for each meteorite disc. In this paper the arithmetic mean of three shock wave velocities of each Jilin ordinary chondrite disc is given due to considerable shock-impedance difference of different kinds of minerals. However, the planar front velocity of each Nandan iron meteorite disc was corrected geometrically [7]. W was measured by the magnet-induced system for flyer velocity [14]. Particle velocity u was calculated by the impedance-match method [15]. The standard material for flyer and driver plate are copper and Ly-12#Al. The Hugoniot parameters of copper are ρ_0=8.930 g/cm^3, C_0=3.940 km/s, λ=1.489 while those of Ly-12#Al are ρ_0=2.70 g/cm^3, C_0=5.328 km/s, λ=1.338 [16].

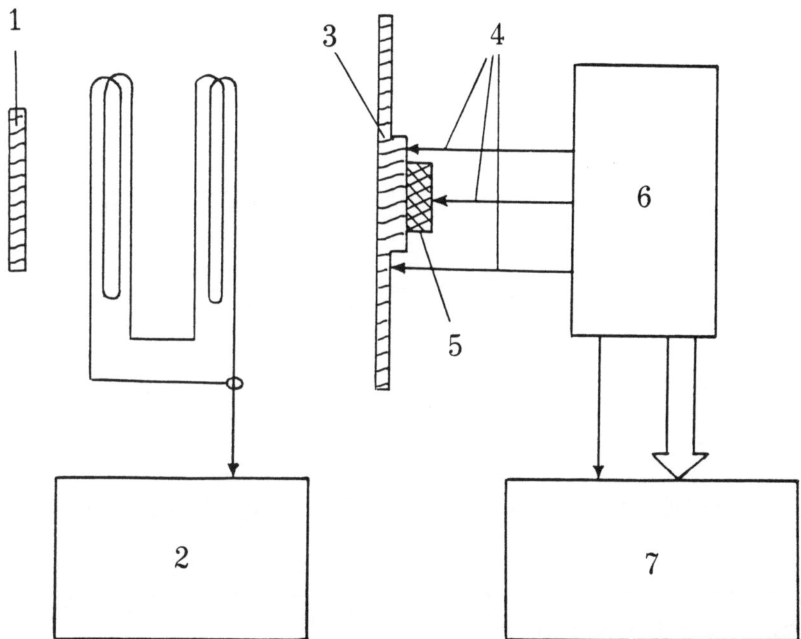

Figure 1. Schematic diagram of shock wave velocity measurement for meteorite disc. 1, Flyer plate; 2, Magnet-induced system for projectile velocity measurement; 3, Driver plate; 4, Electric-contact pins; 5, Sample disc; 6, Pulse circuit network; 7, Time interval recorder.

Results

Hugoniot data for Jilin ordinary chondrite are listed in Table 1, and $D(u)$ Hugoniot is plotted in Fig. 2.

As shown in Fig. 2, in the experimental pressure range (10–60GPa), the D *versus* u of Jilin ordinary chondrite can be fitted to be in linear relation with the square-least method,

$$D=3.895(\pm 0.252)+1.175(\pm 0.114)u \quad (0.5\ \text{km/s}<u<2.6\ \text{km/s}) \qquad (1)$$

where the number in parentheses is one standard deviation uncertainty. Based on mass and energy conservation equations of the compressed and uncompressed medium loaded by parallel planar shock wave [15], and *eq.* (1) is substituted into them, the pressure-volume relation of Jilin ordinary chondrite can be expressed as

$$P=\rho_0 C_0^2 \eta/(1-\lambda\eta)^2 \quad (2)$$

here $\eta=1-V/V_0$, representing relative compressibility. $C_0=3.895(\pm 0.252)$ km/s, approximating to zero-pressure bulk sound velocity. $\lambda=1.175(\pm 0.114)$, which is experimental constant.

shot No.	flyer-driver	W_f (km/s)	ρ_0 (g/cm^3)	d (mm)	t_{1i} (ns)	t_1 (ns)	t_2 (ns)	D (km/s)	u (km/s)	P (GPa)
1	Al-Al	2.862	3.397	3.130	240 204 198	214	767	5.660	1.447	27.82
2	Al-Al	3.733	3.489	2.790	190 170 201	187	646	6.078	1.869	39.64
3	Al-Al	2.138	3.498	2.918	218 241 167	209	769	5.211	1.074	19.57
4	Al-Al	1.061	3.491	3.131	263 332 348	314	1038	4.325	0.552	8.33
5	Al-Al	4.626	3.500	2.990	145 152 150	149	600	6.630	2.300	53.38
6	Cu-Al	3.772	3.448	3.024	196 140 91	142	596	6.661	2.614	60.04
7	Al-Al	3.831	3.535	2.975	191 203 180	191	651	6.468	1.863	42.60

Table 1. Hugoniot data for Jilin ordinary chondrite.
W_f represents projectile(flyer) velocity; ρ_0 represents initial density; d represents thickness of sample disc; t_{1i} represents the time when shock front triggers three electric-contact pins on the front surface of sample disc; $t_1=(t_{11}+t_{12}+t_{13})/3$; t_2 is the time when shock wave front reaches the back surface of sample disc; D represents shock wave velocity; u represents particle velocity; P represents shock pressure.

It should be pointed out that Hugoniot data points of which shock pressures are over 70GPa deviate from the linear band, and scattered seriously. After performing several conditional experiments, we believe that such a scattered phenomenon or unstable state is not occasional, possibly result from thinness of specimen discs.

Hugoniot data for Nandan iron meteorite measured by the same method are given in Table 2, and $D(u)$Hugoniot is plotted in Fig. 3.

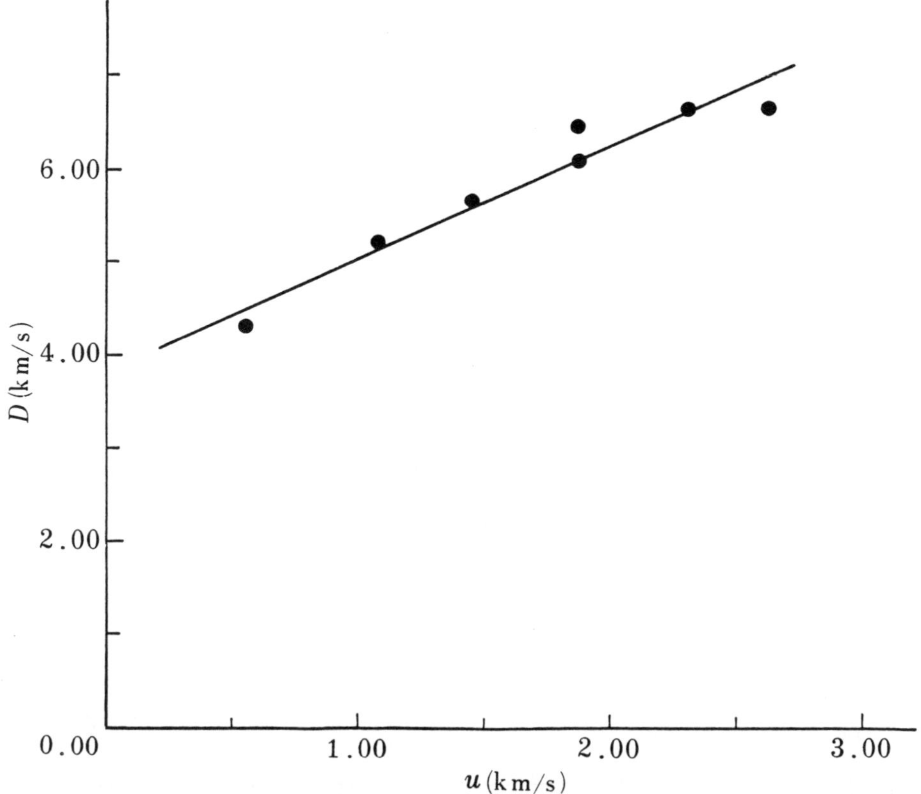

Figure 2. $D(u)$ Hugoniot of Jilin ordinary chondrite.

As shown in Fig. 3, D *versus* u of Nandan iron meteorite is a binomial, which can be fitted to be

$$D=3.884(\pm 0.006)+1.840(\pm 0.005)u-0.12267(\pm 0.00105)u^2 \qquad (3)$$

where the number in parentheses is one standard deviation uncertainty. In the same way, the pressure-volume relation of Nandan iron meteorite can be expressed as

$$P=\{\rho_0 C_0^2\, \eta/[(1-\lambda\eta)^2-2C_0\lambda'\eta^2]\}$$

$$\{1+C_0^2\lambda'^2\, \eta/[(1-\lambda\eta)^2-2C_0\lambda'\eta^2]+\cdots\} \qquad (4)$$

here ρ_0=7.780 g/cm^3, $\eta=1-V/V_0$, P is shock pressure, ρ is density, C_0=3.884(±0.006) km/s, λ=1.840(±0.005), λ'=–0.12267(±0.00105)(km/s)$^{-1}$ and subscript 0 represents initial state.

EQUATION-OF-STATES

Jilin Ordinary Chondrite

Equation-of-states possibly fit for describing adiabatic compression of Jilin ordinary chondrite in the experimental pressure range are universal, density polynomial, finite stress, and Murnaghan forms. If choosing the universal form [12],

$$P_S=3K_{0S}(1-x)\,x^{-2}\exp[(3/2)(K'_{0S}-1)(1-x)] \quad (5a)$$

and

$$\Delta E_S=9\{K_{0S}V_0/[(3/2)(K'_{0S}-1)]^2\}$$

$$\{1+[(3/2)(K'_{0S}-1)(1-x)-1]\exp[(3/2)(K'_{0S}-1)(1-x)]\} \quad (5b)$$

where $x=(V/V_0)^{1/3}=(\rho_0/\rho)^{1/3}$. K_{0S} and K'_{0S} are zero-pressure bulk modulus and its pressure derivative along Hugoniot curve. ΔE represents the change of internal energy between compressed and uncompressed state. Subscript S represents adiabatic compression. $V_0=1/\rho_0=0.2883$ cm^3/g. Parameters to be evaluated only are K_{0S} and K'_{0S}. According to their definitions,

$$K_{0S}=-V(dp/dV)_{p\to 0}=\rho_0 C_0^2\,[(1-\eta)(1+\lambda\eta)/(1-\lambda\eta)^3]_{\eta\to 0}=\rho_0 C_0^2 \quad (6a)$$

and

$$K'_{0S}=(dK/dp)_S=4\lambda-1 \quad (6b)$$

Substituting $\rho_0=3.469$ g/cm^3, $C_0=3.895(\pm0.252)$ km/s, $\lambda=1.175(\pm0.114)$ into *eqs*. (6a) and (6b),

$$K_{0S}=52.80(\pm0.22)\text{GPa};\ K'_{0S}=3.70(\pm0.46).$$

The standard deviation σ_P between measured pressure-volume data and $P_S(x)$ curve is 3.71GPa, suggesting it is suitable on the whole.

No. of shots	flyer-driver	W (km/s)	D (km/s)	u (km/s)	P (GPa)	ρ (g/cm^3)	$\theta°$
1	Cu-Cu	2.592	6.160	1.347	64.2	9.950	1.03
2	Cu-Cu	3.332	6.684	1.723	91.2	10.432	5.20
3	Cu-Cu	4.206	7.240	2.223	123.0	11.272	3.85
4	Cu-Cu	4.215	7.297	2.198	125.5	11.133	4.62
5	Cu-Cu	4.206	7.341	2.193	127.0	11.094	2.75
6	Cu-Cu	4.879	7.594	2.564	151.9	11.746	2.81
7	Cu-Cu	4.875	7.957	2.546	155.5	11.440	3.20
8	Cu-Cu	4.908	7.848	2.634	158.0	11.710	5.53
9	Cu-Cu	5.164	7.895	2.754	162.0	12.206	3.00
10	Cu-Cu	5.591	7.802	2.984	181.0	12.598	0.21
11	Cu-Cu	5.956	8.393	3.115	204.0	12.372	2.90

Table 2. Hugoniot data for Nandan iron meteorite
θ is the angle included between the shock front(planar) and sample surface, of which calculation is referred to reference [7].

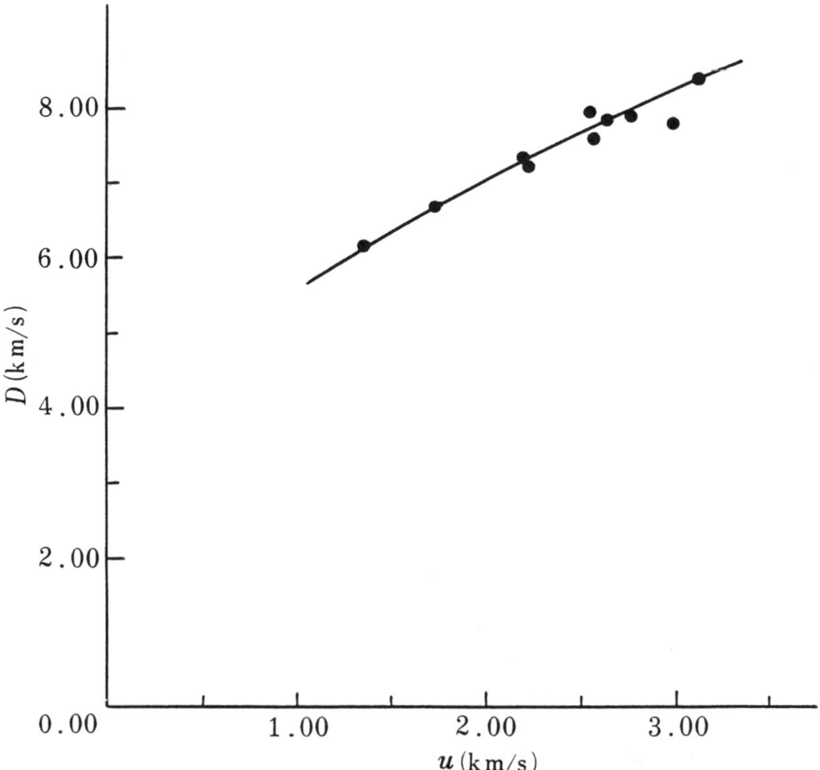

Figure 3. $D(u)$ Hugoniot of Nandan iron meteorite.

If choosing the density polynomial form [11],

$$P_S = K_{0S}(w-1)\{1+[(K'_{0S}-1)/2](w-1)\} \quad (7a)$$

and

$$\Delta E_S = K_{0S}V_0\{\ln w + 1/w - 1 + [(K'_{0S}-1)/2](w - 1/w - 2\ln w)\} \quad (7b)$$

here $w=V_0/V$. The standard deviation σ_P Between measured pressure-volume data and $P_S(w)$ curve is 4.31GPa.

If choosing the finite strain form [13],

$$P_S = 3K_{0S}f(nf+1)^{(n+3)/n}(1+a_1 f) \quad (8a)$$

and

$$\Delta E_S = 9(K_{0S}V_0/2)[1+2(a_1/3)f]f^2 \quad (8b)$$

where $f=(1/n)[(V_0/V)^{n/3}-1]$; $a_1=(3/2)(K'_{0S}-n-2)$. If choosing Eulerian stress reference, n=2, $f=(1/2)[(V_0/V)^{2/3}-1]$. The standard deviation σ_P between measured pressure-volume data and $P_S(f)$ curve is 4.22GPa.

If choosing Murnaghan form [9], $P = K_0/K'_0 [(V_0/V)^{K'} - 1]$. P_{SM} curve describing P-V relation is obviously deviated from measured pressure-volume points over about 40 GPa shock pressure. Demonstrated by calculated standard deviation σ_P above or shown in Fig. 4, the universal equation-of-state is the fittest one.

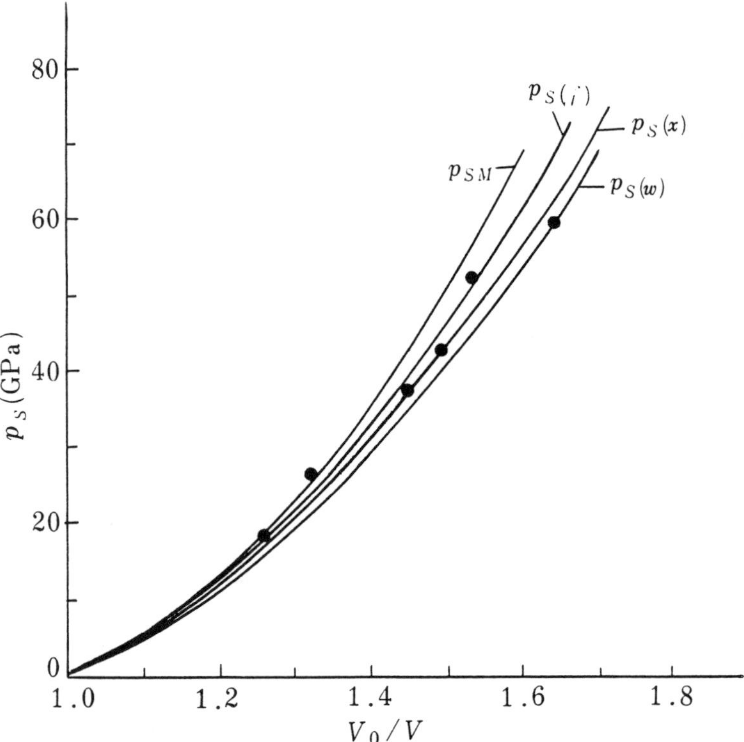

Figure 4. Comparison between equation-of-state curves describing P-V relation and measured pressure-volume points for Jilin ordinary chondrite. $P_S(x)$, $P_S(f)$, $P_S(w)$ and P_{SM} represent the universal, finite strain, density polynomial and Murnaghan forms equation-of-states for Jilin ordinary chondrite.

Nandan Iron Meteorite

Choosing the three-term form of equation-of-state [17],

$$P(V, T) = p_x(V) + [(3\gamma + z)/(1+z)](\rho RT/\mu) f(\Theta/T) + (1/4)\rho\beta_0 (V/V_0)^{1/2} T^2 \quad (9a)$$

and

$$E(V, T) = E_x(V) + [(2+z)/(1+z)](3RT)/(2\mu) f(\Theta/T) + (1/2)\beta_0 (V/V_0)^{1/2} T^2 \quad (9b)$$

where the first term $P_x(V)$ and $E(V)$ are the contributions to $P(V, T)$ and $E(V, T)$ at zero Kelvin respectively. The second terms are heat contributions of lattice vibration to $P(V, T)$ and $E(V, T)$ respectively. The third terms are heat contributions of free electrons to $P(V, T)$ and $E(V, T)$. $\gamma = \gamma(V)$ is Gruneisen coefficient. z is the parameter of vibration deviation of solid. Θ is Debye temperature. T is Kelvin temperature. μ is mass per mole. R is universal gas constant. β_0 is specific heat coefficient of electrons at ambient pressure. Function $f(\Theta/T)$ is

$$f(\Theta/T) = 3/x^3 \int_0 y^3 \, dy/(e^y - 1),$$

$x = \Theta/T$. For Nandan iron meteorite, μ=56. 14g/mol. Θ and β_0 can be substituted by those of pure iron(Θ=457K, β_0=1. 939×10^{-5}J/g·K^2) [17] respectively. In fact, at relatively lower temperature (T<10^5K). It cannot cause considerable error because Θ and β_0 play less important effect on pressure and internal energy. Seen from *eqs*. (9a) and (9b), determining z and γ are the key to determining the heat contributions of lattice vibration.

The parameter of vibration derivation of solid z can be expressed as [7]

$$z = lRT / \mu C_x^{\ 2} \tag{10a}$$

where l is non-harmonic vibration parameter. The other parameters are the same as above in physical meaning. C_x is sound velocity at zero Kelvin,

$$C_x = [(1/\rho_{0K})dP_x/d\delta]^{1/2} \tag{10b}$$

where $\delta = \rho/\rho_{0K}$, representing shock compression ratio. ρ_{0K} is zero-pressure density at zero Kelvin. For Nandan iron meteorite, ρ_{0K} =7.95 g/cm^3, l = 8 (substituted with that of pure iron [17]).

Gruneisen coefficient $\gamma(V)$ is defined as below [7],

$$\gamma(V) = (\alpha/2 - 2/3) - (V/2) \cdot \frac{d^2(P_x V^{\alpha})/dV^2}{d(P_x V^{\alpha})/dV} \tag{11}$$

If choosing $\alpha = 0$, $\gamma(V)$ is Slater formula; if choosing α=2/3, $\gamma(V)$ is Dugdale-MacDonald formula; if choosing α=4/3, $\gamma(V)$ is free volume formula. α value chosen in *eq*. (11) should be in agreement with Hugoniot data. For Nandan iron meteorite, α value is chosen to be zero as a result of the fact that γ_{S0} at zero pressure is fit for the empirical formula λ=(γ_{S0}+2/3)/2.

$P_x(V)$ and $E_x(V)$ of Nandan iron meteorite can be expressed in Born-Meyer potential [17],

$$P_x(V) = Q\delta^{2/3}\{\exp[q(1 - \delta^{-1/3})] - \delta^{2/3}\} \tag{12a}$$

and

$$E_x(V) = 3(Q/\rho_{0K})\{(1/q)\exp[q(1-\delta^{-1/3})] - \delta^{1/3} - (1/q-1)\} \tag{12b}$$

where ρ_{0K} and δ represent the same physical meaning as above. Q and q are two parameters representing matter properties. Q and q are only two parameter to be evaluated. Q and q are determined by the two-dimensional optimum seeking method [17]. Q=41.23531GPa, q =12.27179, which are determined by optimizations and substitutions in *eqs* (4), (9a) and (9b).

The pressure-volume curves of $P(V, T)$ equations for Nandan iron meteorite is plotted in Fig. 5. In the experimental pressure range (64-204GPa), the standard deviation σ_P between measured pressure-volume points and $P(V, T)$ curve of the three-term form of equation-of-state is 15.89GPa. If choosing the density polynomial form of equation-of-state, the standard deviation σ_P =17.59GPa. If choosing the universal form of equation-of-state, the standard deviation σ_P =21.07GPa. The finite strain and Murnaghan form of equation-of-state are considerably deviated from measured data points. Demonstrated by the calculated standard deviation σ_p above or seen from Fig. 5, the universal form of equation-of-state is the fittest one.

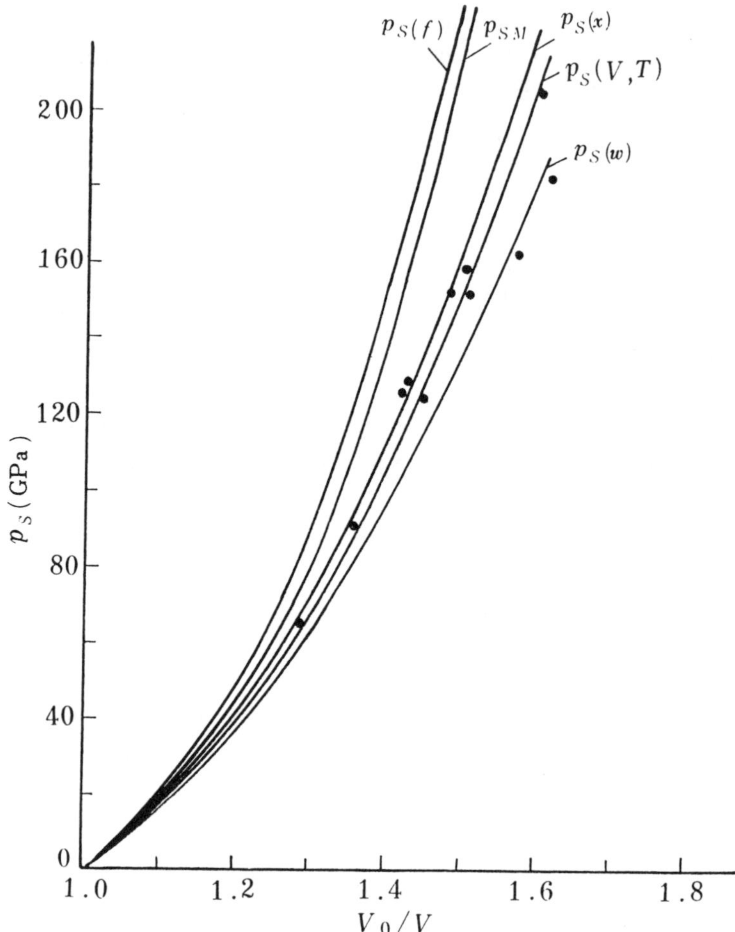

Figure 5. Comparison between equation-of-state curves describing *P-V* relation and measured pressure-volume points for Nandan iron meteorite. $P(V, T)$, $P_S(x)$, $P_S(f)$, $P_S(w)$ and P_{SM} represent the three-term, universal, finite strain, density polynomial and Murnaghan forms of equation-of-states, respectively.

Three-term form of equation-of-state is based on quantum mechanics and Debye model. That is to say, only under the circumstances of known crystal structure and values of parameters will it be possible to define their mathematical expressions of three-terms. The three-term form of equation-of-state is only fit for describing the compression of element and chemical compound solids. Therefore, it is impossible to define the three-term form of equation-of-state of Jilin ordinary chondrite due to different mineral assemblage. Even if Nandan iron meteorite is a Fe-Ni alloy, values of some physical parameters of pure iron were used in the course of defining its three-term form of equation-of-state. However, we are still satisfied with the result. The universal form of equation-of-state is based on a model of interatomic bonding energies. So it is fit for describing *P-V-E* relation of Nandan iron meteorite. The density polynomial form is simply derived from an expansion of the pressure in powers of relative density change. So it is perhaps fit for iron meteorites and ordinary chondrites. Similarly, the finite strain form is derived from an expansion of the pressure in powers of relative specific volume change(η=1−V/V_0). It is fit for describing the *p-V-E* relation of Jilin ordinary chondrite at relatively low pressure. But it causes considerable derivation to define that of Nandan iron meteorite at very high pressure, possibly due to the plastic deformation or liquidation. Murnaghan equation is based on the assumption that the bulk modulus can be expressed as a series in $P(K=K_0+K_0'P)$, of which the upper pressure limit is not more than 50GPa. Therefore, it is within our expectation that Murnaghan equation is unfit for describing the compression of iron meteorite and ordinary chondrite to high pressure.

SUMMARY

It is not beyond our expectation that Jilin ordinary chondrite and Nandan iron meteorite belonging to two basic types of meteorites respectively are fit for describing their shock compression to high pressure by different forms of equation-of-states as a result of the fact that Nandan iron meteorite is a Fe-Ni alloy on the whole while Jilin ordinary chondrite is an assemblage of different minerals, and their structures, porosities as well as other physical properties are different. Generally speaking, the adiabatic compression of ordinary chondrites can be described by universal, finite strain and density polynomial forms of equation-of-states. However, that of iron meteorite can be described by three-term, density polynomial and universal forms of equation-of-states. Practically, we can choose several expression of equation-of-states which are expectedly fit for meteorite we studied, evaluate values of parameters of those equation-of-states, compare the *P-V* or *P-E* relation curve of those equation-of-states with measured pressure-volume point, and then choose the form of equation-of-states of least standard deviation as the fittest one.

In the experimental pressure range, the adiabatic compression of Jilin ordinary can be best described by the universal form of equation-of-state, of which parameters are zero-pressure bulk modulus K_{0S} =52.80(±0.22)GPa, its pressure derivative K'_{0S}=3.70(±0.46). That of Nandan iron meteorite can be best described by the three-term form of equation-of-state, of which parameters of matter are Q=41.23531GPa, q =12.27179. Their application will be discussed in another paper.

Acknowledgements

We thank Chen Pansen, Huang Yue, Dong Shi, He Hongliang, Hu Jinbiao, Hu Qiang in National Laboratory of Shock Wave and Detonation, Chinese Academy of Engineering Physics for their excellent technical assistance in carrying out the shock experiments. We gratefully acknowledge the constructive suggestions by Prof. Jing Fuqian. This work is supported by grants from National Natural Science Foundation of China (NSFC49274201).

REFERENCES

1. Lin Wenzhu. Shock wave compression of Jilin ordinary chondrite and pyrolite model, *Chinese Journal of Space Science* **4:4**, 338-345 (1984).
2. T. Matsui, K. Kani and Y. Matsushima. Hugoniot of some ordinary chondrites. *Papers Presented to 11th Symposium on Antarctic Meteorites* 150-151. NIPR, Tokyo (1986) (abstract).

3. R. T. Schmitt, A. Deutsch and D. Stoffler. Calculation of Hugoniot curves and post-shock temperatures for H- and L-chondrites. *Lunar and Planet. Sci. Conf. XXV* 1209-1210 (1994) (abstract).
4. R. G. McQueen, S. P. Marsh and J. N. Fritz. Hugoniot equation of state of twelve rocks. *J. Geophys. Res.* **72:20**, 4999-5036 (1967).
5. Wang Daode, Chen Yongheng, Li Zhaohui, Lin Yangting, Dai Chengda, Hu Ruiying, Yi Weixi, Liu Jingfa and Huang Wankang. *An Introduction to Chinese Meteorites*. Science Press, Beijing, 434-451(1993) (in Chinese).
6. T. J. Shankland, O. L. Anderson and D. Young. Iron workers Convention I: Workshop on physics of iron, *J. Geophys. Res.* **95:B13**, 21689-21690 (1990).
7. Jing Fuqian, Li Dahong and Zhu Lichang. *An Introduction to Experimental Methods of Equation-of-states of Solids*. Science Press, Beijing (1986) (in Chinese).
8. P. W. Bridgman. *The Physics of High Pressure*. Bell and Sons, London (1958).
9. F. D. Murnaghan. The compressibility of media under extreme pressure, *Proc. Natl. Acad. Sci. U. S. A.* **30,** 244-247 (1944).
10. J. C. Slater. *Introduction to Chemical Physics*. McGraw-Hill, New York (1939).
11. S. Eliezer, A. Ghatak and H. Hora. *An Introduction to Equation of State: Theory and Application*. Cambridge University Press, New York (1986).
12. P. Vinet, J. Ferrante, J. R. Smith and J. H. Rose. A universal equation of state for solids, *J. Phys. C* **19**, L467-L473 (1986).
13. R. Jeanloz. Shock wave equation of state and finite strain theory, *J. Geophys Res.* **94:B5**, 5873-5886 (1986).
14. Shi Shangchun, Chen Pansen and Huang Yue. Velocity measurement of magnet induced system for projectile, *Chinese Journal of High Pressure Physics* **5:3**, 205-214 (1991).
15. D. Stoffler. Density of minerals and rocks under shock compression. In; *Landolt-Bornstein-Numerical Data and Functional Relationship in Science and Technology. New Series, Group V, Geophys. Space Res. , Vol. sub. a.* K. H. Hellwege (Ed). pp. 120-183. Springer-Verlag, Berlin, (1982).
16. R. G. McQueen, S. P. Marsh, J. W. Taylor, J. N. Fritz and W. J. Carter. Chapter VII. In: *High-Velocity Impact Phenomena.* R. Kinslow (Ed). Academic Press, New York and London (1970).
17. Xu Xishen and Zhang Wanxiang. *An Introduction to Theorey of Equation-of-states*. Science Press, Beijing (1986) (in Chinese).

Proc. 30th Intern. Geol. Congr., Vol. 26, pp. 67-74
Wang *et al.* (Eds)

The 1908 Tunguska Explosion: Discovery of Iridium and Other Element Anomalies

HOU QUANLIN* MA PEIXUE
Institute of High Energy Physics, Chinese Academy of Sciences, P.O. Box 2732, Beijing 100080, P. R. China
* *present address: Laboratory of Lithosphere Tectonic Evolution, Institute of Geology, Chinese Academy of Sciences, Beijing 100029, P. R. China*

Abstract

The 15 peat samples collected from the layers affected by the 1908 Tunguska explosion in Tunguska area of central Siberia, Russia, were analyzed by NAA to determine the contents of Ir, REE, Ni, Co, Fe and Sb. The analytical results indicate that Ir concentrations in and above the event layer were 0.54~0.24 ppb, about 10~20 times higher than that in layers below, and other elements were all enriched by a factor of 2 to 5. In addition, the variation of Ni concentration was closely related with Ir in the explosive layer, and the patterns of chondrite-normalized REE are almost flat with a ratio of about 1. Hence, it can be inferred from the characteristics of elemental geochemistry that the explosion likely was associated with extraterrestrial material. Due to its explosion before reaching the ground, the density of the explosion body must have been quite low, the chemical composition of which was probably similar to the carbonaceous chondrite (C1). In terms of the Ir flux in the explosion area, it can be estimated that the celestial body exploded weighed about 1.9×10^{10} tons, corresponding to 1~1.5km in radius, and might have released energy equal to ~10^4 megatons.

Keywords: 1908, Tunguska explosion, elements anomaly, NAA

INTRODUCTION

On 30 June 1908 at 7:17 a.m. local time (12:17 U.T.), a great explosion occurred in the sky over the basin of River Podkamennaya Tunguska in Central Siberia (60° 55′ N, 101° 57′ E). Since then, many expeditions have been undertaken to the site and many books and reports published [1-21]. No crater has ever been discovered during any of these expeditions. The explosion was splendid and observed over a radius of 600km to 1000 Km in central Siberia, sunlight reflected from the debris lit up the night sky for several days over Europe and western Asia, and trees were blown down over an area of several hundred square kilometers near the epicenter.

The explosion, the latest one in modern history, has inspired many exotic explanations. Antimatter [4], a small black hole [8] and, inevitably, an exploding saucer [1], have all been proposed as means of liberating tens of megatons of energy in the atmosphere without cratering the Earth surface. Quantitative explanations in terms of a less exotic object have often suggested that it must have been a low-density comet ($\rho_m \approx 10^{-3}$ - 10^{-2}g cm^{-3}) to have exploded before reaching the ground [10, 11, 20]. Compared to the density of roughly 0.6g cm^{-3} - 1.0g cm^{-3} reported for comet Halley [18], however, this would make the Tunguska object decidedly unusual [15]. Proponents the cometary explosion theory could point to the possibility of sunlight reflected from the parts of the comet tail that missed the Earth, but other meteorites do not possess tails. Ganapathy thought that the exploded body weighted more than 7 million tons, was more than 0.16 kilometer in diameter, and may well have been a stony meteorite, and that the sunlight reflected from the debris from the explosion lit up the night sky [6]. The debris from the explosion was also discovered in a South Pole ice core [6]; this discovery indicates that the Tunguska object exploded and vaporized in the atmosphere with subsequent stratospheric injection and transport of the debris all over the Earth. Chyba *et al.* pointed out that the Tunguska event represents a typical fate for stony asteroids tens of meters in radius entering the Earth's atmosphere at common hypersonic velocities, and that Comet and carbonaceous asteroids of the appropriate energy disrupt too high, whereas typical iron objects reach and crater the terrestrial surface [3]. The energy released from the explosion has been estimated from tens magetons [3] to 10^3 megatons [20]. In contrast, Lyne *et al.* suggested that a carbonaceous chondrite was the most probable cause of the Tunguska event [13].

sample No.	dry weight (mg)	dry ratio (%)	ashed weight (mg)	ashed rate (%)
N4-1	617.11	15.40	12.47	2.02
N4-2	694.74	17.37	9.78	1.41
N4-3	901.47	22.54	7.54	0.84
N4-4	813.56	20.34	7.09	0.87
N4-5	425.05	10.63	4.66	1.10
N4-6	700.20	17.51	5.15	0.74
N4-7	940.49	23.51	7.39	0.79
N4-8	741.98	18.55	7.3	0.98
N4-9	559.96	14.00	4.84	0.86
N4-10	383.45	9.59	4.19	1.09
N4-11	685.36	17.13	7.06	1.03
N4-12	842.06	21.05	9.90	1.18
N4-13	437.7	10.94	7.11	1.62
N4-14	1083.61	27.09	15.95	1.44
N4-15	948.26	23.71	14.22	1.50

Table 1. The related parameter in pretreatment.

Korina and the cooperators [10] analyzed the 15 peat samples near the event layer in Tunguska using neutron active analysis (NAA), and discovered a little Ir anomaly (17.2 ppt) at the bottom of the event layer, but no at the other site in the event layer. The anomalous Ir was 3~4 times higher than the background value (Ir=3.8 ± 1.8 ppt), and was considered to be caused by the Tunguska explosion [10]. In fact, such a low Ir anomaly at the bottom of the event layer was unnecessarily caused by the explosive body, because the 17.2 ppt Ir is still within the range of the upper crust (average 20 ppt Ir in the upper crust [22]). The elemental anomaly in strata caused by extraterrestrial event usually forms a few or tens years, even more, behind the event, because the debris produced by extraterrestrial event, e. g. the Tunguska explosion, can eject to the atmosphere even stratosphere, then gradually fall down to the Earth's surface. Therefore, the elemental anomaly caused by the Tunguska explosion, if present, should be at or above the event layer, rather than at the bottom.

The Tunguska explosion was the latest unusual event associated with extraterrestrial material. It has great significance to study the elemental geochemistry of the Tunguska explosion, not only for itself, but also for the extraterrestrial events taken place in the Earth's history, such as the K—T boundary event occurred 65Ma ago and undergone a long process of geological reformation [23, 24].

In this work, 15 peat samples collected from a core across the event layer in Tunguska area, the explosion site, in central Siberia, Russia, were analyzed using NAA to determine the contents of Ir, REE, Ni, Co, Fe and Sb. The aim is to investigate the characteristics of elemental variation in the upper part across the layer and to provide new data for interpretation of the Tunguska explosion.

EXPERIMENTS

Samples collection

The samples used in this work were collected from a core of the peat, having a section of 100 cm^2 and depth of 61 cm, at the epicenter of the Tunguska catastrophe. The peat age was determined with a method by Muldiyarov *et al.* in the paper [10]. The peat layers in the studied interval were found to be deposited from 1875 to 1953 [10]. The peat bed of 1908, i. e. the event layer, was estimated to be in the interval of 43cm-46cm (Fig.1). The collected 15 peat samples were from 45 cm to 0 cm (Fig. 1).

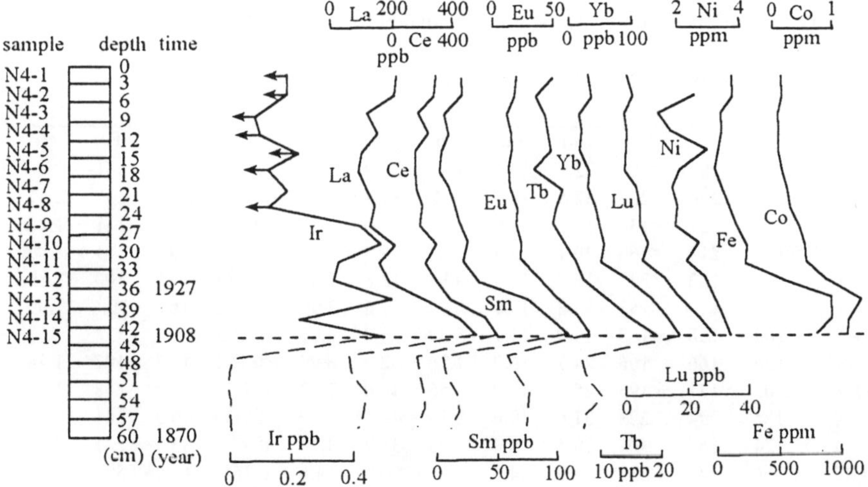

Figure 1. The elements abundance of the peat core at the epicenter of the Tunguska catastrophe — by Korina *et al.* [10].

Preparation

The pretreatment process of the 15 peat samples can be described as following three steps: (1) dry at 100℃ for 2 hours, (2) carbonize at 200 ℃ for 2 hours, and (3) ash at 500℃ for 5 hours. The dry and ash rates are listed in Table 1. The ashed samples were irradiated at a neutron flux of 5.2×10^{13}n cm^{-2} s^{-1} in a reactor of the Institute of Atomic Energy Science, Beijing, China, for 16 hours. They were counted after cooling 8 days and 17 days, respectively, using the big volume HPGe detectors, whose energy resolution and relative efficiency were about 1.85 KeV (1332.5 KeV for

^{60}Co), and 28%. A DD-8000 spectrometer made by Tsinghua University, China, was used to collect and process the radioactive data.

RESULTS AND DISCUSSION

The analytical results for the 15 peat samples are listed in Table 2. It can be seen that the Ir contents in the depth of 45cm-24cm (samples N4-15~N4-9) are 0.54 ppb-0.24ppb (Table 2, Fig. 1), much higher than that in the samples below the event layer (less than 0.02ppb [10]), while that in the samples above the depth of 24cm is less than the detection limit. The Ni content ranges from 3.47 ppm to 2.47 ppm, showing a similar changing tendency with Ir and anomaly near the event layer(Table 2, Fig.1). Besides, the REE, Co and Fe also present enrichment to a different degree. In one word, the Ir, REE, Ni, Co and Fe are all anomaly near the event layer. The main anomaly of Ir lies at a depth from 45 cm to 24 cm, and that of other elements, *e.g.* REE, Ni, Co and Fe, lie in relative lower horizons, from 45 cm to 30 cm depth, which can be interpreted that different elements have variable resident time in the atmosphere or the stratosphere.

The Origin of Element Anomalies

The reasons for the elements anomalies near the event layer are not beyond the possibilities as follow: (1) sedimentary rate decrease; (2) meteoric ablation increase; and (3) extraterrestrial material accretion associated with the Tunguska explosion. According to Korina *et al.*[10], the 60 cm depth should correspond to 1870, the 46cm~43 cm to 1908, and the 36cm to 1927. The sedimentary rates from 1908 to 1927 can be calculated to be 0.3~0.5cm/y, nearly same as that from 1870 to 1908. This indicates that the sedimentary rate is roughly stable at least from 1870 to 1927, and the elements anomalies unlikely resulted from the sedimentary rate decrease.

Sample No.	La	Ce	Nd	Sm	Eu	Tb	Yb	Lu	Ir	Sb	Fe	Ni	Co
N4-1	214	282	<286	32.9	19.4	4.93	29.0	3.97	<0.237	81.1	241	n	0.149
N4-2	208	276	<227	32.5	18.1	<1.42	34.4	6.09	<0.232	1111	237	2.47	0.143
N4-3	115	152	<136	18.5	10.9	2.98	12.0	2.68	<0.130	72.7	121	1.28	0.081
N4-4	143	213	<166	20.1	13.4	3.33	16.5	3.29	<0.145	106	141	1.87	0.080
N4-5	90.9	122	<241	14.5	12.3	3.82	12.8	3.00	<0.236	177	100	2.89	0.086
N4-6	78.8	110	<153	13.3	10.1	0.94	27.6	2.99	<0.162	164	90.6	1.99	0.113
N4-7	102	138	<150	19.9	11.1	5.39	23.0	6.01	0.222	96.1	138	1.62	0.200
N4-8	126	166	<166	24.3	16.9	4.26	32.7	6.99	<0.162	45.7	187	1.78	0.219
N4-9	105	131	<195	25.5	15.0	3.55	40.0	7.32	0.452	38.2	218	1.70	0.382
N4-10	181	244	<223	31.6	19.6	5.10	46.1	8.79	0.511	40.0	281	2.43	0.442
N4-11	124	167	105	29.5	19.4	7.08	38.2	7.76	0.371	32.3	271	2.08	0.463
N4-12	166	240	<232	48.9	34.5	8.25	80.9	14.6	0.355	42.9	596	2.69	0.793
N4-13	276	334	255	75.9	46.3	12.0	117	19.2	0.539	52.0	954	2.90	1.320
N4-14	382	539	332	97.0	63.8	15.5	140	25.5	0.240	54.8	996	n	1.140
N4-15	443	610	363	113	72.9	20.4	160	30.6	0.513	51.5	850	3.47	1.090
an.fa.	2	2	n	3	3	3	4	4	2	n	4	1.4	5

Table 2. The element content determined by NAA in and above the event layer of the 1908 Tunguska explosion. note: La-Sb, ppb; Fe-Co, ppm; n = no data; an. fa., anomalous factor: Ir = N4-9~N4-15/other samples, other elements = N4-11~N-15/other samples.

By comparing the distribution of Ir and Ni in the main event layer with that of ablation spheres separated from 2 kg of red clay sediment from the mid-Pacific Ocean [6], it can be found that there is no correlation between Ir and Ni in the 11 black and shiny metallic ablation spheres (Fig. 2a). Although the Ir content in these ablation spheres varies by a factor of 1000, the Ni content varies by less than a factor of 10 (Fig. 2a). The lack of correlation between Ir and Ni in the ablation spheres may be due to two factors: (1) these spheres come from in composition different iron meteors whose Ir concentration varies much more than their Ni concentration and this collection represent all the products over thousands of years, and (2) chemical processes associated with the melting of ablation products could vary the relative proportion of these two elements. In contrast, the correlation between Ir and Ni in the main event layer of the Tunguska explosion and their cosmic Ni/Ir ratio (about 5.8×10^3) indicate that the Ir and Ni come from a single object, *i.e.* the Tunguska explosive body, that contained the nonvolatile elements in cosmic proportion (Fig. 2b). As for the Ni/Ir ratio in the main event layer is 4 times lower than that in C1 (2.3×10^4), it may be due to that the Ni was more volatile than Ir during explosion.

The patterns of chondrite-normalized REE appear two characteristics: (1) The curves are all nearly flat with a radio of < 1 for samples of N4-1~N4-10 (Fig. 3 a, b), and ≈ 1 for samples N4-11~N4-15 (Fig. 3c); and (2) Eu shows slightly positive anomaly. These indicate that the REE has been rarely undergone chemical fractionation. The REE of terrestrial sediments was usually undergone chemical fractionation, and the curves of chondrite-normalized REE have these features: (1) LREE enrichment, HREE depletion; (2) Eu negative anomaly; and (3) total content of REE much higher than that of C1. These imply that the REE at the depth above 46 cm, especially from 46 cm to 30 cm, mainly come from the extraterrestrial matter, the Tunguska explosive body, which can be similar to C1 in chemical composition. Because the REE in chondrite generally is 104 times higher than that in iron asteroid, the Tunguska explosive body can not be typical iron object. In addition, the Sb/Ir ratio of 50 in the main event layer (depth of 46~30 cm) of Tunguska is between 4 and 76 of the metal spheres in the same event layer, which were proved from common stony meteorite by Ganapathy [6], although it is higher than that in C1 (0.29). Therefore, the Tunguska explosive body in 1908 must have been carbonaceous or common stony meteorite, rather than iron meteorite.

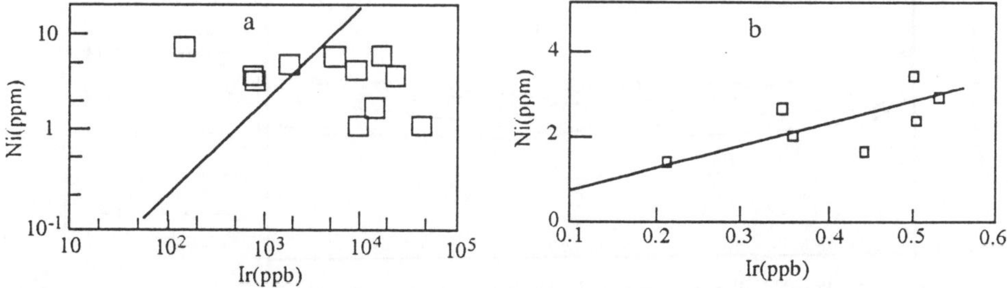

Figure 2. Correlation between iridium and nickel. (a) Metallic spheres from daily, meteoric ablation [6], do not show any such correlation, mainly because they are products from in composition different meteors over thousands of years. (b) In contrast, the correlation between Ir and Ni in the Tunguska main event layer indicates that they come from the same celestial body.

As for whether the Tunguska explosive body was a comet, so far, there is no chemical evidence yet. We agreed to the views of Chyba [3], *i.e.* that short-period and long-period comets with appropriate energies exploded far too high in the atmosphere to account for the observations.

Estimation of the Explosive Body

Some of meteoritic debris and a slightly Ir anomaly were discovered in ice horizons of 1909~1911 at a South Pole ice core [6], which indicates that the Tunguska object exploded and vaporized in the atmosphere with subsequent stratospheric injection and transport of the debris all over the Earth. The deposition rate at the poles, however, is expected to be lower than the global average for two reasons: (1) the poles have an unusually low precipitation rate and (2) the atmospheric circulation pattern preferentially concentrates stratospheric matter in mid-latitude rather than near the poles. This preferential enrichment factor is somewhere between 3 and 5 [25].

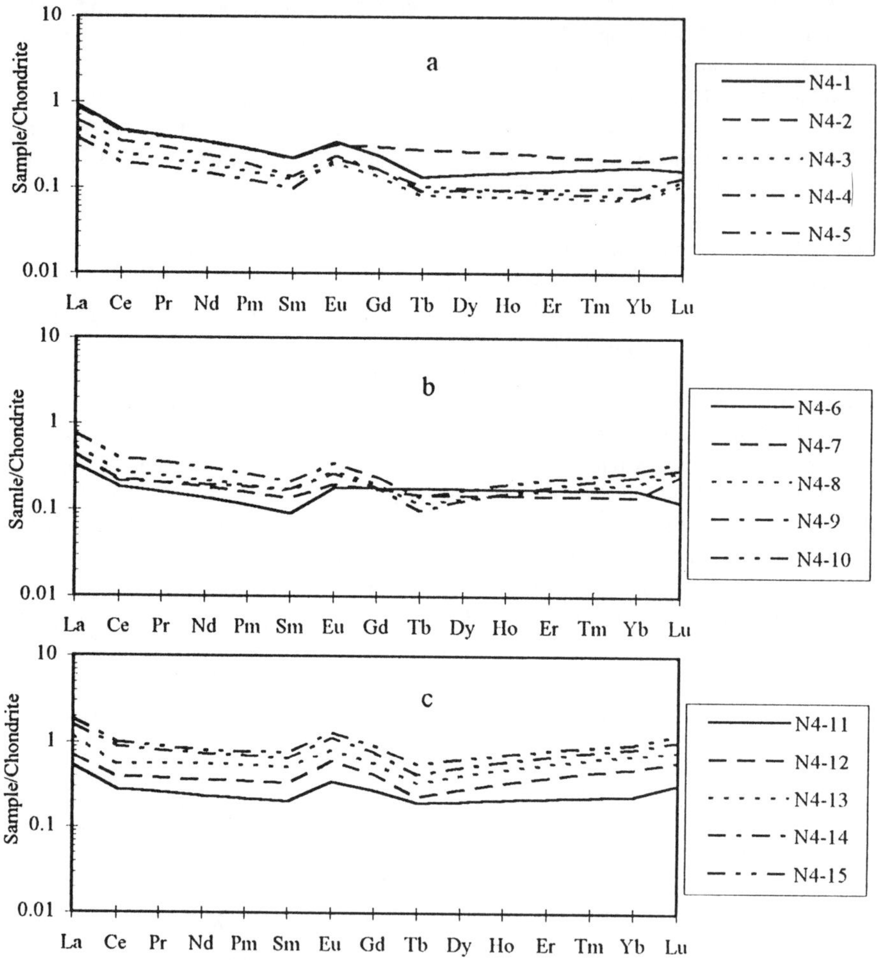

Figure 3. Patterns of chondrite-normalized REE of the 15 peat samples in Tunguska area.

On the basis of Ir contents in the peat layers from the Tunguska explosion site and the density of dried samples (0.12 g cm^{-3}), the flux of excess Ir in Tunguska area can be estimated to be 4×10^{-9} g cm^{-2}. It can be expected that the Ir flux at Tunguska should be much higher than that at Poles. According to Ganapathy data [6], the Ir flux associated with the Tunguska explosion event at the South Pole was about 1.6×10^{-13}g cm^{-2}. As a first-order approximation, using the average value ($\sim 1.2 \times 10^{-9}$g cm^{-2}) of Ir flux between the Tunguska area and South Pole, we can estimated the global fallout of Ir resulted from the Tunguska explosion to be about 6.8×10^{3} tons. Because it exploded before reaching the ground, the density of the explosive body must have been quite low (1.1g cm^{-3} [6]). As mentioned above, additionally, the chemical composition was probably similar to carbonaceous chondrite (C1). Hence, it can be estimated that the explosive body weighed about 1.6×10^{10}tons, corresponding to 1~1.5km in radius. The Tunguska object was usually thought to enter the atmosphere at common hypersonic velocities(20~30 km s^{-1}) [3, 6], and that the energy released by the explosion was only equal to the 1% of the total energy of entering the atmosphere [20]. So the explosive energy can be estimated to be about $\sim 10^{10}$ t TNT, *i.e.* $\sim 10^{4}$ megatons, which is slightly higher than that estimated by Turco *et al.* [20] (10^{3} megatons). Our estimation is higher than that of Ganapathy [6] (0.16km in diameter, weight 7×10^{6} t), and that of Chyba *et al.* [3] (30 m in radius, released energy 10~20 megatons), but they thought that their estimations all were minimum.

CONCLUSIONS

1. The elements anomaly of Ir, REE, Ni, Co and Fe in the main event layer must have been caused by the Tunguska explosion in 1908.

2. The curves of chondrite-normalized REE in the main event layer are relative flat, and the ratio about 1. These indicate that the Tunguska object must have been composed of material similar to carbonaceous chondrite, rather than a typical iron chondrite or comet.

3. The celestial body that exploded over Tunguska in 1908 weighed about 1.6×10^{10} tons, corresponding to a radius of 1~1.5km, and released energy about $\sim 10^{4}$ megatons.

Acknowledgements

The authors are grateful to Dr. E. M. Kolesnikov, Moscow State University, Russia, for providing all the samples, Prof. J. L. Li and H. H. Chen, X. H. Zhou, Institute of Geology, Chinese Academy of Sciences, and Dr. Zhou Yiaoqi, University of Petroleum, China, for helping discussion with authors. Thanks should also be given to the National Natural Science Foundation of China, China Postdoctoral Science Foundation, and President Foundation of the Chinese Academy of Sciences, for their financial support.

REFERENCES

1. J. Baxter and T. Atkins. *The Fire Came by*. Doubleday, Garden City NY (1976).
2. W. H. Beasley and B. A. Tinsley. Tunguska event was not caused by a black hole, *Nature* **250**, 555-556 (1974).
3. C. F. Chyba, P. J. Thomas and K. J. Zahnle. The 1908 Tunguska explosion: atmospheric disruption of a stony asteroid, *Nature* **361**, 40-44 (1993).
4. C. Cowan, C. R. Atluri and W. F. Libby. Possible anti-matter content of the Tunguska meteor of 1908, *Nature* **206**, 861-865 (1965).
5. P. Florensky Kirill. Did a comet collide with the earth in 1908? *Sky and Telescope* **26**, 268-269 (1963).
6. R. Ganapathy. The Tunguska explosion of 1908: Discovery of meteoritic debris near the explosion site at the South Pale, *Science* **220**, 1158-1161 (1983).
7. V. Gentry Robert. Anti-matter content of the Tunguska meteor, *Nature* **211**, 1071-1072 (1966).

8. A. A. Jackson and M. P. Ryan. Was the Tunguska event due to a black hole? *Nature* **245**, 88-89 (1973).
9. G. H. S. Jones. High explosive analogue of the Tunguska event, *Nature* **267**, 605 (1977).
10. M. I. Korina, M. A. Nazarov, L. D. Barsukova, I. V. Suponeva, G. M. Kolisov and E. M. Kolesnikov. Iridium distribution in the peat layers from area of Tunguska event, *Eighteenth Lunar and Planetary Science Conference*, Houston, TX. Mar. 16-20, 1987, 501-502 (1987).
11. E. L. Krinov. *Giant Meteorites.* Pergamon Press, Oxford (1966).
12. B. Yu Levin and V. A. Bronshten. The Tunguska event and the meteors with terminal flares, *Meteoritics* **21**, 199-215 (1986).
13. J. E. Lyne and M. Tauber. Origin of the Tunguska event, *Nature* **375**, 368-369 (1995).
14. H. J. Melosh. Airblast scars on Venus, *Nature* **358**, 622-623 (1992).
15. H. J. Melosh. Tunguska comes down to earth, *Nature* **361**, 14-15 (1993).
16. M. A. Nazarov, M. I. Korina, G. M. Kolesov, N. V. Vasil'ev and E. M. Kolesnikov. The Tunguska event: Mineralogical and geochemical data, *Eighteenth Lunar and Planetary Science Conference*, Houston, TX. Mar 16-20, 1987, 548-549 (1987).
17. R. Rochia, M. de Angelis, D. Boclet, Ph. Donte, C. Jehanno and E. Robin. Search for the Tunguska event in the Antarctic snow, global catastrophes in earth history, *An Interdisciplinary Conference on Impact, Volcanism, and Mass Mortality*, Snowbird, UT, Oct. 20-23, 1988, 156-157 (1988).
18. R. Z. Sagdeev, P. E. Elysberg and V. I. Moroz. Is the nucleus of comet Halley a low density body? *Nature* **331**, 240-242 (1988).
19. Spall Henry. The Tunguska comet or great Siberian mystery explosion of 1908, *Earthquakes and Volcanoes* **18**, 93-97 (1986).
20. R. P. Turco *et al.*. An analysis of the physical, chemical, optical, and historical impact of the 1908 Tunguska meteor fall, *Icarus* **50**, 1-52 (1982).
21. G. L. Wick and J. D. Isaacs. Tunguska event revisited, *Nature* **245**, 139-140 (1973).
22. S. R. Taylor and S. M. McLennan. *The continental crust: Its composition and evolution.* Blackwell Scientific Publications (1985).
23. L. W. Alvarez. Extraterrestrial cause for the Cretaceous-Tertiary extinction, *Science* **208**, 1095-1107 (1980).
24. L. W. Alvarez. Experimental evidence that an asteroid impact led to the extinction of many species 65 million years ago, *Proc. Natl. Acad. Sci. USA* **80**, 627-642 (1983).
25. J. L. Barker and E. Anders. Accretion rate of cosmic matter from iridium and osmium contents of deep-sea sediments, *Geochim. Cosmochim. Acta* **32**, 627-645 (1968).

Proc. 30th Intern. Geol. Congr., Vol. 26, pp. 75-86
Wang *et al.* (Eds)

High-Frequency Carbonate Cycles of the Middle Cambrian Oolitic Shoal Sequence Set on the North China Platform: Evidence and Significance of Milankovitch Controlled Events

MENG XIANGHUA GE MING LIU YONGQING DENG CHENGLONG
Institute of Sedimentary Basin Research, China University of Geosciences, Beijing, P. R. China

Abstract

A series of syngenetic and petrodynamical high-frequency asymmetric meter-scale cyclic sequences of oolitic limestones and microfacies in oolitic barrier shoal systems of the Cambrian in North China are identified. By applying the Fischer Plot method of accommodation temporal analysis, 2 III-order, 10 IV-order, 19 V-order, 76 VI-order, 187 VII-order cycles in the Middle Cambrian oolitic shoal formations are found on the basis of meter-scale cycles and their stacking patterns. Generally, the thickest oolitic shoal formation in a complete III-order cycle is formed in long-term regression, and deposited on average 2.5-3.5 m for 0.1 Ma, and 0.5-0.7 m for 0.02-0.04 Ma. Through comparison with Milankovitch long (0.4 Ma) and short (0.04 Ma) cycles, it was found that the III-order may correlate to the Oort cycle (3-4 Ma), and the V-order and VII-order cycle to the Milankovitch long cycle and Milankovitch short cycle respectively. There are three epochs for the maximum thickness values of oolitic shoal deposits in the III-order regression, which are concordant with the IV- or V-order sea-level changes, occurring roughly from the minimum sea-level to the following rising epoch. On the contrary, oolitic shoal reaches a maximum thickness from falling sea-level to the minimum point. But the rising sea-level, especially the maximum sea flooding, usually terminates or restricts the development of oolitic shoal. Very extensive oolitic shoal environment is often formed on platforms accompanied by shoal island, back-barrier lagoon, and tidal environment. The formation of asymmetric high-frequency sea-level cycles in the Middle Cambrian oolitic shoal sequences is the control of Milankovitch events. It may provide a method for making short-term isochronous lithofacies-palaeogeographic maps and for study of detailed sedimentological evolution history through the long Milankovitch cycle (0.4 Ma) correlation.

Keywords: high-frequency cycles, carbonate platform, Milankovitch event, Middle Cambrian, North China

INTRODUCTION

The North China Platform (Fig. 1) provides an ideal research area of the cosmo-earth system of deposition, as it covers a huge area of around 1500 (E-W) and by 1000 (N-S) km, including the major parts of more than 7 provinces. The Middle Cambrian cyclic sequence consists predominantly of oolitic shoal system with high-frequency carbonate cycles, and tidal flat, mid-outer and mid-inner ramp depositional systems [1, 2, 3, 4, 5].

The general subdivision of the Middle Cambrian in North China is shown in Table 1. In general, the III-order sequence III 4 and III 5 consist of the transgressive system tract (TST) and the highstand system tract (HST). The oolitic shoal sequences are well developed in each system tract, especially in HST. The Ⅳ- and Ⅴ-order cycles may correspond respectively to parasequence set and parasequence as commonly used.

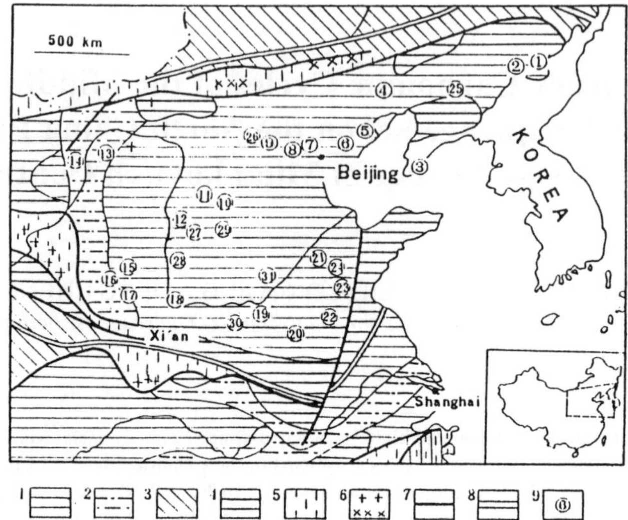

Figure 1. Schematic geological and structural map of the North China region, showing the principal localities of Cambro-Ordovician sections.

Localities: 1. Changbai; 2. Hunjiang; 3. Jinxian; 4. Pingquan; 5. Qinhuangdao; 6. Tangshan; 7. Dingjiatan and Xiaweidian, Beijing; 8. Qingbaikou, Beijing; 9. Hunyuan; 10. Jingjing; 11. Taiyuan; 12. Lishi-Liuling; 13. Zhuozishan; 14. Suyukou, Helanshan; 15. Mizhi; 16. Pingliang; 17. Tangwangling; 18. Ruicheng; 19. Xinji, Lushan; 20. Fengtai; 21. Laiwu; 22. Suxian; 23. Jiagou; 24. Gushan; 25. Benxi; 26. Kouquan; 27. Zhongyang; 28. Hongdong; 29. Zuoquan; 30. Mianchi; 31. Fengfeng, Handan.

Legends: 1. post-Ordovician uplift of epeiric seas; 2. post-Ordovician uplift of basinal areas; 3. marginal basin; 4. post-Cambrian uplift of epeiric seas; 5. post-Silurian uplift of basinal areas; 6. granite and ophiolite; 7. major faults; 8. sutures; 9. section studied.

Chinese Stages		Stratigraphic Unit	Ⅲ—order Sequence		Higher Orders Sequences		
Middle Cambrian	Zhangxian Stage	523**	523	Ⅲ5	Ⅳ5	Ⅴ9—10	187
		Zhangxia Fm.			Ⅳ4	Ⅴ7—8	
		Damesella	Zhangxia				
		Amphoton—Taitzuia			Ⅳ3	Ⅴ5—6	
		Crepicephalina					
		Liaoyangaspis			Ⅳ2	Ⅴ3—4	
		527**	527.5*		Ⅳ1	Ⅴ1—2	
	Xuzhuangian Stage	Xuzhuang Fm.	Xuzhuang	Ⅲ4	Ⅳ5 | Ⅳ1	Ⅴ9 | Ⅴ1	80
		Bailiella					
		Sunaspis					
		530**	531.5*				

Table 1. Subdivision of the Middle Cambrian in North China.
* Data results calculated by Meng *et al.* (1996)
** According to Wang *et al.* (1996)

THE MILANKOVITCH EVENTS AND ITS CONTROL ON SEDIMENTATION

It is well known that the Milankovitch curve [6] represents the orbital eccentricity of the earth changes. In the last 20 years, the study of the impact motion on the earth has greatly improved research on sequence stratigraphy and cyclostratigraphy, and has made important achievements in stratigraphy, sedimentology and oceanography [7, 8, 9, 10]. Many scholars have published papers dealing with the relation of cosmo-earth system with sedimentation and climate.

Recent research has shown that impact on the Earth is characterized by asymmetric cycles which is a peculiarity of Milankovitch cycles. It seems that the sea-level highs may represent a greenhouse effect episode, and the relation between the Earth's orbital change and climate is evident [11]. This kind of asymmetric cycles results probably from lag deposits of climate-controlled sea-level changes, and at the same time by lag carbonate sedimentation owing to sea-level changes. It is therefore important to study the time-space distribution and superimposition of the basic meter-scale cycles before investigating the cosmological effect on the Earth.

METER-SCALE CYCLES AND THEIR STACKING PATTERNS

The depositional systems of oolitic shoal and the types of lithofacies in the Middle Cambrian of North China were shown in Table 2 [12], as based on the modern carbonate sedimentary environments and lithofacies designation.

General features of the meter-scale cycle sequence

In the Cambrian of North China, the meter-scale oolitic cyclic sequence is well developed and is mainly distributed in the HST, *e.g.* in the mid-upper part of Xuzhuang and Zhangxia formations (Fig. 2, 4). The general features of a typical meter-scale cycle of the carbonate platform in North China is summarized as follows.

The bottom unit (Fig. 2, unit a) is usually composed of an upward-shallowing lithofacies set including from lower upwards: deeper shallow carbonate facies, submarine shoal facies, shallow water shoal, deeper shallow water shoal, and shelf deposits. Deposits of the middle and lower shallow-ramp and shallow basin facies, including deep-ramp biolimestone and deep-ramp storm deposits in a deep-ramp condensed section may also be developed. The lithofacies characteristics of the bottom unit are consistent with the sea-level changes formed in the meter-scale cycles.

As the sea-level starts to rise rapidly, the platform will submerge and results in punctuated non-deposition intervals, covered by deeper water sediments of hardground and thin, starved laminations below the photic zone, as shown in Fig. 2, unit b. The lower unit (unit b) usually consists of the upward-shallowing lithofacies, ranging from continued rising of sea-level to slowly rising sea-level with increasing carbonate productivity. The upper unit (unit c) is composed of high-energy dolominization or of exposed lithofacies. When the sea-level begins to fall, carbonate deposition will keep equilibrium with the sea-level changes, causing the depositional surface to be exposed, and a series of exposed structure marks of supratidal bank islands, reef islands will be developed, and fresh water seepage zone and karst weathering crust will also appear (Fig. 2, unit c).

The various sequences stacked up by meter-scale cycles

The characteristics of the basic meter-scale cycles are described above, including lithology or lithofacies, sedimentary indicators and vertical successions, are varied in different palaeogeographic settings in North China, including inner-shelf, mid-shelf and outer-shelf. Consequently a variety of meter-scale cycles are produced. It is essential to recognize these carbonate meter-scale cycles in different geological periods in the study of sequence stratigraphy and cyclostratigraphy.

	Subfacies		Microfacies Types and Association	Characteristic Microfacies Symbols
back ramp	Tidal flat	TF	A(paleokarst mud) B(sandy, muddy limestone dolomite and aragonite)	1. bearing terrigenous quartz sand and silt 2. usually red or greyish green 3. usually with dry ripple marks 4. with cryptalgae lamination
	Lagoon and inter-tidal	LI	C(semi-pellitic stromatolite limestone) D(oolitic micritic limestone)	1. showing rock salt pseudomorphs 2. with semi-pelletic algae sheet 3. with convolute bedding 4. algae ooide, algae clot
shallow ramp	Up—marine surface shoal facies	BI	E(radiating oolitic micritic limestone) F(fresh-water leaching oolitic limestone) G(concentric oolitic sparite)	1. dolomitization 2. monocrystal ooide, solution ooide 3. erosional marks 4. oolitic with vadose cement
		BC	H(resedimentary oolitic clastic limestone)	1. lag deposits 2. double orientational cross bedding 3. assymmetric ripple marks and interference ripple marks
		BP	I(caliche mud—cemented and oolitic limestone) J(SH-stromatolite limestone)	1. algae clot, columnar algae sheet 2. algae patch
	Sub—marine shoal facies	SM	K(oolitic limestone with herringbone cross-beddings) L(resedimentary intraclastics) M(carlith algae limestone and bioherm) N(bioclastic limestone)	1. vertically crossing double-orientational water-flow distribution 2. erosional and intersecting bioherm structure 3. large— and medium-scale assymmetric ripple marks of sandy slopes
deep ramp	Inner deep ramp	ID	O(biolimestone and calcilutite)	1. glauconite present in oolitic shoal 2. bioclastic limestone and micritic
	outer deep ramp and basin	OD	P(calcilutite)	1. storm intraclastic 2. storm fine graded bedding 3. storm scoured mark 4. hummocky cross—bedding
			Q(calcilutite mudstone) R(hardground+ glauconitic mudstone)	1. thin—bedded limestone 2. marine shale 3. muddy-micritic limestone 4. with glauconite

Table 2. Typical identification symbols of various subfacies and microfacies(modified after Meng [12], 1996). BI: barrier island; BC: on barrier drainage channel; BP: on barrier pond and mound.

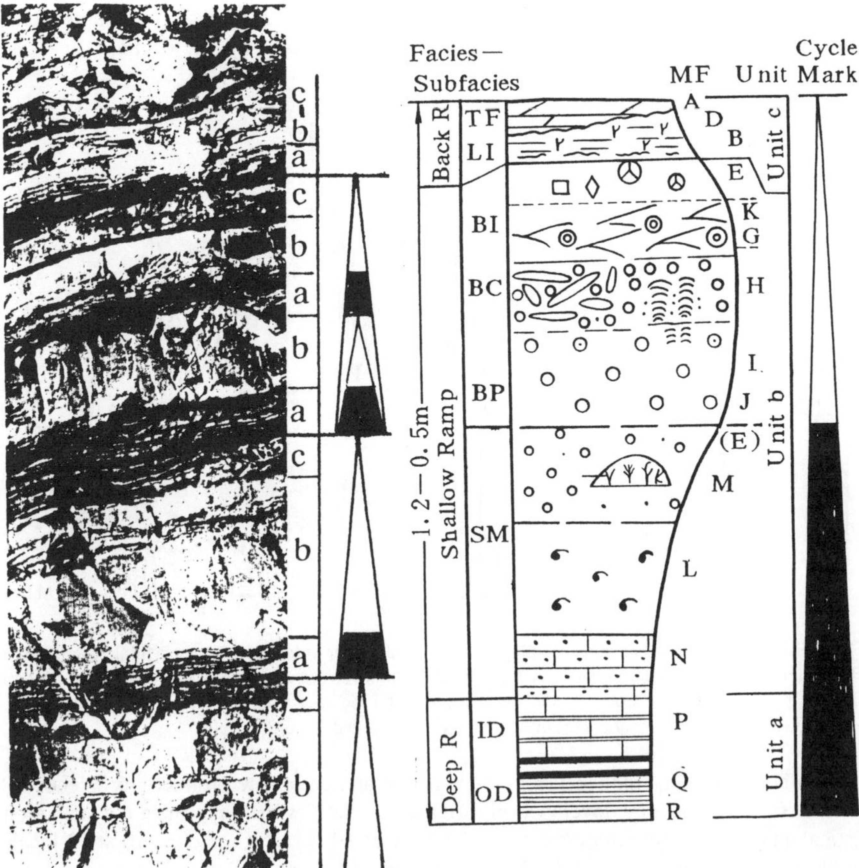

Figure 2. The general features of meter-scale cycles (identified symbols of microfacies see Table 2) and the typical meter-scale cycle composed of the mud-banded limestone (unit a), the mid-thick-bedded oolitic limestone (unit b) and the dolominization is developed at the top of the oolitic limestone (unit c) of Zhangxia Formation (Middle Cambrian) in the Western Hills of Beijing. There are hardgrounds and glauconite at the bottom of the mud-banded limestone. The pen is 15 cm long.

Owing to the complex interaction of the various orders of high-frequency and low-frequency sea-level changes and composite eustatic fluctuations, numerous parasequences or parasequence sets with different stacking patterns are produced, which are common in carbonate stratigraphy of the Lower Palaeozoic of North China.

Carbonate cycles, meter-scale cycles and their stacking patterns into different orders of sequences in North China may include different types. The quadri-rhythmic parasequence set may be exemplified by V-/VI- orders cycles and penta-rhythmic sequence in HST and LST may be exemplified by III-/IV- order cycles. Incompletely developed parasequences (late HST, early LST) and submerged parasequences of upward-shallowing and upward-thinning sequences are usually formed in transgression and CS-type parasequence set usually in maximum flood and are often met with in mid-shelf, the slope and lower part of inner shelf.

It is essential to recognize meter-scale cycles within continuous outcrop sections and their stacking patterns, which are based on meter-scale cycles and measured thickness in the sections, Fischer plots are helpful in defining long-term sea-level changes. Fig. 3 shows an example verified by correction of compaction and water-depth in the Middle Cambrian of North China.

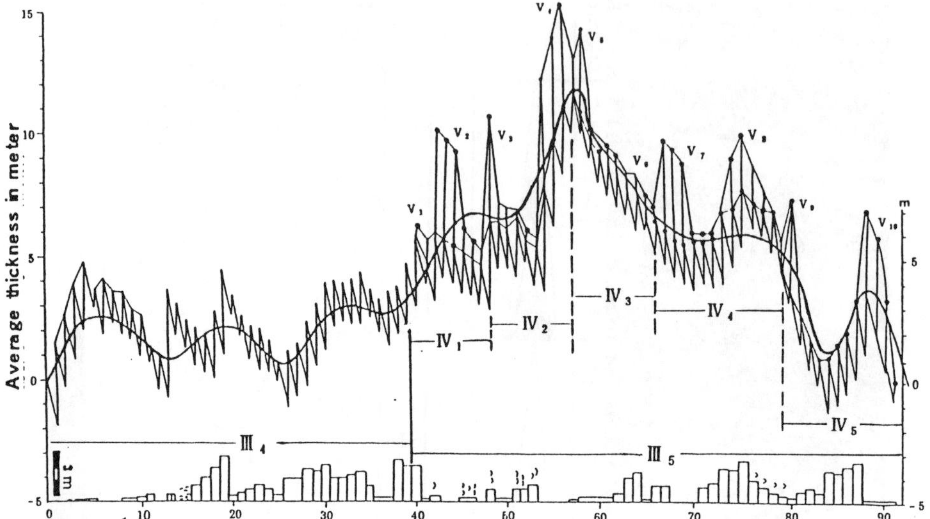

Figure 3. The synthetic sea-level changes in III-, IV-, V-, and higher order cycles in the Middle Cambrian in North China.(Plotted on the basis of meter-scale cycles and Fischer Plots corrected in terms of compaction and water-depth. The columns in the lower part represent the thickness distribution of the oolitic and biolimestone in the cycles probably attributable to the VI-order cycles or incompletely developed VII-order cycles).

The relation between high-frequency sea-level change and carbonate productivity is shown by the cycle thickness and frequency developed in different intervals, which appear in a distribution ratio. A statistical analysis of cycle frequencies on the basis of 1008 meter-scale cycles at 12 localities of the Middle-Upper Cambrian in North China shows that the cycle thickness include two ranks, 0.9-1.2 m and 1.5-2.1 m which may correspond to Milankovitch cycles.

ISOCHRONOUS CORRELATION AND DURATION OF CYCLES

Isochronous correlation

While applying the principles and methods of analysis to the meter-scale cyclic sequence, which forms the basic unit of research, it was found that each III-order sequence in the Middle Cambrian consists of five upward-shallowing IV-order sequences and 10 or 9 V-order sequences in the area (Fig. 4). The high-resolution correlation of stratigraphy could be made in terms of the Fischer Plots, which suggest that different cycles, *e.g.* the IV- and V-order sequences, are isochronous sedimentary units across the whole platform and are probably controlled by the global cosmo-earth system.

The internal components and the duration of the sequences

As there is no isotopic dating for the sequence surface, we can only use statistical method of the continuously measured cyclic sequences and meter-scale cycle sequences (Fischer Plots). In this way, the IV-, V- and VI-order sequences can be recognized, and the time interval of each order of sequence can be calculated.

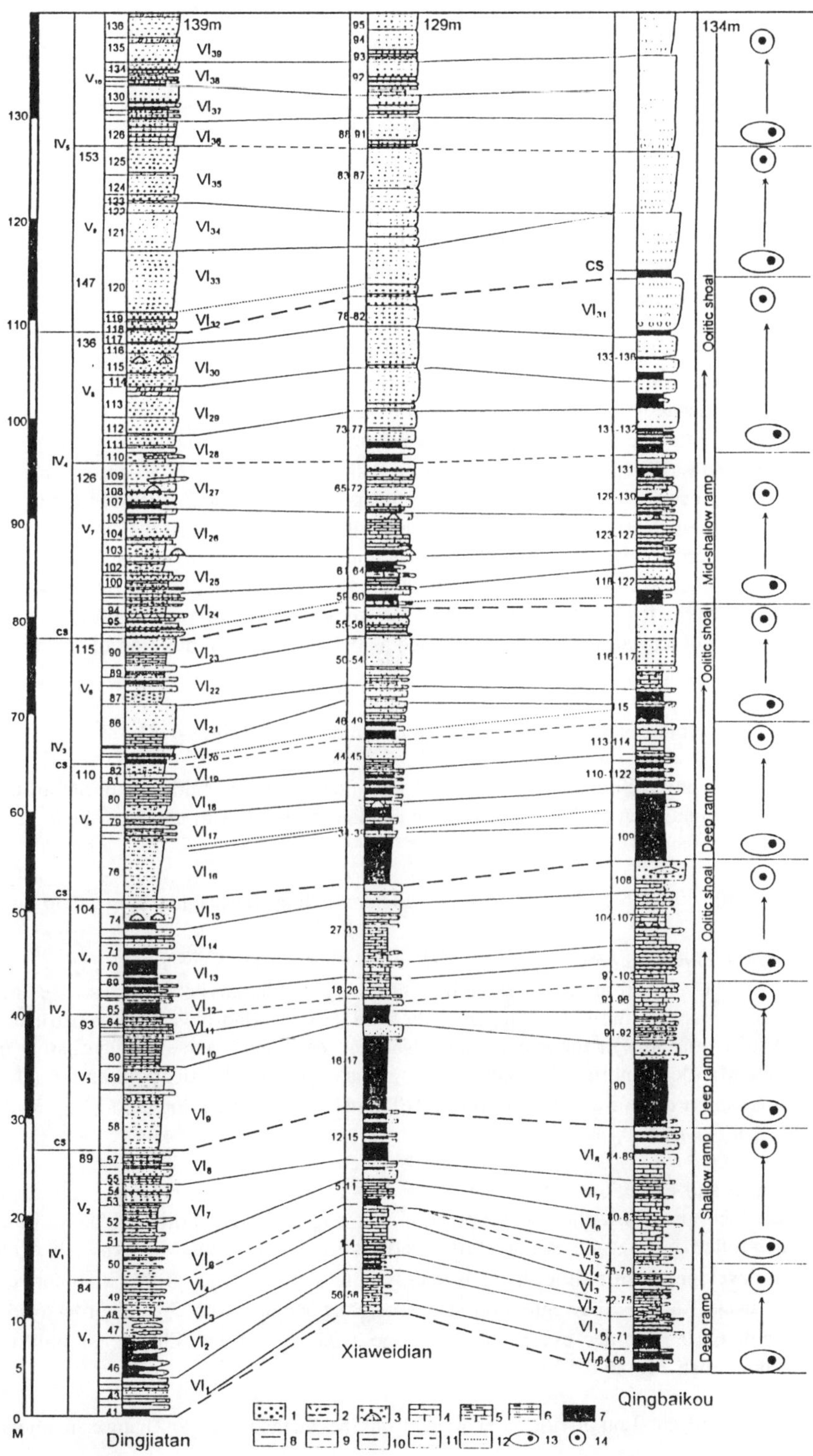
139m
129m
134m
Dingjiatan
Xiaweidian
Qingbaikou
Deep ramp
Shallow ramp
Deep ramp
Oolitic shoal
Deep ramp
Oolitic shoal
Mid-shallow ramp
Oolitic shoal
CS

Figure 4. Sequence column correlation diagram of the III-order cyclic sequence (III 5, Zhangxian Formation) in Western Hills of Beijing.
Legends: 1. oolitic limestone; 2. storm calcirudite; 3. biohermal oolitic limestone; 4. micrite; 5. mud-banded micrite; 6. muddy calcareous siltstone; 7. silty calcareous mudstone; 8. boundaries of VI-order cyclic sequences; 9. boundaries of V-order cyclic sequences; 10. boundaries of IV-order cyclic sequences; 11. boundaries of III-order cyclic sequences; 12. condensed sections (CS) 13. maximum eccentricity; 14. minimum eccentricity.

By calculating the probable thickness of each order of cycles, and drawing the Fischer Plot of meter-scale cyclic sequences and correlating the regional sections, we have studied the components and the number of each order of sequences within the III-order sequence III 4 and III 5, and calculated their time intervals. The results show that the two III-order sequences may be interpreted representing the Oort cycle, which is approximately 3-4 Ma in duration.

By using the optimum-seeking method to correlate the sequence stratigraphy sections and calculating their time interval, we were able to diminish the influence of lacking cyclic records caused by local palaeogeographical conditions and eroding factors in deposition. This method is to choose a section with maximum number of cycles in the III-order sequence with best cyclic correlation, such as the Kouquan section of Shanxi. Although the IV- and V-order cycles can be well correlated between the Xuankongsi section of Shanxi and the Western Hills section of Beijing, the cycle number are different. Take the Xuzhuangian (III 4) as an example, the V-order cycles in the Xuankongsi section and in the Dingjiatan section are 42 and 80 respectively. Thus, the Dingjiatan and similar sections (Xiaweidian and Qingbaikou section) in the Western Hills of Beijing were chosen, as the standard to calculate the cycle numbers of various orders. If the time interval of each V-order parasequence is estimated as 0.4 Ma, the total duration of sequence III 4 and III 5 are 3.6 Ma and 4.0 Ma respectively.

To conclude, the acceptable and reasonable way of calculation is as follows:

1. An estimate by Wang (SSLC report, 1996), the time interval of the Middle Cambrian of China is 11 Ma, including Zhangxian (523-527 Ma), Xuzhuangian (527-530 Ma) and Maoahuangian (530-534 Ma).

2. 10 or 9 long Milankovitch cycles and 40 short Milankovitch cycles just make up an Oort cycle.

Construction of isochronous palaeogeography using long Milankovitch cycle (0.4 Ma)
According to the isochronous correlation of long Milankovitch cycles, sedimentary palaeogeographic maps of individual stages in the Middle Cambrian have been constructed. The lithofacies distribution area in these maps has also been calculated, and the quantitative change diagrams of the Middle Cambrian lithofacies and their palaeogeographic distribution ratio drawn in Fig. 5. The maximum of oolitic shoal is up to 394000~411000 square kilometers (*e.g.* V 2, V 6 and V 7 of Zhangxian).

The development and evolution history of the Middle Cambrian oolitic shoals are represented on the diagram, which also show the superimposition and migration regularities of the oolitic shoal sequences controlled by eustatic sea-level fluctuations (including long Milankovitch cycle, 0.4 Ma).The best development of oolitic shoals is mainly in regression stages of the III-order sea-level changes Oort cycle, such as the regression stages of Xuzhuangian with 5-8 V-order cycles and Zhangxian with 6-10 V-order cycles. The average thickness of the Zhangxian oolitic shoal sequences is more than 20m in 0.4 Ma.

The oolitic shoal configuration obviously changed from the Xuzhuangian to Zhangxian, and several stages may be recognized in terms of the long Milankovitch cycle in Fig. 6. Stage A: Embryonic shoal pattern represents the beginning to the maximum transgression in the long Milankovitch cycle within the III-order sea-level fluctuation; Stage B: H-shaped shoal pattern is in the early HST stage of the III-order sea-level fluctuation, and Stage C: Listric shoal pattern represents the keep-up rate stage when the oolitic shoal growth-rate is equals or exceeds the sea-level rising. It is suggested that the formation of the funnel-shaped shoal pattern [13] on carbonate platforms is caused by the higher carbonate productivity and diagenesis of the platform margin than in the interplatform depressions (basins).

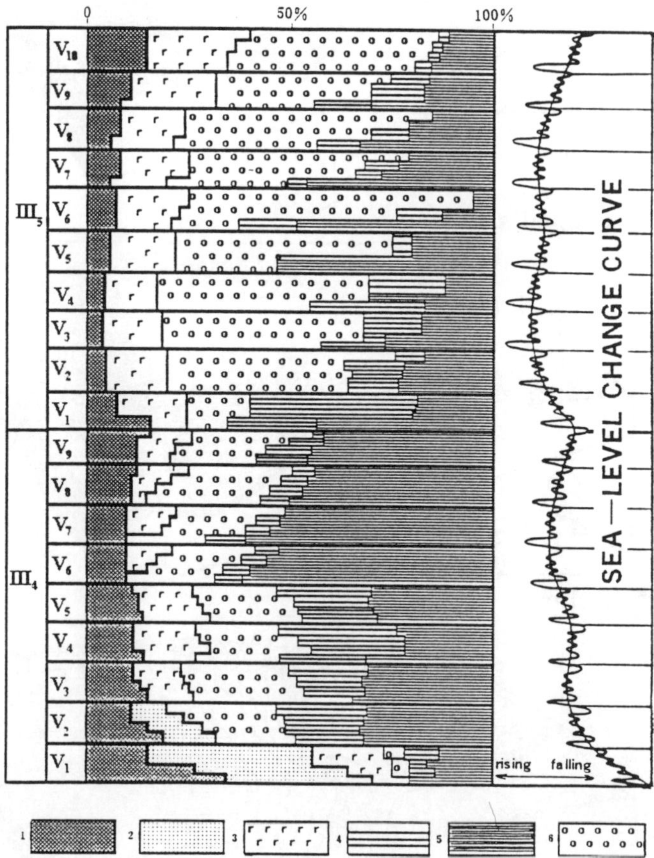

Figure 5. The quantitative diagram of the facies area ratio based on the isochronous lithofacies palaeogeographic map for the V-order cycle (long Milankovitch cycle) in the Middle Cambrian (sequence III 4 and III 5).
Legends: 1. oldland; 2. clastic flat of back ramp; 3. carbonate tidal flat back ramp; 4. inner deep ramp; 5. outer deep ramp and basin; 6. oolitic shoal of shallow ramp.

CONCLUSIONS

1. The high-frequency oolitic shoal carbonate cyclic sequences in the Middle Cambrian of North China are well developed, and resulted from is composite superimposition of various ranks of asymmetric cyclic sequences (PAC) controlled by the Milankovitch events.

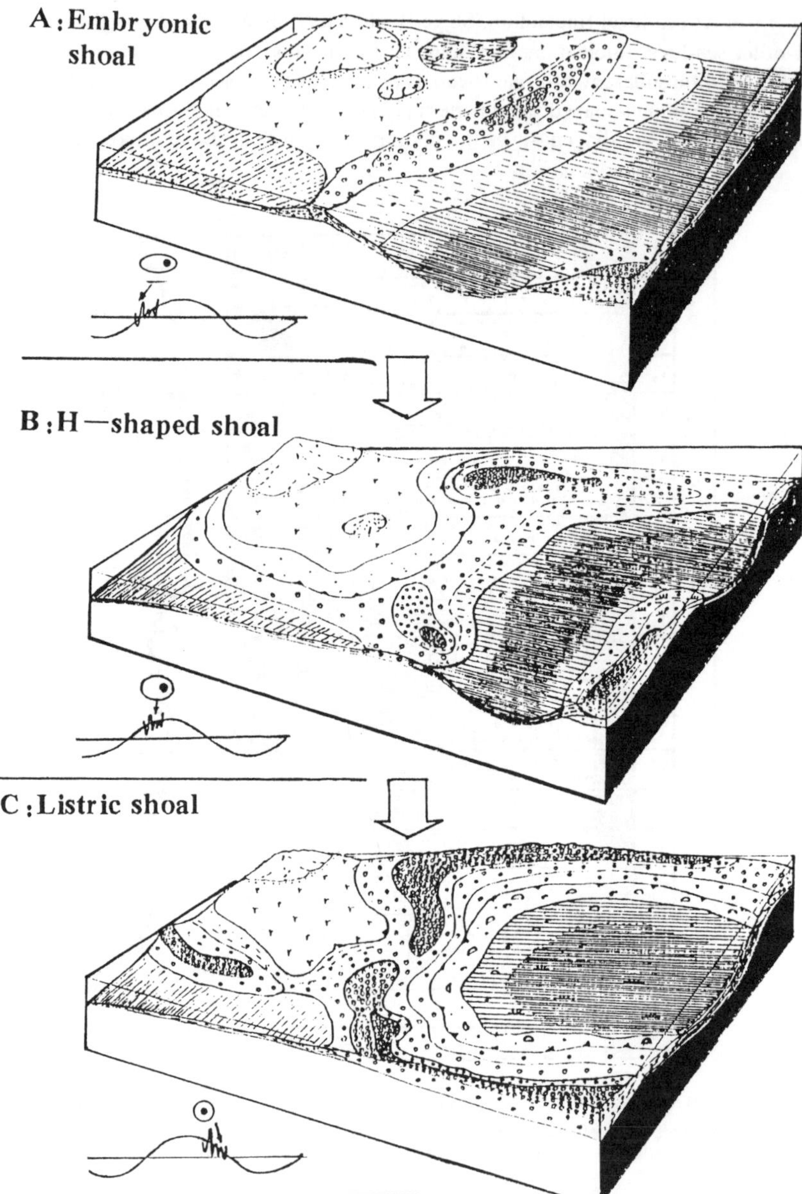

Figure 6. Evolutional stages of the oolitic shoal models in North China (Sedimentation controlled by the long Milankovitch cycle, legends as in Fig. 5).

2. On the basis of recognition of the cyclic sequences and their typical succession in the field, and by frequency analysis and Fischer Plot, the different ranks of cycles are correlated, and their individual duration is estimated, using the time interval of the Milankovitch long cycle as 0.4 Ma and the short cycle as 0.04 Ma. It was found that the III-order sequence may correspond to the Oort

cycle (3-4 Ma).The V-order to the Milankovitch long cycle and the VII-order to the Milankovitch short cycle.

3. The different ranks of asymmetric cycles show generally a 1/5 to 1/4 stacking pattern exemplified by III/IV and V/VI ratios.

4. The Middle Cambrian high-frequency sequences of North China permit of isochronous correlation, and can be used for mapping quantitatively the lithofacies palaeogeography and determining the sea-level changes.

5. The Milankovitch long cycle is the main cause for formation of the oolitic shoals and controls their form and distribution. The III-order cycle (Oort cycle) also plays a role in controlling and strengthening them.

Acknowledgements

We are grateful to the National Natural Science Foundation of China (NNSC) and the Royal Society of London for financial support. Professor Wang Hongzhen kindly read the manuscript and made constructive criticisms.

REFERENCES

1. Meng Xianghua *et al.* Study on Sea - level Fluctuation Geodynamics of Carbonate Depositional Cycles in the North China Platform, *Acta Sedimentologica Sinica* **14:2**, 29-40 (1996).
2. Ge Ming, Liu Yongqing and Meng Xianghua. *Field trip T313 guidebook of the 30th International Geological Congress: The depositional sequences and the evolutionary history of the North China carbonate Platform of Early Palaeozoic.* The Geological Publishing House, Biejing (1996)
3. Meng Xianghua, Ge Ming and Liu Yongqing. *Field trip T224 guidebook of the 30t International Geological Congress: The Cambrian deposition facies, sequence stratigraphy, and the high-frequency cyclic sequences of the carbonate platform at the Western Hills, Beijing.* The Geological Publishing House, Beijing (1996).
4. Ge Ming, Meng Xianghua and M. E. Tucker. The Changshanian maximum sea flooding event and its global correlation, *Chinese Science Bulletin* **40**, 818-821 (1995).
5. R. K. Goldhammer *et al.* Depositional cycles, composite sea - level changes, cycle stacking patterns and the hierarchy of stratigraphic forcing: Example from Alpins Triassic platform carbonates, *Geol. Soc. Am. Bull.* **102**, 535-552 (1990).
6. M. Milankovitch. Kanon der Erdbestrahlung und seine Anwendung auf das Eiszeitproblem, *Belgrade, Acad. Royal. Serbe, edns. spec.* **133**, 633 (1941).
7. W. F. Koerschner, III and J. F. Read. Field and modeling studies of Cambrian carbonate cycles, Virginia Appalachians, *Journal of Sedimentary Petrology* **59**, 654-687 (1989).
8. M. R. House. A new approach to an absolute time scale from measurements of orbital cycles and sedimentary microrhythms, *Nature* **316**, 721-725 (1985).
9. D. A. Osleger and J. F. Read. Relation of eustasy to stacking patterns of meter - scale carbonate cycles, Late Cambrian, U. S. A., *Journal of Sedimentary Petrology* **61**, 1225-1252 (1991).
10. F. B. Van Houten. Search for Milankovitch patterns among oolitic ironstones, *Palaeooceanography* **1**, 459-466 (1986).
11. G. M. Friedman, J. E. Sanders and D. C. Kopaska - Merkel. *Principles of sedimentary*

deposits — stratigraphy and sedimentology. Macmillan Publishing Company, New York; Maxwell Macmillan Canada, Toronto; Maxwell Macmillan International, New York, Oxford, Singapore, Sydney (1992).

12. Meng Xianghua and Ge Ming. *Sinian – Ordovician palaeogeography, cyclicity - rhythm and sedimentary events of China*. International Academic Publishers, Beijing, China (1996).
13. G. St. C. Kindall and W. Schalager. Carbonates and relative changes in sea – level, *Marine Geology* **44**, 181-212 (1992).

Proc. 30th Intern. Geol. Congr., Vol. 26, pp. 87-95
Wang *et al.* (Eds)

On The Extraterrestrial Impact and Plate Tectonic Dynamics: A Possible Interpretation

WAN TIANFENG* YIN YANHONG** ZHANG CHANGHOU*
* *China University of Geosciences, Beijing 100083, P. R. China*
***Institute of Marine Geology, MGMR Qingdao 266071, P. R. China*

Abstract

The plate tectonics theory arose in the sixties, and its geometry and kinematics have been rather well solved in recent years. However, the dynamics of plate tectonics remains still a controversial problem. According to recently discovered data, the stress orientation of the lithosphere plates varied remarkably in an interval of 10 million years in the Meso-Cenozoic. Sometimes the plates even show radial movements, and the tectonic stress (differential stress) decreases rapidly with the depth under the earth's surface. The mega-impact center and microtektite about 1 Ma ago was situated near the triple junction of the southern Indian plates, and those 34 Ma in age was near the triple junction of the eastern Pacific plates. The velocities of plate motion are well known as 2-12 cm/a, but the velocities of mantle convection are believed to be only several mm/a. Many data show that the hypothesis of mantle convection driving the lithosphere plate motions is difficult to be established., and the influence of extraterrestrial impact on plate motions should not be ignored. Depending on the limited data, we consider that the mega-impact and microtektite impact could produce craters with hundreds of kilometers in diameter and tens of kilometers in depth, thus forming partial debility in the upper part of the lithosphere, especially in the oceanic part. This would bring about the upward surge of the mantle material through isostatic compensation and lead to horizontal spreading of the oceanic plates, resulting in subduction, collision or shearing between them. It seems that in the Meso-Cenozoic the extraterrestrial mega-impact may be the main inducing factor for plate motions and the partial upward surge of the mantle material may have driven the plates apart.

Keywords: Microtektite, Impact, Plate tectonics, Geodynamics, Meso-Cenozoic

INTRODUCTION

The theory of plate tectonics rose since sixties and was a great contribution for the development of geosciences, as was generally recognized by most geoscientists. However, some problems exist in the plate tectonics, although the essential data and view points have been repeatedly confirmed.

The subdivision of the main plates on the globe and their kinematics were described in the plate tectonics theory, which basically cioncided with fact. However, there are many problems to be discussed in regard to the evolution and the formation of continental plates before Cenozoic, especially the dynamics of plates. To solve the above problems is obviously the frontier task of modern geosciences.

In this respect, we are inclined to base our discussion on the research of geometry and kinematics of global plates in Mesozoic and Cenozoic, and then to discuss preliminarily their dynamics [1, 2], in which the Earth is treated not as an isolated system, but as a constituent part of the solar system, and even of the whole cosmic system. We consider that the motions of lithosphere plates were induced by the giant meteorite-microtektite impact events, as a working hypothesis for research.

In the first place, in the discussion of dynamics of plate tectonics, we will briefly review the recent new progress of plate tectonics research, and then try to develop or revise the existing ideas.

ON THE KINEMATICS AND DYNAMICS OF THE PLATES

The view-points generally recognized in recent years are that the lithosphere plates are in continuous motion, which is shown by migration of the lithosphere plates with different velocities (2-12 cm/a) and orientations, amounting to an extent of several hundreds to thousands of kilometers in displacement. In the seventies, the above conclusions were determined by the palaeomagnetic data, with the reference system of hotspots [3]. In the last several years, 1122 comprehensive datailed data of palaeomagnetism and slip vectors of earthquake, more precise sketches of recent motion vectors for global plates were compiled (NUVEL-1) [4]. Thus the reliability of plate kinematics was confirmed by the similarity between the above two sets of results.

In the last twenty years, according to the intraplate deformation and palaeomagnetism research, the palaeogeographic and palaeotectonic map of Meso-Cenozoic in many areas were reconstructed, and the motion orientations of some palaeoplates and maximum compressive principal stress direction were determined. Smith *et al.* [5] were the first to carry out this project, who compiled the global palaeogeographic maps in Meso-Cenozoic showing the position and orientation changes of the continents. Letouzey *et al.* [6], based on the micro structural data, discovered the changes of palaeostress orientation around the Mediterranean, where the maximum compressive stress were twice NW trending (in Late Cretaceous and Late Miocene), twice near N-S trending (in Eocene and Quaternary), and once in a NE trending (in Late Oligocene to Early Miocene). One of authors [7, 8], while studying the rock deformation and palaeomagnetism in different tectonosynthems of China and neighboring regions, discovered that since Triassic they were three times under the influence of the northward migration of the blocks dispersed from Gondwanaland in the time interval of 250-208 Ma, 135-52 Ma, and 23.3-0.73 Ma respectively, and at last caused the Indian plate to move to the north of equator and collided with the Eurasian plate. They were also twice under the influence of westward subduction and compression of oceanic plates (Izanagi, 208-135 Ma; Pacific, 52-23.3 Ma). Since Middle Pleistocene (0.73 Ma-) an isostatic state has existed for the Eurasian, Pacific, Philippine Sea and Indian-Australian plates. Moore [9] studied the position and moving direction of the palaeoplates surrounding the Pacific area(Fig. 1) and found that there occurred radiating plate motions in the Late Jurassic, when the Izanagi plate migrated to NW, the Farallon plate to NE, the Pacific plate to SW and the Phoenix plate to SE. However, in the Late Cretaceous, the main oceanic and continental plates on the globe had probably all migrated northwards. At the end of Eocene, the Pacific plate suddenly changed its direction of motion from NNW trending to WNW trending, the Nazca plate migrated to the east and the Indian-Australian plate move to the north. After that, the mentioned above plates basically kept their own motion directions and essentially coincided with the calculated results of recent stress orientation [10]. The results of studies in eastern Asia by Moore [9], Zoback [10] and Wan [7, 8] are on the whole coincident with each other.

The above results show that the migration orientations of plates had changed many times in Mesozoic and Cenozoic. This recognition was an obvious progress in the seventies, when the plate tectonics theory was based only on the oceanic magnetic anomalies, and the orientation of plate motions was regarded as essentially unchanged since the Jurassic.

Based on the research results of palaeotectonic stress magnitudes in the last twenty years, many researchers have recognized that the rock deformation caused by the differential stress are concentrated in the lithosphere, especially in the upper crust. The differential stress magnitudes in geological history may reach 100-200 MPa. At a depth of 50-60 km below earth surface, they may decrease to 8 MPa in the continental rift zone, to 45-5 MPa in the continental extension zone. At a depth of 200 km under the earth's surface, the differential stresses are generally reduced to 15-5 MPa [7, 11, 12, 13], which are almost inconsiderable. However, the confining pressure may reach 2700-5500 MPa, indicating that the density of rocks increased obviously, and rock deformation becomes very weak at that depth.

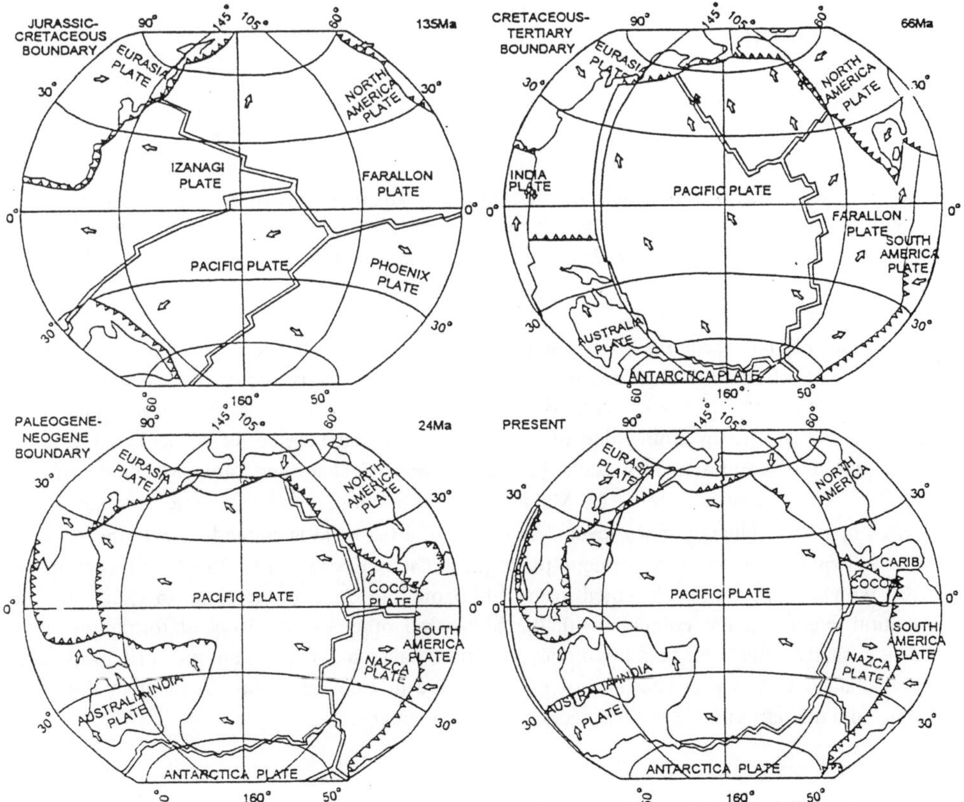

Figure 1. Palaeogeographic maps showing the palaeoplate movements around the Pacific Ocean area (after Moore [9]).

Recently, there are accumulated about 20-30 MPa of differential stress in the interior plate of the upper lithosphere [14, 15, 16] and bigger than that at the boundary between the plates. Forsyth and Uyeda [15], and Turcotte and Schubert [17] considered that the negative buoyancy of plate subduction zone reaching 1 GPa, is most important, and is a main driving force of plate motion.

Zang and Ning [18], considering the temperature and density parameters in various depth during the different subducting processes, have calculated out the effective tensile stress driven by the negative buoyancy in subducting zones, which is only about 40-290 MPa. This result seems to be more reasonable. It is generally recognized that the push force of recent oceanic spreading zone is about 30 MPa [14, 19].

The above results were all made on the presumption that the deformation of lithosphere and the horizontal motion of plates are formed in the upper part of lithosphere and their differential stress magnitudes vary from tens to hundreds MPa. It is difficult to be explained by the Earth expansion and contraction or the variation of the rotation velocity of the Earth, when such big differential stress magnitudes exist in the upper lithosphere, and such orientation change occur in the migration of plates.

THE PERIODICITY OF PLATE MOTIONS AND INTRAPLATE DEFORMATIONS

The periodicity of plate motions and intraplate deformations is also related to the idea of progression by stage in tectonic evolution [7, 20], which was described as the periodicity of pulse of orogeny [21, 22]. Up to present, some geoscientists would still disregard the ideas of catastrophic and progression by stages in tectonic evolution [23, 24]. However, much data seem to point to the periodicity and the progression by stages in tectonic evolution. Of course, it is no denying of the rheology in the tectonic evolution [7, 25], and does not mean unitary orogeny all over the globe [25, 26].

Recently, Wan [7] has recognized that there is periodicity in global tectonic evolution with the characteristics of alternating long time interval stable and short active epoch (catastrophic epoch). The periodicity of global tectonic evolution is various in different regions and is irregular in the same region, and is different from palaeomagnetic inversion, mega meteorite impact event, or biotic mass extinction. Rampino and Stothers [27, 28] proposed that there are two kinds of long periodicities: 33 ± 3 Ma and 260 ± 25 Ma in biotic extinction, palaeomagnetic inversion, giant meteorite impact and change of sea level. The period of 33 ± 3 Ma is coincident with the period of the solar system's crossing of the galactic plane [29]. The period of 260 ± 25 Ma is a cosmic year [29, 30], which is approximately equal to the Mesozoic and Cenozoic in time. In China the rock deformation events do not coincide with the biotic and other events. Rock deformation on the continents usually come later than the biotic extinction, palaeomagnetic, giant meteorite impact and sea level change events [7]. The data of plate tectonic motions are more frequent in the Meso-Cenozoic, and the period of 33 ± 3 Ma seems also to be remarkable.

THE MICROTEKTITE IMPACT EVENTS

Based on the data of Deep Sea Drilling Project (DSDP) and Ocean Drilling Program (ODP), Glass [31] first summarized two macro-scale microtektite impact events on the Earth (Fig. 2). One microtektite impact event is called "Australasian". The microtektites with diameter less than 1 mm where strewn in an area of cycloid shape in the Indian Ocean, southern Asia and Australia, which covered nearly one tenth of the globe. The formation age of microtektite, determined by K-Ar and fission-track methods, is between 0.7-0.83 Ma, *i.e.* Early-Middle Pleistocene. The total weight of microtektites is estimated at about 1×10^8 tons. North American microtektites were strewn in the Caribbean, Latin America extending oblique across the Pacific Ocean to southeastern Asia with a zone distribution of ENE to WSW orientation. The dating result of the meteorites are between 34.6-37.5 Ma, near the boundary between Eocene and Oligocene. The total weight reached $(1\text{-}10) \times 10^9$ tons. The time interval between the above two microtektite impact events is about 33-36 Ma, which is well coincided with the periodicity of 33 ± 3 Ma of giant impact, biotic extinction event, change of sea level and basalt eruption on the globe.

The chemical contents of these two microtektites are characterized by high Cr, Ni, Co, Ni/Fe ratio and MgO, low volatile elements (Cu, Ge, Sn, Pb, etc.), Low Fe_2O_3 / FeO ratio, low alkali (Na_2O and K_2O), low water content (< 0.01%) and very high SiO_2 (> 60 %). They are not only different from the common volcanic glass on the Earth, but also distinct from other meteorites or lunatic rocks. Glass [31] and his colleagues considered that the microtektites were mixed and remelted with the meteorite and oceanic clay during the impact processes through detailed analysis. Their results were reasonable and recognized by most researchers.

The two microtektite impact events were studied rather meticulously and the position of the impact centers and microtektite crater are still not clear. Some researchers have inferred that the crater of Australasian microtektite may have been situated in Zhamanshin, southern Siberia, but the evidences are weak. The crater of the North American microtektite remained uncertain [31].

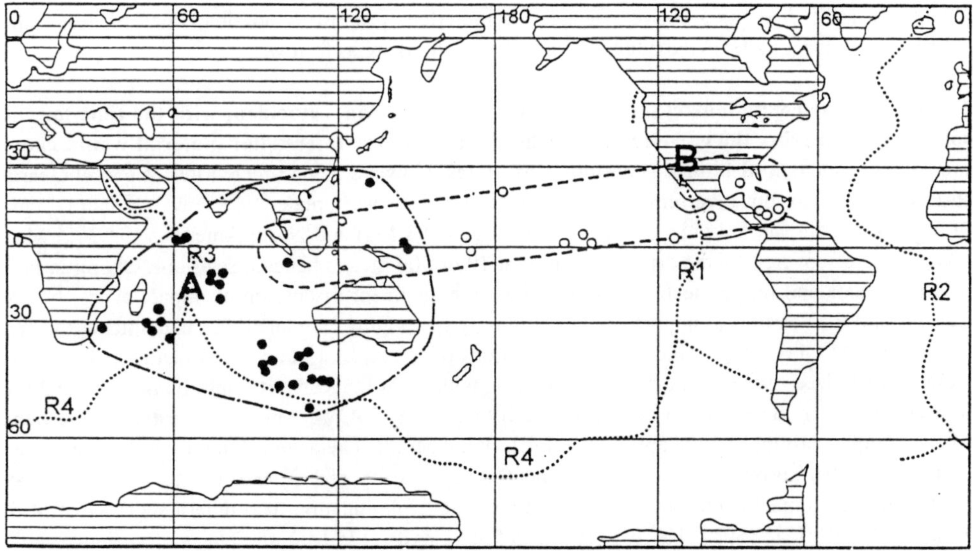

Figure 2. The Australasian (A) and North American (B) microtektite strewnfields and oceanic ridges (modified after Glass [31]. The location of cores containing Australasian and North American microtektites are indicated by solid circles and open circles respectively. R1, East Pacific Ocean Ridge; R2, Mid-Atlantic Ocean Ridge; R3, Mid-Indian Ocean Ridge; R4 Circum-Antarctic Ocean Ridge. Circular faults are on the oceanic floor are indicated on California coast.

In regard to the North American microtektite event, in recent years, Yin *et al.* [32] have made comprehensive analysis of the biotic mass extinction and geochemistry of sediments, and pointed out that the circular faults, south of California shown on the sketch of tectonic setting and ages of basement for the eastern equatorial Pacific Ocean [33], may be the site of impact crater. The diameter of the circular faults rim reaches 1500 km or more, and the two third of the rim had subducted under the North American plate. Wan [34] first pointed out that the impact event might have caused a sudden change of motion direction (from NNW to WNW) of the Pacific plate, and also induced the giant dextral strike-slip of the San Andreas fault zone in western North America. Yin and Wan [35] studied the cause of motion direction change of the Pacific plate at the end of Eocene, which is necessary for increase of a SW 245-trending force, and made calculation in related mechanics. The force direction pointed out [35] just coincides with the strewnfield trending of the North American microtektites [31]. This is a further proof of the possible cause of motion direction change of the Pacific plate originated in the depth.

The impact crater of Australasian microtektites can not possibly be situated in Siberia, but may be in the northwestern part of strewnfield, near the triple junction of the Mid-Indian Ocean Ridge (67°E, 25°S). The Australasian impact may have induced the northward migration of the Indian plate, which moved northwards several hundreds of kilometers in the Quaternary and caused the uplifting and crustal shortening of Qinhai-Xizang (Tibet) plateau.

MICROTEKTITE IMPACT INDUCED THE PLATE MOTION: A POSSIBLE INTERPRETATION

On the basis of data cited above, we have ventured to propose a possible interpretation –the microtektite impact induced the plate motion. The essential are as follows:

1. The giant meteorite impact has a periodicity of 33 ± 3 Ma, coincident with the time for the solar system to cross the galactic plane, where more intense interstellar material would cause meteorite impacts and induce the motion of plates. However, small meteorites impact the earth surface more often, but give limited influences to the lithosphere plate. Nowadays substantial data of microtektite impact are only the Australasian (0.7-0.83 Ma) and North American (34-37.5 Ma), which were discovered by the drilling samples at 10 or 500 meters beneath sea floor. Owing to lack of drilling data and limited techniques, data are not enough for discussion of impact events before Eocene, and the position of impact center could only be roughly estimated. For example, at the end Cretaceous (67 Ma) catastrophic epoch, the motions of the bulk of plates, continental and oceanic, were characterized by nearly northward migration with great range. So it might be inferred that the impact center may be situated near the Weddell Sea (65°S, 35°W), Antarctica, instead of near the Mexico Gulf as proposed by Shapton *et al.* [36]. In the Middle Cretaceous (about 100 Ma), when an important global anoxic event occurred, and the Atlantic opened from south to north, we may predict that the impact center may be situated near the triple junction of southern segment of Mid-Atlantic Ocean Ridge (50°S, 10° W). At the end of Jurassic (135 Ma), the plates of ancient Pacific region were characterized by radiating migration [9], and we may estimate the impact center as near the equator about 130°W in longitude. It is more difficult to estimate the position of impact center for eariler stages.

2. The angle and orientation of meteorite impact on the earth surface will influence the direction of plate motion. If the meteorite impact is vertical to the earth surface, the triple junction of extensional faults (*i.e.* mid-oceanic ridge) and the radiating plate motions will occur around the impact crater. If it is oblique to the earth surface, owing to the influence of the incident orientation of the meteorite, the orientation of plate migration may have a notable change for instance in the case of the North American microtektite impact [31, 32, 35].

3. The giant meteorite will produce a great crater and induce the partial material debility in the upper lithosphere. However, most of meteorites were not discovered, being destroyed by various geological processes or buried by sediments. According to calculated statistic data of meteorite impacts on the continents [37, 38], a meteorite with 2 km in diameter is able to form a crater about 200 km in diameter, 50-60 km in depth and widespread rock fractures and faults in thousands of kilometers in distance. Thus a partial material debility in the upper lithosphere will occur to keep isostasy of the lithosphere. The partial rise of mantle material, and the heat transfer will affect the upper mantle and change the model of mantle convection [39, 40], sometimes forming eruption of basaltic magma or hotspots [28, 41, 42]. Under the influence of giant meteorite impact, the spreading center and triple junction of lithosphere plates will occur around the crater and form radiating horizontal motion.

4. Because of the thin oceanic lithosphere (50-60 km), the impact of giant meteorite is most likely to destroy the lithosphere and change the orientation of migration of oceanic plates and the model of mantle convection [40]. However, the giant meteorite impact on the much thicker continental lithosphere (100-400 km), will generally not cause the plates to disperse, and the crater will induce some partial tectono-magmatization, which was often destroyed by later geological processes. The breakup of Gondwanaland might have been caused by meteorite impacts [43], but it is a controversy in the geoscientific sphere [44, 45, 46].

To sum up, the research on the relation between the dynamics of plate tectonics and the microtektite impact event has just begun. Although we have tried to offer an interpretation for it, it is evident that much more work needs to be done in this respect.

Acknowledgements

We thank professors Wang Hongzhen, Yin Hongfu, Liu Benpei and Zhuang Peiren for helpful discussions. We are grateful to The Laboratory of Lithosphere Tectonics and Dynamics, supported by MGMR, for facilities of research.

REFERENCES

1. The Laboratory of Lithosphere tectonics and its Dynamics. *Annual Report for1994, The Laboratory of Tectonics and its Dynamics, China Univ. Geosci. (Beijing).* Seismological Press, Beijing (1995).
2. The Laboratory of Lithosphere Tectonics and its Dynamics. *Annual Report for 1995, The Laboratory of Tectonics and its Dynamics, China Univ. Geosci. (Beijing).* Geological Publishing House, Beijing (1996).
3. D. Taring and M. Taring. *Continental Drift.* Doubleday Anchor Books (1971).
4. C. Demets, R. G. Gorden and D. F. Argus *et al.* Current plate motions, *Geophys. J. Int.*, **101**, 425-478 (1990).
5. A. G. Smith and J. C. Briden. *Mesozoic and Cenozoic Palaeocontinental Maps.* Cambridge University Press, Cambridge (1977).
6. J. Letouzey and P. Tremolieres. Palae-stress fields around the Mediterrenean since the Mesozoic

from micro-tectonics comparison with plate tectonic data, *Rock Mechanics*, Supple, **9**, 173-192 (1980).

7. Wan Tianfeng. *Intraplate Deformation, Tectonic stress field and their Application for Eastern China in Meso-Cenozoic.* China University of Geosciences Press, Wuhan (1994).
8. Wan Tianfeng and Zhu Hong. Study on the Kinematics of Eastern Asian continental blocks in Meso-Cenozoic, *Annual Report for 1995, The Laboratory of Lithosphere tectonics and its Dynamics, China Univ. Geosci. (Beijing)*, 21-29. Geological Publishing House, Beijing (1996).
9. G. W. Moore. Mesozoic and Cenozoic palaeogeographic development of the Pacific region, *Abstract 28th International Geological Congress, Washington D. C. USA* **2**. 455-456 (1989).
10. M. L. Zoback and K. Burke. Lithosphere stress patterns: A global view, *EOS.* **74**, 609-618 (1993).
11. Wan Tianfeng. *Palaeotectonic Stress Field.* Geological Publishing House, Beijing (1988)(in Chinese).
12. J. G. Mercier. Magnitude of the continental lithosphere stresses inferred from rheomorphic petrology, *J. Geophys. Res.* **85 B,** 6293-6303 (1980).
13. P. England and P. Molnar. Inferences of deviatoric stress in actively deforming belts from simple physical models, *Phil. Trans. R. Soc. London,* **A 337**, 151-164 (1991).
14. M. H. P. Bott and N. J. Kusznir. The origin of tectonic stress in the lithosphere, *Tectonophysics* **105**, 1-13 (1984).
15. D. Forsyth and S. Uyeda. On the relative importance of the driving forces of plate motion, *Geophys. J. R. Astrom. Soc.* **43**, 163-200 (1975).
16. M. L. Zoback and M. Magee. Stress magnitudes in the crust: Constraints from stress orientation and relative magnitude data, *Phil. Trans. R. Soc. London,* **A 337**, 181-194 (1991).
17. D. L. Turcotte and G. Schubert. *Geodynamics; Applications of Continuum Physics to Geophysical Problems.* John Wiley & Sons, New York (1992).
18. Zang Shaoxian and Ning Yuanjie. Progress and problem for the study on the global geodynamics, *Proceedings of Recent Geodynamics,* Oct. 6-9 1993, Beijing, 13-18 Seismological Press (1994).
19. B. H. Hager and R. J. O'Conell. A simple global model of plate dynamics and mantle convection, *J. Geophys. Res.* **86**, 4843-4867 (1981).
20. Wang Hongzhen. The Main stages of crustal development of China, *Earth Science — Jour. Wuhan College of Geology* **3**, 155-177 (1982).
21. W. Bucher. *The Deformation of the Earth's Crust.* Princeton University Press (1993).
22. J. H. F. Umbgrove. *The Pulse of Earth.* Nihoff, The Hague (1947).
23. K. J. Hsu. Time and space in Alpine orogenesis, *Geo. Soc. London Spec. Pub.,* **45**, 421-443 (1989).
24. A. M. C. Sengor. Timing of orogenic events: A persistent geological controversy. In: *Controversies in Modern Geology: Evolution of Geological Theories in Sedimentology, Earth History and Tectonics.* D. W. Mulle, J. A. Mckenzie and H. Weissert (Eds), pp. 405-476. Academic Press, London (1991).
25. Li Jiliang. Time and space for orogenic lithosphere tectonics evolution, *1989-1990 Annual report, Laboratory of Lithosphere Tectonics Evolution, Institute of Geology, Academica Sinica,* 119-123. Chinese Science and Technology Press (1991) (in Chinese with English abstract).
26. Wang Hongzhen. Retrospect of the study on global tectonics, *Earth Science Frontiers (China University of Geosciences, Beijing)* **2**, 37-42 (1995) (in Chinese with English abstract).
27. M. R. Rampino and R. B. Stothers. Geological rhythms and commentary impacts, *Science* **226**, 1427-1431(1984).
28. M. R. Rampino and R. B. Stothers. Flood basalt volcanism during the past 250 Million years,

Science **241**, 663-668 (1988).
29. K. A. Innanen. The sun's orbit in a mass model of the galactic system, *Zeits. Astrophysics* **64**, 457-459 (1966).
30. J. Steiner. The sequence of geological events and the dynamics of the Milk Way galaxy, *Jour. Geol. Soc. Australia* **14**, 99-131 (1967).
31. B. P. Glass. *Introduction to Planetary Geology*. Cambridge University Press (1982).
32. Yin Yanhong and Zhou Moqing. The catastrophic event of the eastern Pacific in the end of Eocene, *Annual Report for 1994, The Laboratory of Lithosphere Tectonics and its Dynamics, China Univ. Geosci. (Beijing)*, 77-86. Seismological Press, Beijing (1995).
33. W. W. Hey. Palaeoceanography: A review for the GSA centennial, *Geological Society of America Bulletin* **100**, 1934-1956 (1988).
34. Wan Tianfeng and Cao Rueping. Tectonic event and stress fields from Middle Eocene to Early Pleistocene in China, *Geosciences* **6**, 275-285 (1992) (in Chinese with English abstract).
35. Yin Yanhong and Wan Tianfeng. The possiblility and dynamics of a microtektite impacted the Pacific plate and caused the change of its moving direction in the end of Eocene, *Annual Report for 1995, The Laboratory of Lithosphere Tectonics and its Dynamics, China Univ. Geosci. (Beijing)*, 122-132. Geological Publishing House, Beijing (1996).
36. V. L. Shapton, G. B. Dalrymlple *et al.* New links between the Chicxulub impact structure and the Cretaceous / Tertiary boundary, *Nature* **359**, 819-821 (1992).
37. R. A. F. Grieve. Terrestrial impact: The record in the rocks, *Meteoritics* **26**, 175-194 (1991).
38. B. D. Dressier, G. G. Morrison *et al.* The Sudbury structure, Ontario, Canada-A review. In: *Research in Terrestrial Impact Structure.* J. Pohl (Ed). pp. 39-68. Friedr, Vieweg & Sohn, Braunschweig (1987).
39. Shi Yaolin. Impact induced thermal effects on the Lithosphere, *Annual Report for 1994, The Laboratory of Lithosphere Tectonics and its Dynamics, China Univ. Geosci. (Beijing)*, 59-72. Seismological Press, Beijing (1995).
40. Shi Yaolin. Possible effects of meteorite impact on mantle convection and plate motion, *Annual Report for 1995, The Laboratory of Lithosphere Tectonics and its Dynamics, China Univ. Geosci. (Beijing)*, 136-145. Geological Publishing House, Beijing (1996).
41. D. Ait *et al.* Terrestrial maria: The origins of large basalt plateaus hotspot tracks and spreading ridge, *Jour. Geology* **96**, 647-662 (1988).
42. M. R. Rampino and K. Caldeira. Antipodal hotspot pairs on the Earth, *Geophys. Research Letters* **19**, 2011-2014 (1992).
43. R. Oberbeck, T. R. Marshall and H. Aggarwal. Impact tillites and the breakup of Gondwanaland, *Jour. Geology* **101**, 1-19 (1993).
44. M. G. Young. Impact, tillitites and the breakup of Gondwanaland: A discussion, *Jour. Geology* **101**, 675-679 (1993).
45. J. P. Le Roux. Impact Tillitites and the breakup of Gondwanaland: A second discussion, *Jour. Geology* **102**, 483-485 (1994).
46. Zhang Shunxin. Did the impact result in the breakup of Gondwanaland? *Annual Report for 1995, The Laboratory of Lithosphere Tectonics and its Dynamics, China Univ. Geosci. (Beijing)*, 95-104. Geological Publishing House, Beijing (1996).

Proc. 30th Intern. Geol. Congr., Vol. 26, pp. 97-108
Wang *et al.* (Eds)

An Unique Terrestrial Occurrence of Extraterrestrial Microspherules

WANG ERKANG WAN YUQIU
Department of Earth Sciences, Nanjing University, Nanjing 210093, P. R. China
Laboratory of Solid State Microstructures, Nanjing University, P. R. China

HU ZHONGWEI
Department of Astronomy, Nanjing University, Nanjing 210093, P. R. China

BI DONG
Department of Geology, University of Alberta, Edmonton, AB T6G 2E3, Canada

Abstract

Microspherules were recently recovered in granite in Suzhou, China, and have been studied by using X-ray diffraction, scanning electron microscope, electron microprobe, and neutron activation analysis. Results show that iron microspherules are composed of magnetite, magnesioferrite, and kamacite; some of them contain α-iron metal core; siliceous microspherules are mainly of ringwoodite, the only identified minerals in siliceous microspherules. The similarities in morphology, mineralogy, and internal structure to spherules of extraterrestrial origin found in other environments, especially the identification of ringwoodite and the unusual chemical compositions of siliceous microspherules suggest that these spheruels from Suzhou granite may be of extraterrestrial origin, and even might be the products of an explosion.

Keywords: Microspherules, Ringwoodite, Extraterrestrial Origin, Granite, Suzhou, China.

INTRODUCTION

Over a century ago, Murray and Renard discovered microscopic magnetic spherules in deep-sea sediments [1]. Finding both "stony" and "iron" types, they suggested that the spheres were produced by melting meteoroids as they entered the atmosphere. Since then, deep-sea spheres have been subjected to a variety of investigations. Besides, magnetic spherules have been found in samples collected from many terrestrial occurrences, including deep-ocean clays [2, 3], the polar ice caps [4, 5, 6], manganese nodules [2], beach sands [7], sedimentary rocks range in age back to the early-Paleozoic, Pleistocene sediments [7, 8] and Recent fluvial sediments [9]. Above studies showed that these spherules can be classified into three categories, iron-, glass-, and silicate-dominated, of which iron-rich spherules are the most common [10].

Microspherules were recently discovered in granite in Suzhou, Jiangsu province, China during an investigation on the placer minerals of Suzhou granite. The purpose of this paper is to present the results of morphological, internal structural and mineralogical studies of these microspherules and discuss the possible origin of such microspherules. First, let us introduce the geological background of Suzhou area.

GEOLOGICAL SETTING

Located in east China, the Suzhou granite suite occurs in the Yangtze-Qiantang Paleozoic Depression Zone, west of the subduction zone of the West Pacific Plate against the China Continental Plate, bordering on the southeastern flank of the giant Changjiang tension fault. In late Triassic, an NE-striking wide synclinorium was formed in response to the Indosinian Movement. Crustal uplifting took place over vast areas in early Yenshanian (180-140Ma), accompanied by magma intrusions. In late Yenshanian (140-70 Ma), tectonic activities, particularly faulting, were extensively developed, leading to volcanic eruptions. The Suzhou granite suite was emplaced into the uplift region during middle-early Yenshanian, at the intersection of the NE-striking Mudu syncline tension fault and the NW-striking fault. The rocks are of anorogenic origin, showing clear-cut intrusive contact with their surroundings.

Strata exposed in the area belong to the Jiangnan System, including Devonian to Triassic sedimentary rocks and Jurassic volcanic rocks. Well-developed magmatic activities in the Suzhou area may be divided into three periods: (1) Indosinian epoch intrusive rocks are mainly of granite-porphyry. (2) Early Yenshanian intrusive rocks are dominant, including granite, quartz porphyry and circularly distributed granodiorite. (3) Late Yenshanian magmatism is presented mainly by intermediate-alkaline volcanic rocks spreading over vast areas northeast of the granite suite. The granite suite occurs as a diapiric intrusion, showing characteristics of shallow emplacement with a polygonal outline on the surface. Geophysical data indicate that the granite suite extends deep downwards in the form of a more or less cylinder-shaped stock, measuring approximately 65 km^2 across with an outcrop of about 11 km^2 [11] (Fig. 1).

MICROSPHERULES FROM SUZHOU GRANITE

Considering that Suzhou granite suite is located at the lower reaches of Yangtze River, the most prospect area of China, with a convenient transportation, since 1924, this granite suite has been studied. Furthermore, because of its color and hardness and lack of fractures, it has been used as the best material for memorial buildings. Within the granite suite, there are many stone quarries distributed, which provide us a good opportunity to get fresh granite samples. We have been engaged in detail investigations on placer minerals of Indosinian period granite-porphyry and Late Yenshanian period granite. The results indicate that microspherules are contained in both granites. The microspherules are distributed randomly in Indosinian period granite-porphyry, but for the Late Yenshanian period granite, they are only contained in its early stage. In heavy mineral fractions of suzhou granite, it is easy for these spheres to be noted due to their highly spherical shapes, magnetic properties and their metallic or pearl luster.

HANDLE METHODS

The fresh granite samples were cleaned with condensed air, and then broken by jaw breaker. After that the broken samples were sieved and elutriated, then were subject to a vibrating table to remove minerals with smaller gravity. These procedures would remove nearly all silicate minerals. The residues were then concentrated further using a high intensive permanent magnet to obtain ferromagnetic fractions. It is necessary to point out that the jaw breaker was the only contact metal tool during the whole handling process, and in every breaking process, the same breaker was used. Sampling and analyzing the jaw breaker shows that its trace elements are entirely different from granite magnetic microspherules.

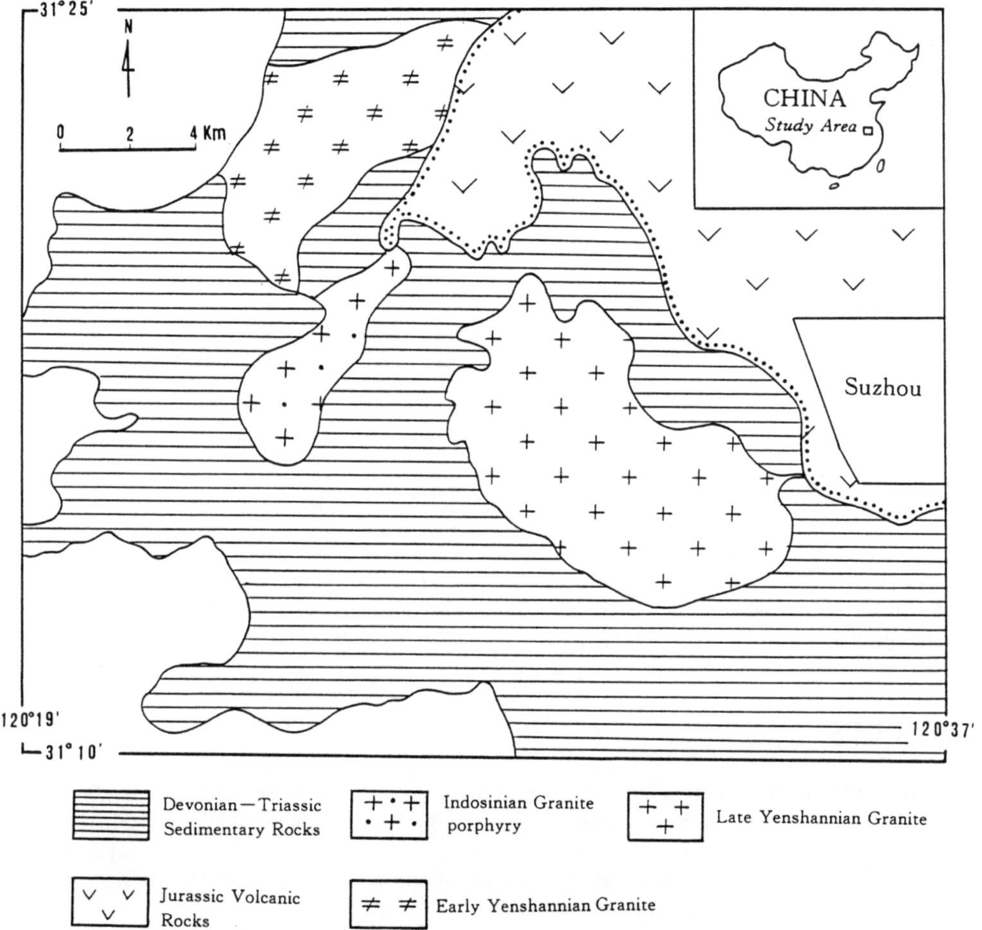

Figure 1. Geological map of Suzhou area, eastern China.

ANALYSIS METHODS

The spherules were picked out manually from the ferromagnetic fractions under a binocular microscope and were cleaned ultrasonically and washed with acetone. In the meantime, spherules were also found in thin sections when we carried out petrogaphical studies. The spherules were then placed on double-sided tape on a copper mount for surface morphology analysis using a X-600 scanning electron microscope equipped with an energy disperse X-ray spectrometer. After SEM analysis some of the spherules were embedded in epoxy resin, and were sectioned and polished to expose a flat surface for determination of compositions using an JEOL JXA-8800 M supper electron probe microanalyzer equipped with four wavelength disperse spectrometers pure metals were used as the standards for Ni, Cr, and Mn, and hematite and quartz as the standards for Fe and SiO_2. All analyses were carried out at 20 kV accelerating voltage. A group of siliceous spherules were crushed for X-ray diffraction(XRD) analysis. Data from the XRD analysis are presented in Table 1.

line	Suzhou Siliceous	granite[a] microspherule	
	d	I	
1	3.35[b]	80	
2	2.85	20	
3	2.46	80	
4	2.10	20	
5	2.04	10	
6	1.92	10	
7	1.81	10	
8			
9	1.54	20	
10	1.486	20	
11	1.442	20	
12	1.335	10	
13			
14	1.195	30	
15	1.120	2	

Table 1. X-ray Powder Diffract Data for Siliceous Microspherules in Suzhou Granite.
a) Determined by Kong Youhua, Institute of Geology and Ore Deposits, Zhejiang Province.
b) Affected by strong lines of quartz attached to surface of microspherules.

Instrumental neutron activation analysis (INAA) of five spherules was performed at the Institute of Atomic Energy, Academic Sinica, Beijing China. The spherules were irradiated for 100h in heavy water reactor.

RESULTS

Based on their distinct mineral and chemical compositions, the microspherules are readily divided into two main groups: iron and siliceous.The sizes of the all accounted microspherules were measured on the SEM screen. Their size ranges are from 40-210μm, mainly in the scope of 60-130 μm, dominated by 60-90 μm, some are larger than 300 μm. The histogram of the size distribution of Suzhou granite micrspherules is shown in Fig. 2.

Iron Microspherule

The iron microspherules are black, shiny and ferromagnetic. Most of the microspherules are nearly spherical and some are ellipsoidal in shape and have a surface with a distinct polygonal pattern or a dendritic pattern. Surface cavities or pits are common. Some spherules have a smaller sphere attached. A few are in the form of fruit core-like in shape. The presence of the dendritic structure indicate that the microspherules were once molten droplets [2]. In polished section, some of the microspherules are solid, some are hollow. The solid spherules tend to be made up of crust and core. The determination of mineralogy using X-ray powder diffraction for microspherules indicates that they are magnetite, magnesiofferite, α-iron, kamacite, and troilite. Most of the solid spherules are composed of Fe-oxide, mainly magnetite. The crust is mainly composed of magnetite and magnesioferrite, and the core mainly of α-iron and kamacite. Figure 3 presents some of the SEM images of iron microspherules.

Siliceous Microspherules

Under binocular, the microspherules are transparent or semitransparent with color from black, brown, some of them are of brownish-green, brownish-yellow and pale white. Most of the siliceous microspherules have a smooth surfaces. From their appearance features it is obvious that these microspherules have undergone an ablation process during atmospheric entry heating, such as ejection, explosion, collision, impaction and finally rapid cooling (Fig. 4).

About one thousand such siliceous microspherules were broken to make a big sample to be photographed by X-ray powder diffraction. The obtained data show that the only mineral hosted in siliceous microspherules is ringwoodite (Table 1) of which X-ray diffraction data is just similar to that of Tenham meteorite [12].

Five randomly chosen siliceous microspherules were detected by INAA. Comparing with CI chondritic abundance, the results show that the refractory elements such as Al Ca,Sc,Ti, V are uniformly enriched by factors of 10-50 × or more, and W,Hf,U,Th display much more magnitude enrichment, up to 200 ×. In addition, REEs also have a high enrichment, about 100 ×. Here we present REE data of Suzhou granite microspherules and the corresponding REE composition of mean Cl chondrite from Anders and Grevesse (1989) (Table 2).

Sample Weight	No. 8-2 22.9ug	No. 3 9.0ug	No. 8-4 4.8ug	No. 5 18.3ug	No. 10 24.7ug	mean-CI* Chondrite
La	55.4	80.9	206	42.1	101	0.235
Ce	130	205	810	112	280	0.603
Nd	74	75	820	29	129	0.452
Sm	8.20	13.7	16.9	6.40	13.3	0.147
Eu	2.97	5.06	11.6	2.65	4.87	0.056
Tb	1.9	6.2	7.4	---	4.4	0.036
Dy	11.1	25.2	41.9	14.5	19.8	0.243
Ho	2.23	4.85	16.9	3.10	4.5	0.056
Er	6.90	---	---	9.9	---	0.159
Yb	13.8	29.1	33.3	21.8	14.7	0.163
Lu	1.79	3.27	5.35	3.23	2.33	0.024

Table 2. REEs in the Siliceous Microspherules of Suzhou Granite (Determined by INAA) (All Data in ppm).
* From E. Anders and N. Grevess (1989).

Based on the 532 EDS spot analyses for more than 100 siliceous microspherules, nine principal composition types can be available. Table 3 is the analytical results about nine composition types of the siliceous microspherules. Their main composition is SiO_2, Al_2O_3 FeO, CaO and K_2O. The general characteristic of composition is rich in Si and poor in Mg. The content of Si, Al oxides varies inversely with the content of Fe oxide. Besides, both Ca and K in some microspherules show very high content.

ORIGIN OF MICROSPHERULES

The characteristic of mineral composition hosted in microspherules is an important criterion to identify whether they are of extraterrestrial or not. For the iron microspherules, kamacite is a common meteoritic mineral, its existence provides us a clue to prove the microspherules to be of extraterrestrial origin.

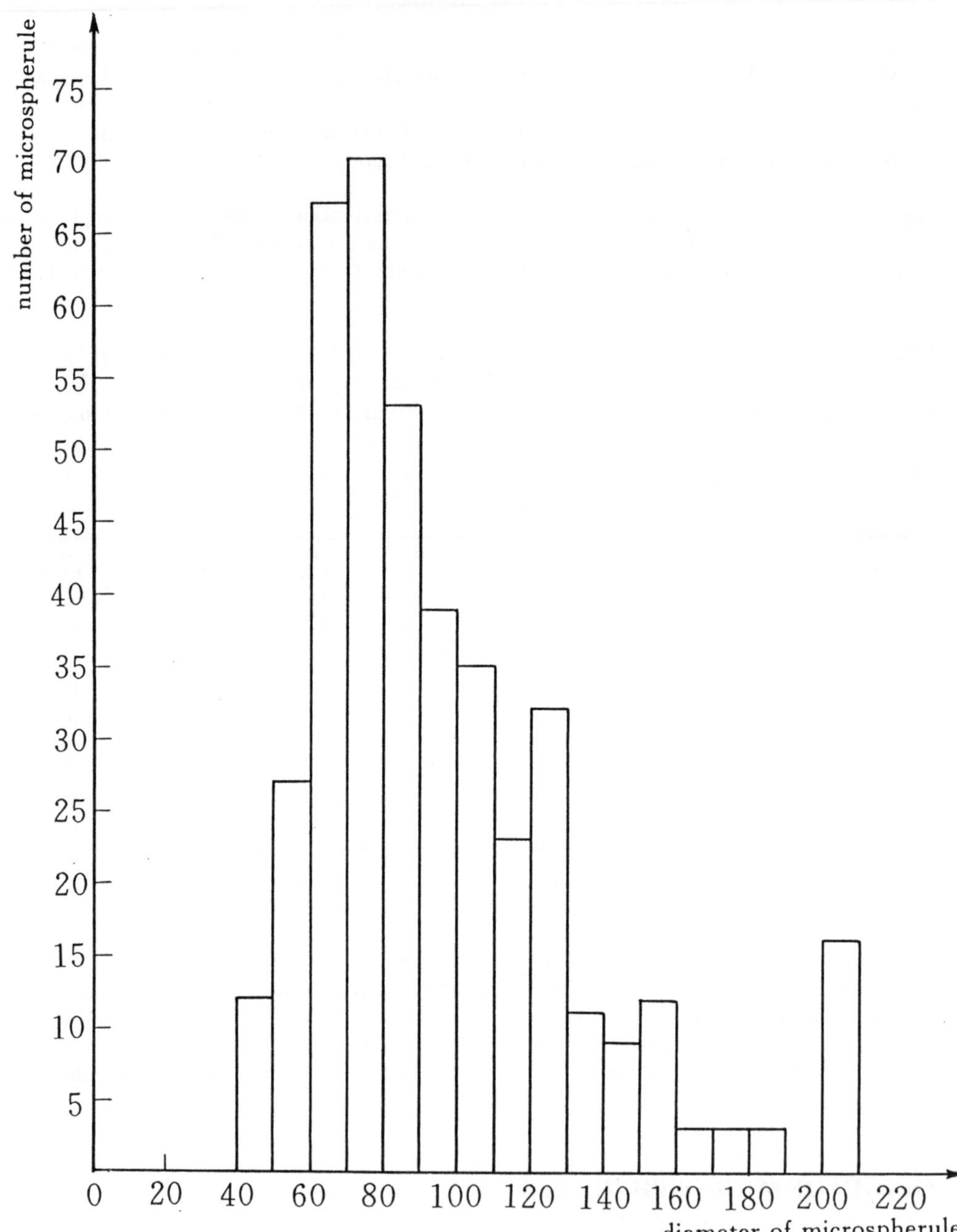

Figure 2. The histogram of size distribution of the microspherules in Suzhou granite.

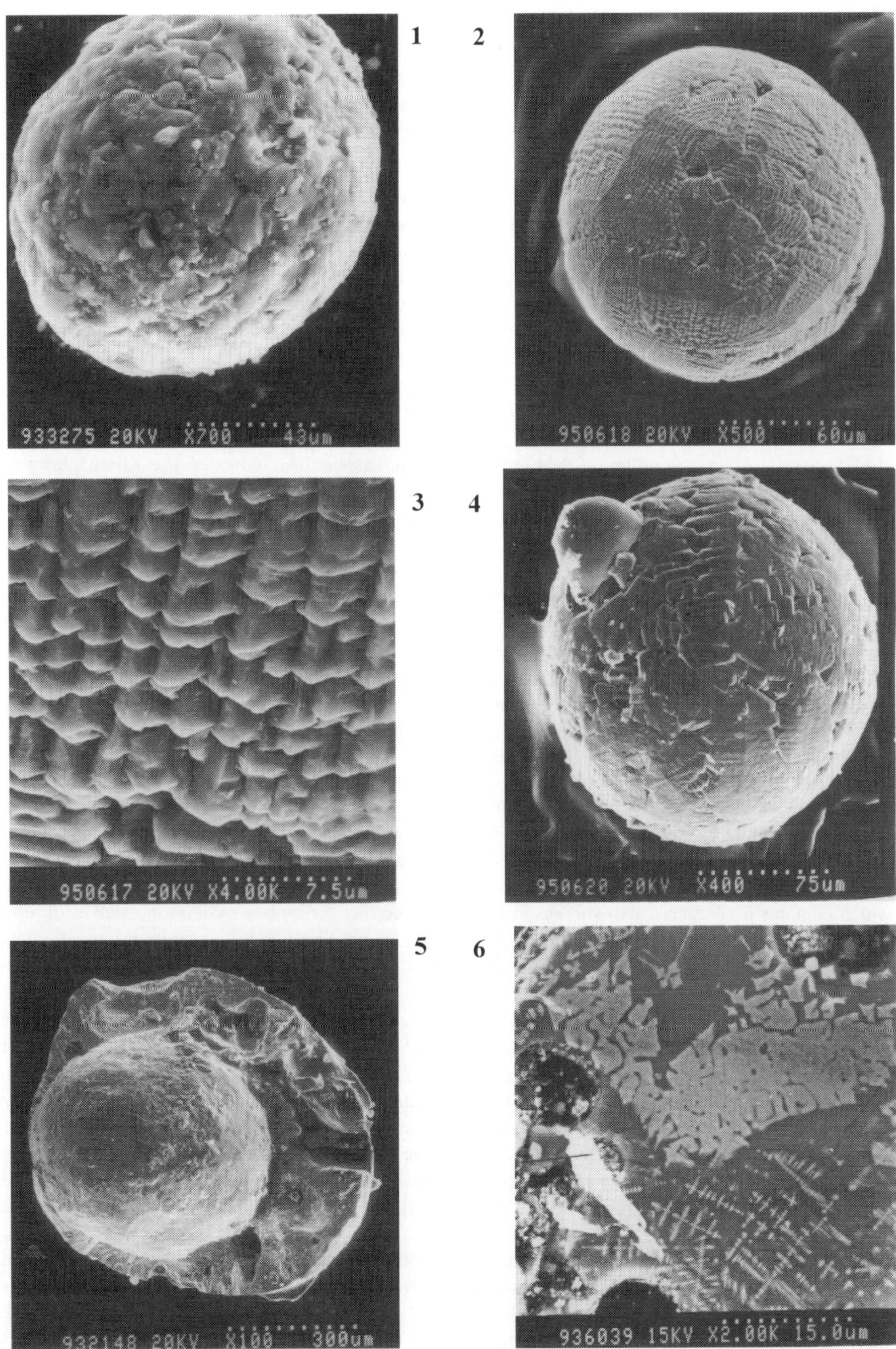
1
2
3
4
5
6
933275 20KV X700 43um
950618 20KV X500 60um
950617 20KV X4.00K 7.5um
950620 20KV X400 75um
932148 20KV X100 300um
936039 15KV X2.00K 15.0um

Figure 3. SEM images of iron microspherules from Suzhou granite.
1. A elliptical microspherule with gas stamp on its surface which shows brain-like wave.
2. A magnesioferrite microspherule, its surface is the aggregate of many small crystals.
3. Closed-up of image 2, magnesioferrite crystal druse can be seen clearly.
4. A magnetic microspherule of complete crystals, its top left with a small sphere attached.
5. A natural section clearly showing crust and metal core, crust is composed of magnetite and core of a-iron.
6. Part image of a polished section showing skeleton and dendritic structures.

For the siliceous microspherules, the presence of ringwoodite is sufficient proof for us identify them to be of extraterrestrial origin. Ringwoodite in natural occurrence was first discovered by Binns [13, 14] in the impact vein of Tenham chondritic meteorite. After that, this mineral was found in Coorara and Catherwood chondrites [15]. Based on the experimental results of A. E. Ringwood [16], ringwoodite can be formed in the condition of a very high temperature and pressure [17]. Only beneath the depth of 500 km of crust can meet such formation condition. So for the terrestrial minerals, ringwoodite only present theoretically not really exist, whereas for the extraterrestrial minerals, it was reported in cosmic occurrence many times and deeply studied [4, 13, 14, 18]. Ringwoodite belongs to the minerals which own great typomorphic implication.

The discovery of ringwoodite in Suzhou granite micrsospherules indicates the microspherules maybe experienced a similar process to Tenham. For such small microspherules and ringwoodite containing in them, in order to meet the very high temperature and a very high pressure, the only explanation, we suggest, is that these microspherules would be formed by particulates of an meteoritic explosion.

For the REE composition of the siliceous microspherules in Suzhou granite (Fig. 5a), it is obvious that total amounts of REE are greatly enriched, and there is an appreciable enrichment of the LREEs relative to HREEs. REE abundance pattern favors a meteorite origin as a roughly flat distribution, namely, REE did not experienced a notable chemical differentiation. Similar REE pattern observed in the famous Bishopville aubrites reported by C. Floss and others [19]. Figure 5 shows the comparison among their distribution patterns.

Extraterrestrial origin spherules have been found in samples collected in many environments. Among them the spherules from deep-sea clay was a representative. Comparing Suzhou microspherules with deep-sea spherules shows that Suzhou microspherules have unusual compositions. Of all the siliceous microspherules in Suzhou granite, there is no one microspherule which contains perfect crystals, whereas in deep-sea spherules, crystals are common, such as olivine and magnetite showing the typical brickwork pattern. So we think Suzhou granite microspherules formed in the condition different from that of deep-sea spherules.

The interesting thing is that there are two mid-air explosion events, Revelstoke and Tunguska events, which produced a large number of iron and siliceous spherules resembling to those of Suzhou granite. Especially for the siliceous spherules. For the chemical compositions of siliceous microspherules in Suzhou granite, they can be priliminarily divided into 9 types, characterized by Fe-rich, Al-rich, K-rich, Ca-rich, respectively. Comparing the chemical compositions of siliceous spherules of Revelstoke detected by Carr. H [20] using EPMA shows that exception for the types of "2","3',"9", the others of all composition types of Revelstoke siliceous microspheruels are in a good match with Suzhou siliceous micro-spherules (Table 3).

The siliceous microspherules of Revelstoke not only dominant the amount of the obtained microspheruls but also present many composition types. These are not the accident, just reflect the conditions of sphere forming. When the explosion of meteorite body occurred in mid-air, from the melting of particles, the forming of droplets to the finally rapid quenching, all these processes completed in a very short time, so preserved the varieties of composition properties at that time.

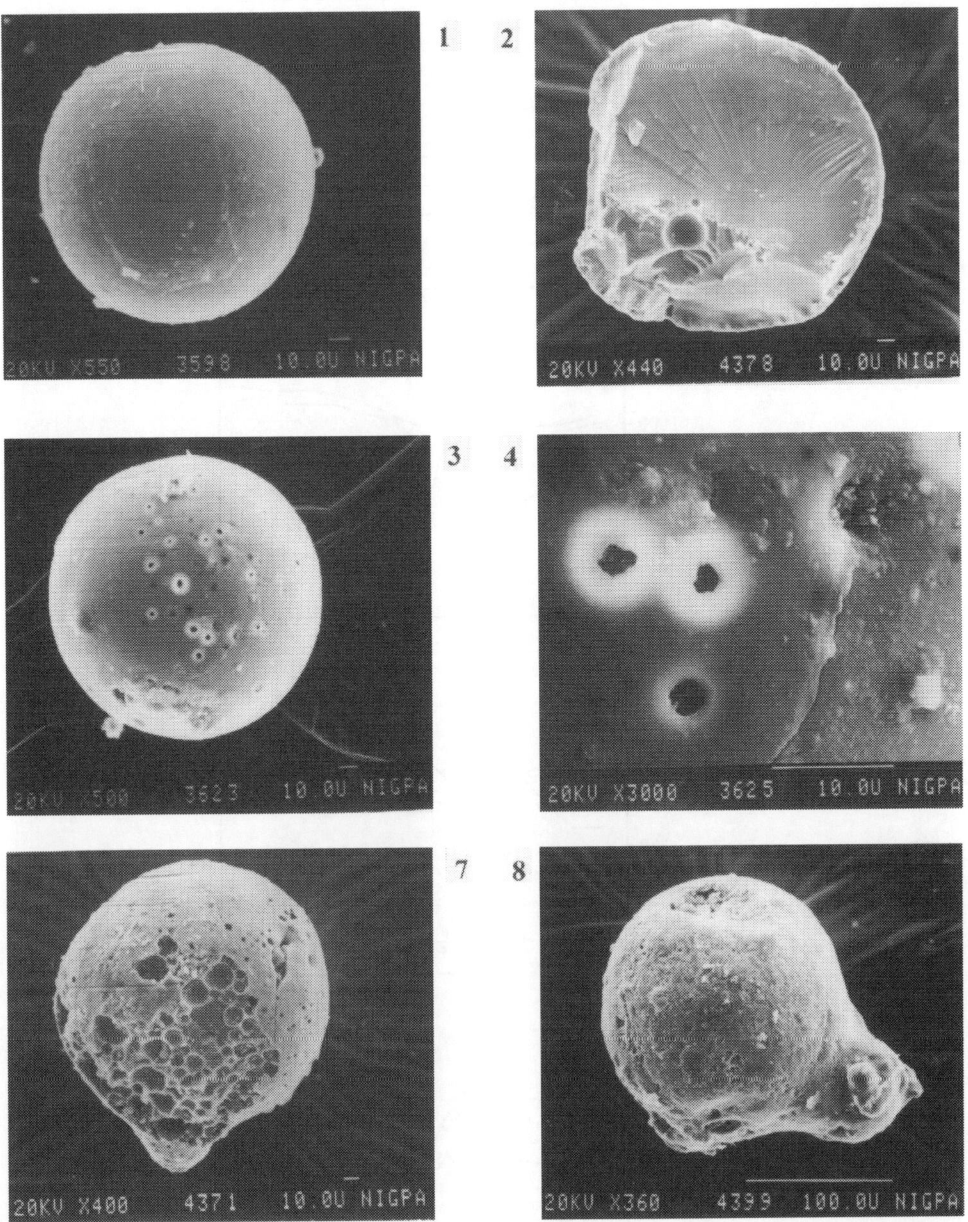

Figure 4. SEM images of siliceous microspherules from Suzhou granite.

1. A siliceous spherule with a smooth surface.
2. Natural section of a siliceous spherule displaying conchoidal fracture.
3. A sphere with a number of small holes in its surface, its left side with a small protrusion, a possible not broken bubble.
4. closed-up of image 3, clearly showing hole surrounded by gas halo.
5. A droplet spherule with many cavities in its bottom side.
6. A compound spherule welded by big spherule and a small spherule, with a pit on its upper side.

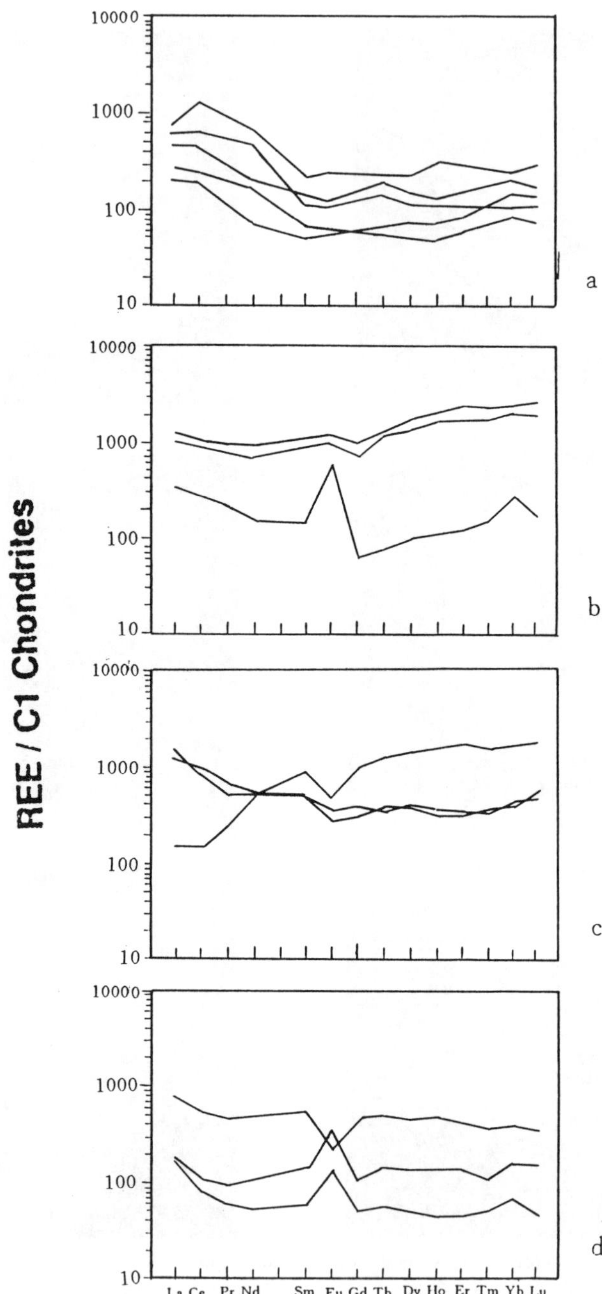

Figure 5. CI chondrite-normalized REE patterns.
a. for microspherules of Suzhou granite.
b, c, d. for seven areas of weathered CaS from Bishopville meteorite.

The similarities of Chemical composition types of siliceous microspherules between Suzhou granite and Revelstoke indicate the possible similarity of their formation processes. The occurrence of ringwoodite in microspherules of Suzhou granite only implies that the energy of the explosion would be much greater. In addition, there is a good comparison in compositional type of siliceous microspherules between Suzhou microspherules and Senzeilles microspherules which is concluded as the product of a meteorite impact [21]. Whereas, deep-sea sediment spherules did not tend to display such varieties of siliceous microspherules, they were selected and preserved only those in a collectable size surviving the atmospheric entry ablation [22].

Type	I	II	III	IV	V	VI	VII	VIII	IX
Sample	124-3-1	125-1-1	124-4-1	126-1-5	126-1-6	126-2-1	124-30-5	121-1-1	121-1-4
SiO_2	28.79	36.25	38.58	57.28	16.57	41.10	38.07	50.35	53.35
Al_2O_3	4.07	12.83	9.70	28.75	9.25	14.73	28.68	17.94	26.08
FeO	62.46	44.56	41.93	9.71	41.27	25.91	29.69	6.69	7.13
CaO	0.17	5.20	0.46	1.08	31.76	17.47	2.71	21.28	1.37
K_2O	1.81	1.16	2.88	2.28	0.88	0.41	0.50	1.59	10.06
Feature	Fe>Si Al<<Si	Fe>Si Al<Si	Fe≥Si Al<Si	Fe<<Si Al<Si	Fe>Si Al<Si Ca-rich	Fe<Si Al<Si Ca-rich	Fe<Si Al<Si	Fe<<Si Al<Si Ca-rich	Fe<<Si Al<Si K-rich

Table 3. EDS analysis of siliceous microspherules in Suzhou Granite.

CONCLUSION

Ringwoodite is an meteoritic typomorphic mineral. The identification of ringwoodite in siliceous microspherules of Suzhou granite indicates that the microspherules are no doubt of an extraterrestrial origin. Considering the formation condition of ringwoodite, these microspherules were more possible the products produced by an explosion of a meteorite at mid-air or an impact once happened.

The microspherules from granite in Suzhou show some similarities in morphology and mineralogy to spherules of extraterrestrial origin from other terrestrial environments such as Tunguska, Revelstoke, and are identical to spherules found in Senzeilles F/F boundary in Belgium. This is not an coincident thing. This is because of the similarities of formation process among them. Whether Suzhou microspherules, Tunguska and Revelstoke microspherules or Senzeilles microspherules all experienced an explosion or impact which is different from the ablation from meteorite surface.

How did the microspherules enter the granite? This is still a enigma. We suggest that all these microspherules may be products of an event once happened. They fell down and were aggregated in some strata and then were captured by the emplacement of granitic magma. About this problem, some corresponding researches are being done.

Acknowledgements

The project was financially supported by the National Natural Science Foundation of China (NNSF), for which we are thankful. We are especially grateful to Prof. Ouyang Ziyuan for his encouragement and support of our work, and we thank Prof. Lou Gufeng and Chen Wu for their help in mineral studies. Thanks are also due to Yang Hua and Sun Deming for their supports for our field work.

REFERENCES

1. S. Murray and A. F. Renards. *Report on the scientific results of the HMS Challenger during the years 1873-76*. vol. 3. Neill and Company, Edinburgh.
2. R. B. Finkelman. Magnetic particles extracted from manganese nodules: suggested origin from stony and iron meteorites, *Science* (Washington, D. C) **167**, 980-984 (1970).
3. H. T. Millard and R. B. Finkelman. Chemical and mineralogical compositions of cosmic and terrestrial spherules from marine sediments, *Jour. Geophy. Res.* **75**, 2125-2134 (1970).
4. R. A. Schmidt. Microscopic extraterrestrial particles from the Antarctic Peninusula, *Annals of the New York Academy of Sciences* **119**, 186-204(1964).
5. E. Thiel and R. A. Schmidt. Spheres from the Antarctic ice cap, *Jour. Geophy. Res.* **66**, 307-310 (1961).
6. F. W. Wright, P. W. Hodge and C. S. JR. Kangway. Studies of particles from extraterrestrial origin: 1. Chemical analyses of 118 particles, *Jour. Geophy. Res.* **68**, 5575-5588(1963).
7. U. B. Marvin, and M. T. Einaudi. Black magnetic spherules from Pleistocene beach sands, *Geochimical et Cosmochimica Acta* **31**, 1871-1884 (1967).
8. D. Bi, R. D. Morton, and K. Wang. Cosmic Ni-Fe alloy spherules from Pleistocene sediments Alberta, Canada, *Geochimica et Cosmochimica Acta* **57**, 4129-4136 (1993).
9. D. Bi, and R. D. Morton. Magnetic spherules from recent fluvial sediments in Alberta, Canada: characteristics and possible origin, *Canadian Journal of Earth Sciences* **32**, 351-358 (1995).
10. M. B. Blanchard, D. E. Brownlee, T. E. Bunch, P. W. Hodge, and F. T. Kyte. Meteoroid ablation spheres from deep-sea sediments, *Earth and Planetary Science Letters* **46**, 178-190 (1980).
11. X. L. Zhang, J. B. Wang and B. C. Shen. A study on A-type granite in Suzhou, China, *Geochemistry* (China) **7**, 29-45(1988).
12. E. K. Wang, Y. Q. Wan, Y. H. Kong, Y. Y. Shao and W. L. Zhang. First terrestrial occurrence of ringwoodite, *Chinese Science Bulletine* **38**, 1652-1655(1993)
13. R. A. Binns, Ringwoodite, nature $(Fe,Fe)_2$ SiO_4 spinel in the Tenham meteorite, *Nature* **221**, 943-945 (1969).
14. A. Putnis and G. D. Price. High-pressure $(Mg,Fe)_2$ SiO_4 phases in the Tenham chondritic meteorite, *Nature* **280:19**, 217-218 (1979).
15. J. V. Smith and B. Mason. Pyroxene-garnet transformation in Coorara meteorite, *Science* **160**, 66-67 (1970).
16. A. E. Ringwood and A. Major. Synthesis of Mg_2SiO_4-Fe_2SiO_4 spinel solid solutions, *Earth Plane. Sci. Lett.* **1**, 241-254 (1966).
17. D. Bi. Magnetic spherules from Pleistocene sediments in Alberta, Canada, *Meteoritics* **29**, 88-93 (1994).
18. M. Cheng, T. G. Sharp, A. El. Goresy, B. Wopeka and X. D. Xie. The majorite-pyrope+ magnesiowustite assemblage: Constraints on the history of shock veins in chondrites, *Science* **271**, 1570-1573 (1996).
19. C. Floss, M. M. Strait and C. Crozaz. Rare earth elements and the petrogenesis of aubrites, *Geochimica et Cosmochimica Acta* **54**, 3553-3558(1990).
20. M. H. Carr. Atmospheric collection of debris from the Revelstoke and Allende fireball, *Geochimica et Cosmochimica Acta* **34**, 689-700(1970).
21. P. J. Claeys, G. Casier, and S. V. Margolis. Microtektites and mass extinctions: Evidence for a Late Devonian asteroid impact, *Science* **19:2**, 109-210(1992).
22. S. G. Love and D. E. Brownlee. Heating and thermal transformation of micrometeoroids entering the Earth's atmosphere, *Icarus* **89**, 426-433 (1991).

Proc. 30th Intern. Geol. Congr., Vol. 26, pp. 109-114
Wang *et al.* (Eds)

Yamato -793605: a newly Idenetified Martian Meteorite from Antarctica

KEIZO YANAI
Department of Enviromental and Planetary Sicences, Faculty of Engineering, Iwate University, 4-3-5 Ueda, Morioka 020, Japan

Abstract

Yamato-793605 is a newly identified Martian meteorite from Antarctica. For a long time, this specimen was believed initially to be one of the type B Yamato diogenites, based on it's general appearance and mineral assemblages; however, it contains intermediate plagioclase, whereas Ca-rich plagioclase is present in all known diogenites. Yamato-793605 is a breccia of an originally cumulate rock, showing typical poikilitic texture and consisting of 50~60% pyroxenes (mostly coarse-grained orthpyroxene with chinopyroxene) and granular olivine with interstitial plagioclase. Pyroxene compositions are Mg-rich, orthopyroxene is En75Fs22Wo3 and clinopyroxene is En15Fs50Wo35. Olivine is Mg-rich(Fo70) with little variation. Interstitial plagioclase is completely maskelynite with an intermediate of approximately An50. composition(An55). Orthopyroxene in this meteorite is more Mg-rich than that of the type B diogenites, and plagioclase is Ca-poor in all known diogenites and eucrite-type meteorites. Yamato-793605 is similar to ALH-77005 shergottite in its mineral assemblages and especially mineral compositions. An oxygen isotopic analysis strongly suggests that Yamato-793605 is a Martian meteorite. Its composition is: $\Delta^{18}O$=4.01, $\Delta^{17}O$=2.3 and $\Delta^{17}O$=+0.21. Yamato-793605 is the eleventh known Martian meteorite in the world and the fifth one from Antarctica.

Keywords: Yamato-793605, Martian meteorite, Antarctica

INTRODUCTION

The Yamato-793605 (Y-793605) meteorite was collected by the meteorite search party of the 20th Japanese Antarctic Research Expedition (JARE-20, 1978-1980) [1] on the bare icefield around the Yamato Mountains (71°-73°S, 34°-37°E) in Queen Maud Land, East Antarctica. Over 3,600 specimens were recovered by the Expedition and they all have been initially processed since 1980. During the processing, the Y-793605 was preliminarily identified and classified as one of the diogenite-type achondrites. Especially Y-793605 was believed to be one of the type B Yamato diogenites (simply type B diogenite as Yamato-75032) together with other type B diogenites, such as Yamato-75032 and Yamato791200, because Y-793605 is very similar to the type B diogenites from its general appearance, mineral assemeblage and pyroxene compositions. Therefore, Y-793605 has been stored with other specimens of the type B diogenites for a long time and was not studied in detail. However, during a recent study of all Yamato diogenites, the author discovered important mineralogical evidence that suggests Y-793605 is quite different from type B diogenitic meteorites. Its plagioclase has a composition of approximately An55, whereas plagioclase in diogenites is generally more Ca-rich. The author asked Professor R. Clayton, University of Chicago, to do an oxygen isotope analysis to help resolve the issue. His analysis confirmed that Y-793605 is Martian meteorites [2]. Below is a preliminary description of the mineralogy, petrology and bulk composition of this newly identified Martian meteorite. Table 1 shows all known Martian meteorits in the world up to 1996.

Shergottites		Nakhlites		Chassignites		Orthpyroxenites	
ALH-77005*	482.5g	G.V.	158g	Chassigny	~4kg	ALH84001	1,930.9g
EETA79001	7,942.0g	Lafayette	~800g				
Y-793605*	-	Nakhla	~40kg				
LEW88516*	13.2g						
QUE94201	12.0g						
Shergotty	~5kg						
Zagami	~18kg						

Table 1. Martian meteorites and their original weight.
ALH : Allan Hills, EET: Elephant Moraine, G.V.: Governador Valadares, LEW: Lewis Cliff, QUE : Queen Alexandra Range, Y: Yamato Mountains. *olivine bearing. Total weight: ~78kg.

YAMATO - 93605 MARTIAN METEORITE; DISCOVERY, INITIAL PROCESSING AND PRELIMINARY IDENTIFICATION

The search party for Antarctic meteorites of the JARE-20, led by the author, revisited the Yamato Mountains during the October 1979 and January 1980 austral summer season. The party recovered the Y-793605 meteorite as one of over 3,600 Antarctic meteorite specimens on the Yamato Icefield in this season. Yamato-793605 was found on the bare ice surface, like most other Antarctic meteorites, and it was collected and kept in a clean teflon bag in the same manner as other Antarctic meteorite specimens.

After the initial processing, Yamato-793605 was stored together with other diogenite specimens and was believed to be a small fragment of type B diogenite for a long time without further detailed studies such as PTS examinations and elemental analyses. After a more detailed examination of a polished thin section (PTS), the author suggested that this small chip differs remarkably from all other diogenites in the Yamato meteorite collections.

The specimen was officially named Yamato-79305 (Y-79305), based on the naming system of the Japanese collections of Ataractic meteorites in the initial processing. Detailed description and photography of the specimen was carried out by the curatorial office.

The specimen is a fragment, partly covered with a black fusion crust. Relatively coarse-grained granular crystals are seen on the exposed interior face. A photomicrograph of a thin section (Fig. 1) shows a brecciated and has typically poikilitic texture. Most olivine is euhedral and plagioclase is interstitial with many crystal fragments in the darker portions. Very coarse-porphyritic pyroxene crystals can be seen in the section. The sample also contains some opaque accessory minerals.

PTS, ANALYTICAL METHODS, BULK CHEMISTRY AND OXYGEN ISOTOPES

A small chip of Yamato-793605 was made into a polished thin section (PTS). The section was studied using petrographic microscope in transmitted and reflected light. Quantitative elemental analysis and elemental mapping of the constituent minerals were carried out using an automated

JEOL JCXA733 and JXA-8800M electron probe microanalyzer (EPMA) with five spectrometers. The analytical procedures of the EPMA are the same as those of Kushiro and Nakamura (1970) [3].

An interior chip was used for standard wet chemical analysis and it was chemically analyzed by H. Haramura. A few ten mg of small fragments was analyzed for oxygen isotopes by Claytons group, University of Chicago [2].

PETROGRAPHY AND MINERAL COMPOSITIONS

Examination of the PTS shows that Y-793605 is a breccia consisting mainly of large pyroxene grains in a black matrix of shock origin with many small crystal fragments. Pyroxene grains show typical poikilitic texture including granular olivine and interstitial plagioclase, and Y-793605 appears originally to be a cumulate rock (Fig. 1). Mineral composition are shown in Table 2.

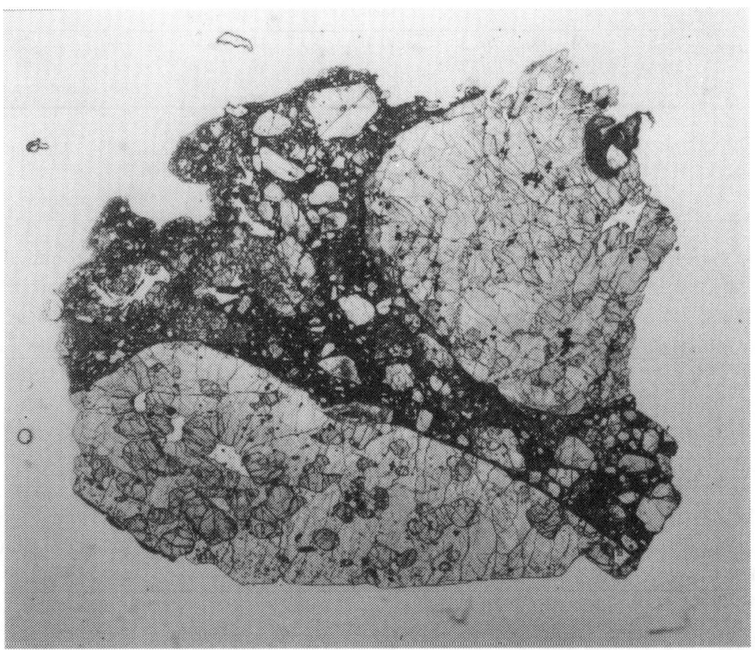

Figure 1. Photomicrograph of a thin section of Yamato-793605, with a field of view of 9 mm. Y-793605 shows a breccia with an originally cumulate texture, showing typical poikilitic texture, consisting mostly of coarse-grained orthopyroxene (Px) with granular olivine (Ol) and interstitial plagioclase (Pl).

Pyroxene

Pyroxene occurs as the main phase in the specimen and large pyroxene grains, almost 1 cm in size, poikilitcally include many olivine grains with plagioclase. Small fragmental pyroxene grains are in the shocked black matrix. Pyroxene compositions are Mg-rich orthopyroxene, average En74.6Fs21.6Wo3.8 and range En763-73.3Fs22.5-20.7Wo3.0-4.4, in the host phase with poikilitic texture. Clinopyroxene compositions average En51.0Fs14.9Wo34.1 and range En52.1-50.0Fs16.7-13.7Wo35.8-31.6, in small grains and some pyroxene grains plot in the pigeonite region of the pyroxene quadrilateral with little compositional variation (Fig. 2a). The pyroxene has a remarkably low FeO/MnO ratio of 26.4 and several grains lie within the range of average proxene from basaltic achondrites, but most are clearly different from those of basaltic achondrites.

		Olivine			Pyroxene				Plagioclase			
		Ol-1	Ol-2		Px-1	Px-2	Px-3		Pl-1	Pl-2	Pl-3	Pl-4
		Main	phase		Mg-rich	Ca-rich	Ca-rich		average			
		Mg-rich	Fe-rich		Opx	Cpx	Pig		(114)	Na-rich	Ca-rich	K-rich
SiO_2		38.19	36.25		56.35	53.02	53.14		53.93	56.15	55.09	54.64
TiO_2		0.02	0.04		0.05	0.27	0.00		0.16	0.10	0.11	0.04
Al_2O_3		0.02	0.03		0.38	1.71	1.36		27.94	27.46	28.91	28.14
FeO		24.28	29.38		13.79	8.25	17.39		0.48	0.63	0.61	0.51
MnO		0.41	0.61		0.50	0.41	0.64		0.07	0.04	0.00	0.04
MgO		38.33	32.62		28.49	16.98	21.66		0.10	0.17	0.14	0.13
CaO		0.21	0.21		1.58	16.79	7.10		11.03	9.83	11.03	10.61
Na_2O		0.00	0.00		0.09	0.41	0.18		4.87	5.68	2.41	5.66
K2O		0.00	0.00		0.00	0.00	0.00		0.22	0.22	0.12	0.40
Cr_2O_3		0.01	0.05		0.47	0.95	0.00		0.02	0.00	0.03	0.00
Total (wt%)		101.47	99.19		101.70	98.79	101.47		98.82	100.28	98.45	100.17
FeO/MnO:		59.2	48.2		27.6	20.1	27.2		-	-	-	-
Mol %:	Fo	73.8	66.4	En	76.3	50.4	59.3	Ab	43.9	50.5	28.1	48.0
	Fa	26.2	33.6	Fs	20.7	13.7	26.7	An	54.7	48.2	71.0	49.8
				Wo	3.0	35.8	14.0	Or	1.3	1.3	0.9	2.2

Table 2. Electron Microprobe analysis of major and minor silicate phases in the Yamato-793605 Martian meteorite (weight percent).

Olivine

Olivine is the major phase in Y-793605. It occurs as euhedral to subhedral and rounded individual grains and some are in aggregates. Most olivine, under lmm accross, is included poikilitically with in large pyroxene grains, but some occur as fragments in the shocked blackened matrix. Microprobe analyses show that olivine is compositionally homogeneous and Mg-rich (average Fo68.8, and range Fo66.4-738, Fig. 2b), similar to those in the ALH-77005 Martian meteorite, some diogenites (for example, Asuka-881377) and some lunar meteorites [4]. Olivine grains associated with the shocked blackened matrix have nearly the same compositions as that of the main olivine.

Plagioclase

Plagioclase occurs interstitially as a minor phase at pyroxene-pyroxene and pyroxene-olivine grain boundaries. Plagioclase is completely maskelynite and its composition is nearly intermediate, average An54.7, with little variation. Its range Ab28.1-50.5An48.2-71.0Or0.7-2.1 (Fig. 2c). Plagioclase of Y-793605 has a very similar composition to those of Martian meteorites, especially shergottites, but it is quite different from those of all other known achondrite groups. Therefore plagioclase is one of the most important and useful indicators in distinguishing Martian meteorites from achondrites.

BULK CHEMISTRY AND OXYGEN ISOTOPES OF YAMATO-793605 AND OTHER MARTIAN METEORITES

The results are presented in Table 3, with data for Y-793605 and other Martian meteorites for comparison. The composition of Y-793605 is generally similar to that of ALH-77005 classified as a shergottite, but it differs markedly from other shergottites such as the Zagami meteorites.

In detail, the composition of Y-793605 is higher in Al2O3, FeO and CaO, but lower in SiO2 and MgO than that of ALH-77005. Y-793605 is higher in FeO and MgO, but lower in SiO2 and CaO, compared with other shergottites. Therefore Y-793605 is similer to ALH-77005, but it might be originally different from other shergottites.

The oxygen isotopic composition of Y-793605 strongly suggests that it is Martian meteorites. Y-793605 has an oxygen composition of d18O=4.01, d17O=2.30 and D17O=+0.21% and it is most similar to the composition of Chassigny (d18O=+3.91; d17O=+2.33) [5].

COMPARISON WITH OTHER MARTIAN METEORITES AND CONCLUSIONS

Yamato-793605 is a breccia of an originally cumulate rock showing typical poikilitic texture and consists of 50~60% pyroxene with olivine and plagioclase. Pyroxene is mostly coarse-grained orthopyroxene including eudral-subhedral granular olivine grains and interstitial plagioclase.

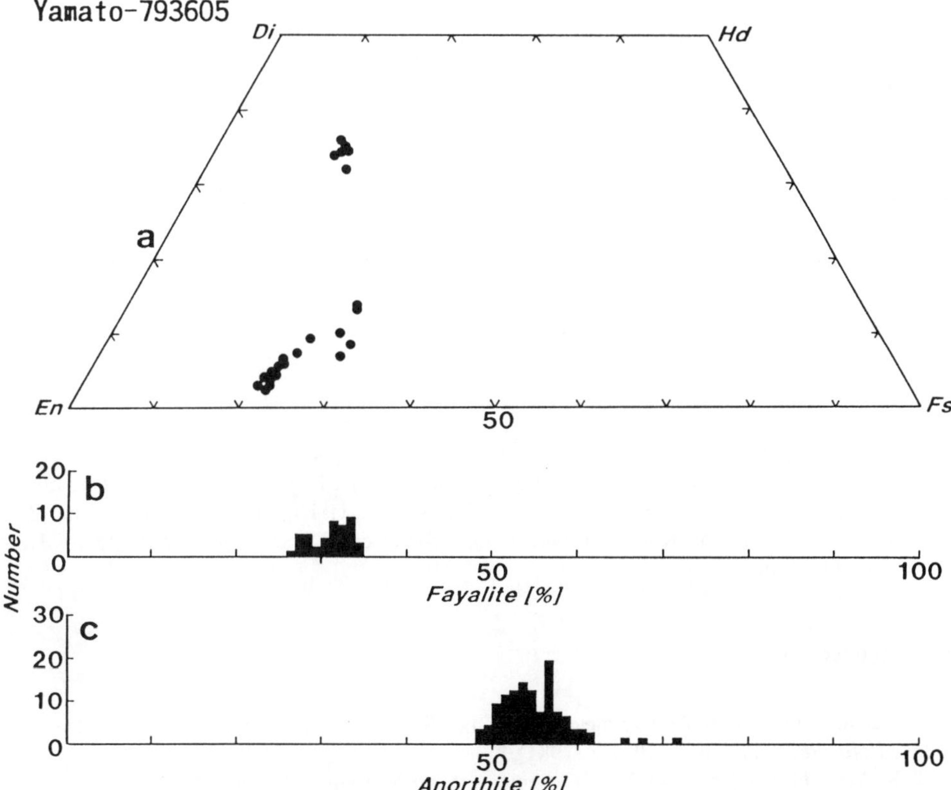

Figure 2. (a) Plot of Yamato-793605 pyroxene compositions on the pyroxene quadrilateral. En is modal Mg/(Ca+Fe+Mg) and Fe is Fe/(Ca+Fe+Mg). The compositions are plot in three regions, as Ca-augite, pigeonite and Mg-orthpyroxene. (b) Distribution of olivine composition show a narrow range of Fe contents. (c) Plagioclase is almost intermediate in composition with little variation.

Yamato-793605 is classified as a shergottite preliminarily based on its mineral assemblages and especially its intermediate plagioclase compositions, texture, mineral compositions, and bulk chemistry. The oxygen isotope data strongly support this classification. Y-793605 is similar to

ALH-77005 in its lithology, mineral comopsitions and especially bulk chemical composition (*e.g.* Table 1 and 3), but the two meteorites are quite different from other shergottites. Therfore Yamato-793605 and ALH-77005 should be separated preliminarily from the main group of shergottites, and temporarily called lherzolitic shergottites. It is proposed that the two lherzolitic shergottites be classified as a new grouplet, but the term requires future consideration and discussion.

	Y-793605	ALH-77005	Zagami
SiO2	42.51	43.02	50.52
TiO2	0.60	0.36	0.84
Al2O3	4.01	2.54	6.27
Fe2O3	0.22	0.38	0
FeO	20.52	18.97	18.03
MnO	0.44	0.45	0.44
MgO	25.98	29.69	12.14
CaO	3.76	2.84	9.57
Na2O	0.24	0.37	0.13
K2O	0.03	0.03	0.08
H2O(-)	0.00	0.00	0.00
H2O(+)	0.3	0.28	0.0
P2O5	0.51	0.39	0.46
Cr2O3	0.82	1.00	0.15
FeS	-	0.25	1.58
Ni%, (ppm)	(913)	0.024	(26)
Co%, (ppm)	(<30)	-	(<30)
S%, (ppm)	(0.56)	-	-
Total	100.50	100.59	100.21

Table 3. Major elemental compositions of Yamato-793605 compared with other Martian meteorites, in weight percent.
Standard wet chemical analysis (H. Haramura).

Acknowledgements

I am particularly grateful to all members of the meteorite search party of the JARE-20 for their great efforts in the search for meteorites in both field seasons. I thank H. Haramura for the standard wet chemical analysis, S. ONO for preparing polished thin section, H. Kojima, M. Naito S. Ikadai (Natl Inst. Polar Res. Tokyo), Y. Kondo (JEOL, Tokyo) for electron microprobe work and M. Noda (Iwate Univ.) for typing.

REFERENCES

1. K. Yanai. Collection of Yamato meteorites in the 1979-1980 field seeson, Antarctica, *Mem. Natl. Inst. Polar Res., Spec. Issue* **20**, 1-8 (1981).
2. T. K. Mayeda, K. Yanai and R. N. Clayton. Another Martian meteorite. *Lunar and planet. Sci. Conf.*, 917 (1995).
3. I. Kushiro and Y. Nakamura. Petrology of some lunar rocks. *Apoll 11 Lunar Sci. Conf.*, 607-672 (1970).
4. K. Yanai and H. Kojima. *Catalg of the Antarctic Meteorites*. Natl Inst. Polar Res., Tokyo (1995).
5. R. N. Clayton and T. K. Mayeda. Oxygen isotopes in eucrites, shergottites makhlites, and chassignites. *Earth Planet. Sci. Lett.* **62**, 1-6 (1983).

GEOLOGICAL EDUCATION

Proc. 30th Intern. Geol. Congr., Vol. 26, pp. 115-121
Wang *et al.* (Eds)

Stimulation of Awareness of Geoscience Knowledge Among the Mass Population of the Society

AFIA AKHTAR
Geological Survey of Bangladesh, 153, Pioneer Road, Segundbagicha, Dhaka 1000, Bangladesh

Abstract

Geoscience is intermingled with our lives. Most of our daily uses, fossil fuels for energy, construction materials for buildings, housing apartments, dams, bridges, roads and highways and minerals for industrial uses, all depend on the contribution of geosciences. The modern civilized world is entirely dependent on mineral resources and their proper utilization, Tremendous landscape changes, huge destruction of property, loss of lives, environmental pollution and consequently the fall of civilization owing to natural disasters are also related to the realm of geoscience. So, geoscience is a vital and growing scientific field that includes vast subject matters. For predicting and mitigating the damage of natural disasters and exploring and exploiting mineral resources, proper geoscientific knowledge is always a necessity.

Geoscience is the only science that plays a vital role in all spheres of life from common to corporate level. Planners and decision makers can utilize geoscientific data from comprehensive reports and maps produced by geoscientists to make right decisions in land-use planning and management. In this regards, there should be good cooperation and coordination between planners, decision makers and geoscientists. It is even the vital duty of geoscientists to persuade the planners and decision makers that there is a need to incorporate geological factors is land-use planning, by providing them documents on case studies. Common people, by acquiring even elementary geoscientific knowledge, can take precaution and protect themselves and their property from damage of hazards, protect natural environment and the valuable commodity water resources from pollution, and thus protect themselves from various diseases. Cultivators may be able to protect soil from erosion, siltation and salination and to increase soil fertility to produce more food to meet the demand of the exploding population, if they could apply geoscience knowledge to their respective enterprises and in their daily lives. As we are dependent on geoscience, it is a need for every citizen to know about it. And it is the prime responsibility of a geoscientist, nay a nation, to stimulate the awareness of geoscience among the overall population of a country. To do so, there must be some propaganda to stress the importance of geoscience through broadcasting media and also there must be given to mass education including an emphasis on geoscience as a compulsory subject or at least a special chapter to the curricula.

Keywords: Geosciences, awareness, mass population, natural disaster, curriculum

INTRODUCTION

Common people of a society do not have any idea about geoscience and its contribution though it is a most vital, exciting and evergrowing scientific field that includes whatever we see around us — the earth as a whole and its natural resources. Moreover, it is the only science that plays an unparalleled role in all spheres of life from common to the corporate level [1].

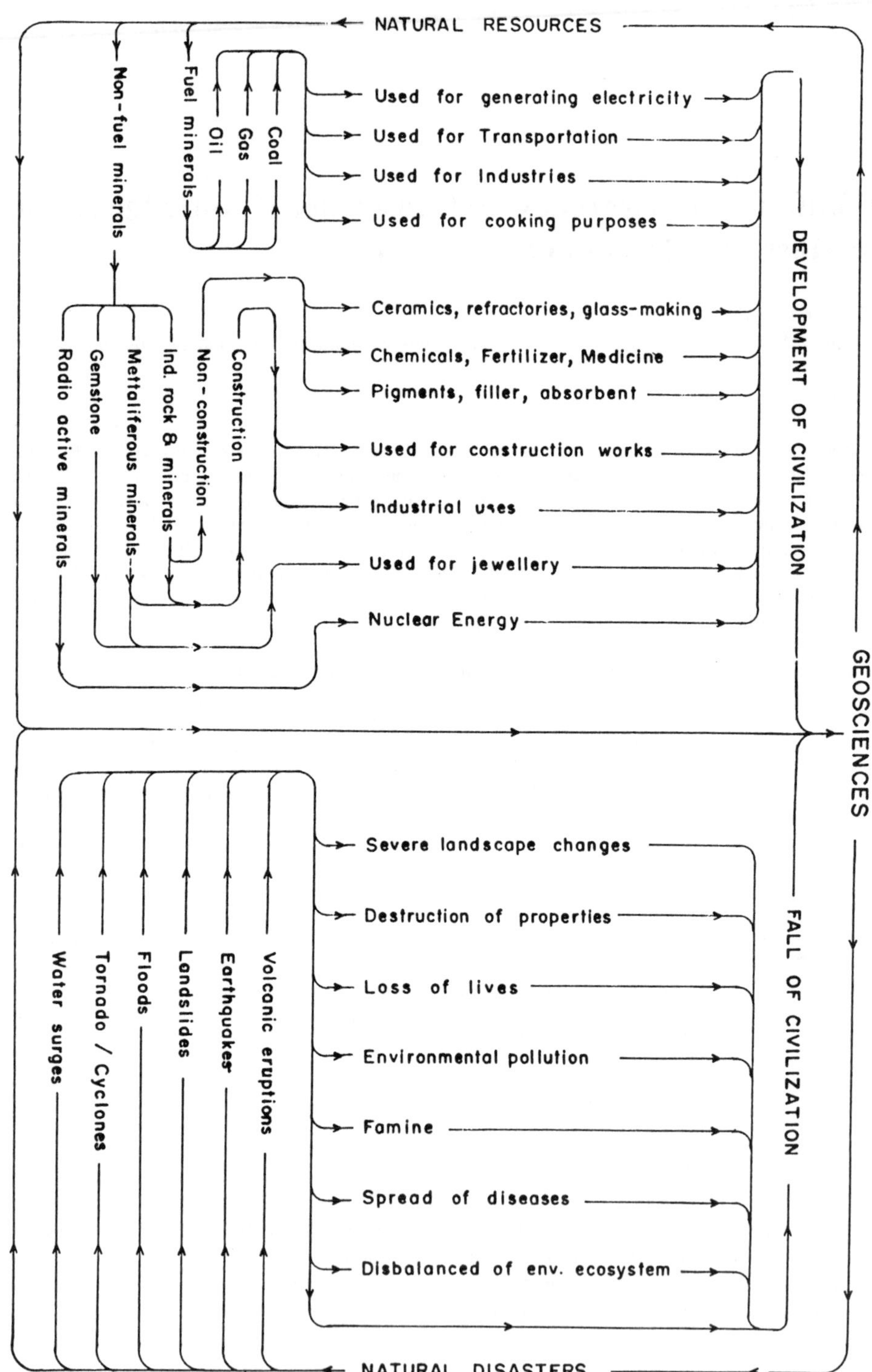

Figure 1. The constructive and destructive effects due to the awareness and unawareness of knowledge about the importance of geosciences and their application on our society.

Varieties of utensils that we use in our daily lives, fuel minerals-oil, gas and coal used for generating electricity, transportation and cooking purposes; non-fuel rocks and minerals, both metallic and non-metallic, for modern industrial and agricultural uses; construction materials for building, housing apartments, major construction works such as dams, bridges, roads and highways; water, the basic ingredient for human survival — all are connected with the contribution of geoscience (Fig. 1).

Natural hazards such as tornadoes, volcanic eruptions, earth quakes, floods, landslides *etc.* that cause tremendous landscape changes, huge destruction of property, death and injury to the people and environmental degradation which in turn is responsible for various disease of living beings can be better understood by acquiring the knowledge of geosciences (Fig. 1 and 2).

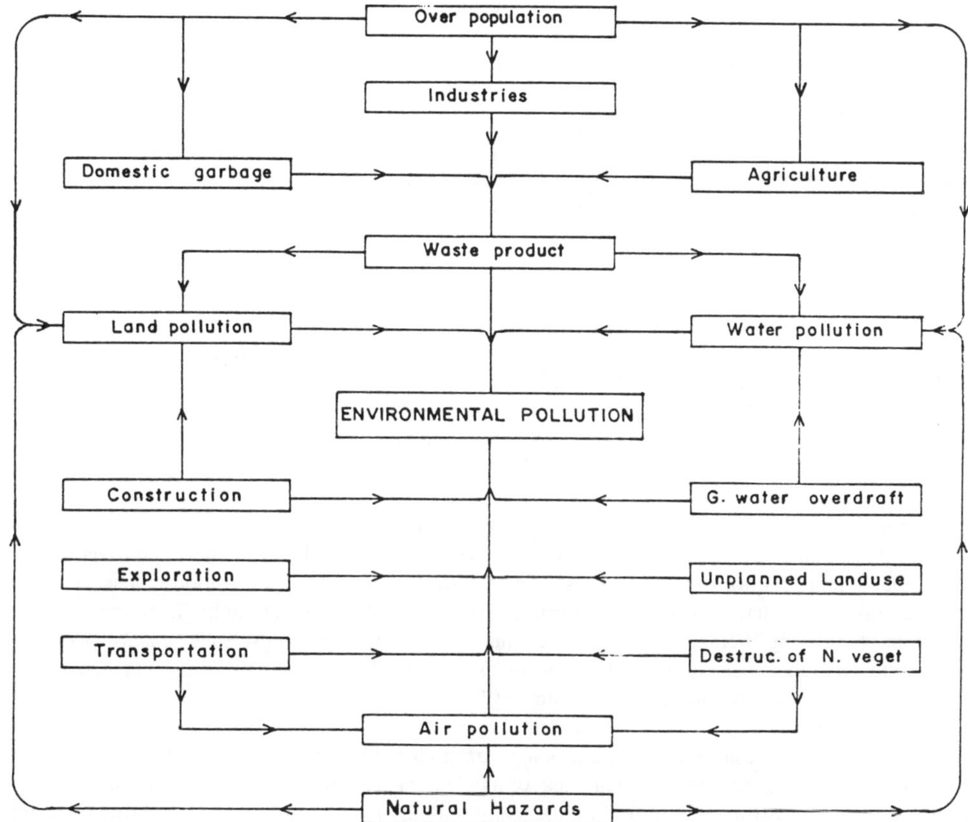

Figure 2. Some major sources of pollution of environmental ecosystem which can be minimised by proper application of geoscientific knowledge.

The modern civilized industrial world in therefore entirely dependent on earth resources and their proper utilization. But at the same time, lack of knowledge of natural hazards may cause the fall of civilization. So, we can not afford to ignore the importance of geoscience as it affects directly of indirectly all of us. But still, geoscience is unknown to the majority of the population of the world. This paper will suggest briefly the ways of creating an awareness and incorporating the geoscientific knowledge into our society.

SOCIETY AND GEOSCIENCES

The standard of living of any society is to a certain extent indicative of how much geoscientific knowledge and techniques is incorporated in our lives by way of its application in exploration and exploitation of mineral resources and their proper utilization [2, 3]. Taking necessary precautions against the destructive effects of natural disasters and protecting environment and water resources from all kinds of pollution are essential for maintaining a balanced healthy and hygienic environmental ecosystem [4, 5]. From these facts, it may be said that geoscience is a part of our lives and our society. And it is only when this visualization comes to our common people, then they will be interested to know about geosciences and their role in improving the quality of life. The following comparative studies show the merits of the awareness of geosciences and the demerits of the lack of knowledge of geosciences.

1) People, because of a lack of geoscientific knowledge, live or build their homes in areas, where severe hazards like landslides, volcanic eruptions, earthquakes, tornadoes, floods, land subsidence *etc.* are the ever present threat. Among these, flood is probably the most frequent of all natural hazards and affects more people and properties than any other hazards [6]. People do not know how destructive it can be and that is why they do not feel the need to consult the concerned authority. As a result, they suffer a lot.

Geoscientifically conscious people, on the other hand, before building their houses or undertaking any construction works in hazardous areas, will think with a geoscientific point of view whether it is safe for the proposed work, and accordingly they will consult the geoscientists of concerned authority to mitigate the damage of possible hazards, for the safety of life and property. Proper geoscientific analyses of the possible hazards may offer predicative model for hazards and hazardous areas. This model can be used for proper national planning, for incorporating the hazardous component adequately, both for preventive and curative measures.

2) As population increases, more land-use planning, land reclamation and site analyses will be needed for safe and effective use of available spaces in future [7]. Unfortunately, geoscientifically unconcious people do not care about these. They destroy their natural vegetation, abuse their land and undertake construction work without proper planning and thus create unhygienic environment, dangerous for living beings. On the other hand, awareness of geoscience makes people to incorporate the geoscientific consideration in proper land-use planning and management for preserving a balanced environmental ecosystems [7].

3) Waste products disposal is a growing problem for modern industrial society. People, especially of developing countries, because of their lack or neglect of geoscientific knowledge, dispose their waste products (industrial, agricultural or household garbage) haphazardly and also throw them into water bodies. As a result, the disease bearing bacteria from the decomposed materials trickle downward and contaminates the ground water. In addition, people build up their industries and toilet facilities close to the water sources, which accelerates water contamination and consequently causes the environmental pollution. Once water became contaminated, it is very dangerous for human health. By drinking such contaminated water, people are easily affected by various diseases such as cholera, dysentery, typhoid, hepatitis, diarrhea, chronic fatigue and even cancer. These will ultimately affect the national economy.

On the contrary, most of the people of the developed world, because of their awareness of geosciences, are cautious and select areas that are geologically and environmentally stable for a long period to dump their waste products. Thus they avoid the danger of polluting water and their environment, and reduce diseases due to water pollution. Moreover, by applying their geoscience knowledge, people can easily produce compost from their waste products and use it as a conditioner to increase soil fertility which in turn increase food production. Even some of the waste products can be reused by reclamation which is economically feasible. Geologically and environmentally conscious people do so by applying proper geoscience knowledge and techniques in order to reduce the volume of the waste products as well as the environmental pollution, and also in an economic way.

4) As common people do not realize how essential ingredient is water for human survival, they misuse this valuable commodity instead of conserving it for future uses. Since most of the village people of the developing world do not have ideas about ground water aquifer, they construct a lot of deep tube wells within a small area [8]. As a result, due to overpumping or excess mining of groundwater, especially in coastal areas, salt water or polluted water easily infiltrates to contaminate the fresh water. Salt water may corrode expensive industrial equipment and may ruin the productivity of the agricultural land. Not only that, over pumping may cause subsidence of the land. These adverse effects will ultimately limit the further economic growth of a nation. But, geoscientifically concious people, as they are aware of the condition of groundwater aquifer, do not commit such kind of mistake in the larger national interest.

5) Soil is another basic necessity for our survival. Soil erosion, siltation and salination are the old problems which affects agriculture. When these problems accumulate to a large proportion, they will influence the economy of the entire nation. Cultivators because of their lack of geoscientific knowledge, do not know how the protect their soil from the above mentioned problems and how to rejuvenate the soil to produce more and more food to meet the demand of increasing population. As soil is a renewable resource, cultivators with geoscience knowledge could increase soil fertility and its ability to withstand erosion, and thus food production will become self sufficient. In addition, plenty of food production and food export will contribute a great deal to the national economy and also lead to a better life.

6) Urbanization is an accompaniment of modern industrial civilization. But it grows in most of the developing world without planning and without proper geoscientific investigations, as there is no cooperation and coordination among the people, the decision makers, the planners and the geoscientists [9]. And it is mainly because geosciences and their proper applications are not well known to the mass population of the society, including the planners and the decision makers. People of the developed world are more or less aware about the geoscientific constraints and as such they plan their urbanization with due consideration to geoscientific factors [1]. As a result they would suffer much less than the people of the developing world.

7) As a whole, people of the developing countries are general not aware of geosciences and their importance in getting a better life. Awareness of geoscience creates inspiration among the people to explore and exploit more and more mineral resources and to utilize these resources properly for the over all development activities of a country to increase the national economy, to elevate the standard of living for the benefit of the common people of the society [10, 11].

CONCLUSIONS AND SUGGESTIONS

People of a society, by using their geoscientific knowledge and techniques and by utilizing the earth resources, can make a great contribution to the overall national development, improve their quality of life and maintain a balanced environmental ecosystem [12, 13]. Planners and decision makers can utilize geoscientific interpretations from comprehensive reports and maps produced by geoscientists to take proper decisions in land-use planning and management and also to take precaution against the destructive effects of disasters. It is the prime responsibility of geoscientists to convince the planners, decision makers and also the general people of a society that there is a need to incorporate geoscientific factors in all spheres of life by providing them documents on case studies. To do so, there must be a good cooperation and coordination among the planners, decision makers, geoscientists and also the mass population of a society. To create such a congenial atmosphere the following steps are suggested:

1) There must be some propaganda to focus the importance of geoscience and its contribution to our society by citing specific examples through broadcasting media such as Radio, TV, daily Newspapers, Magazines and also Audio-Vedio presentation of the disastrous results caused by neglecting geoscientific factors.

2) Vocational training can be arranged for mass population to make them understand about what type of difficulties we are facing due to our unawareness and neglect of geosciences.

3) The public is in general interested to visit museums. Earth science or geoscience museum, in this sense, will be the best place, from where people of all level and of all ages can easily gather ideas about the importance of geosciences and how they effect our lives and society. Proper attention should also be given in this respect to develop the public awareness in the field of geosciences through museum, which is an effective, through informal educational system.

4) Urban areas suffer from a series of serious environmental pollution and hazardous waste disposal. In these circumstances, there should be sufficient public participation [4]. It is a vital duty of geoscientists to motivate people so that they should not undertake any construction works or should not dump their waste products without prior geoscientific investigation or without consultation with concerned authority.

5) Wide publicity through attractive posters is needed to create an awareness among the common people about geoscience and its contribution to improving the quality of life, so that they can take precaution and protect themselves and their property from natural hazards and other dangers caused by geological processes. Such knowledge is also needed to protect natural environment and the water resources from possible pollution and thus to protect themselves from various diseases.

6) Agricultural cultivators are the backbone of a nation as they play a critical role in the national development. Necessary steps should also be taken to make them conscious of geoscientific knowledge, so that they can protect their land better and can increase soil fertility by to produce more grains to improve food condition to the benefit of world economy.

7) Most of the people of the developing world misuse natural resources as they do not have ideas about the limitation of these resources and it is mainly because of lack of proper geoscientific knowledge. To minimise the use of non-renewable resource such as oil, gas, coal, minerals [11, 14] and to protect and conserve renewable resources such as forests, grass lands, farm lands, water *etc.* [15], all necessary steps should be taken to make them conscious of what will be the result of these misuses.

The steps mentioned above can not be implemented successfully until the people have basic education, without which they will not be able to realise the actual situation. So, a first step should be taken to increase the overall literacy rate with emphasis on geoscience as a compulsory subject or at least a special chapter to the curricula, especially in case of the developing world [16]. But above all, to make the mass population literate, there has to be a greater national will and a determined programme.

Acknowledgements

Thanks to M. N. Hasan of the Geological survey of Bangladesh, Dhaka, Bangladesh and S. A. Bilgrami of Corporate Office, Karachi, Pakistan, for their kind review of the manuscript. Thanks is also due to the Director General of the Geological Survey of Bangladesh for his kind permission to publish it outside of Bangladesh.

REFERENCES

1. F. Jr. Betz (Ed). *Environmental Geology*. Dowden Hutchinson and Ross Inc. Benchmark papers in Geology, **28** (1975).
2. D. R. Coates. *Environmental Geology*. John Wiley and Sons, New York (1981).
3. F. A. Keller. *Environmental Geology*. Charles E Merril Pub. Co. , Columbus, Toronto (1979).
4. P. Revelle and C. Revelle. *The Environment: Issues and Choices for Society*. PWS Publishers/Willard Grant Press, Boston (1981).
5. A. O. Strahler and A. H. Straller. *Environmental Geosciences: Interaction between Natural System and Man*. John Wiley and Sons Ins. (1973).
6. R. Altman. *The Complete book of Home Environmental Hazards*. Facts on File, New York (1990).
7. A. R. Geyer and W. G. Meglade. *Environmental Geology for Landuse Planning, Env. Geol. Report 2*. Commonwealth Penn. Dept. of Env. Res. (1972).
8. A. Akhtar. Impacts of groundwater contamination on society and its remedy — a review. *Reg. workshop on environmental aspect of groundwater Development* **IV**, 53-58. Kurukshetra, India (1994)
9. D. D. Chiras. *Environmental Science — A Framework for Decision making*. The Benjamin Cumming Publishing Co. Inc. , 2nd Edition (1988).
10. A. Akhtar and M. N. Hasan. Fossil Fuels in Bangladesh. *Export Oriented Development of Mineral Resources* and *Mineral Based Industries, Proc. 2nd SEGMITE Intern. Conf.* , pp. 79-84. Karachi, Pakistan (1994).
11. A. Akhtar and M. N. Hasan. *Mineral Deposits and Mineral-based industries in Bangladesh. Proc. Intern. Symposium and field workshop on Phosphorites and other industrial minerals*. Abbotabad, Pakistan (1995) (in press).
12. A. S. Boughey. *Reading in Man, The Environmental and Human Ecology*. Macmillan Publishing Co. Inc. , New York (1973).
13. T. C. Jr. Foin. *Ecological Systems* and *the Environment*. Houghton Miffin Co. , U. S. A. (1976).
14. S. Y. Mathers and A. J. G. Notholt (Eds). *Industrial Mineral in Developing countries*. AGIP report series, Geoscience in Int. Development, Nos. 18, B. G. S. (1994).
15. O. S. Owen. *Natural Resources Conservation - An Ecological Approach*. Macmillan Publishing Co. Inc. , New York (1980).
16. A. Akhtar. The importance of geoscience in secondary and higher secondary education. In: *Geoscience Education and Training*. A. V. Stow and G. J. H. McCall (Eds). pp. 67-70. A. A. Balkema Publisher, Rotterdam, the Netherlands (1996).

Proc. 30th Intern. Geol. Congr., Vol. 26, pp. 123-138
Wang *et al.* (Eds)

The Influence of Geology Teaching on the Image of Geosciences

ALFREDO BEZZI
Earth Sciences Department, Genoa University, Viale Benedetto XV, 5 - 16132 Genoa, Italy

Abstract

Science educators consider students' perceptions of science a fruitful area of research for improving the teaching and learning. To explore both students' and teachers' images of (geo)sciences, in the present study the Kelly's repertory grid technique, according to his Personal Construct Theory, was used. The main purposes of this investigation are to outline the image of some geological disciplines constructed by University students, and to verify whether the teaching of Geology could affect the construction of such an image. The subjects who took part in this research were University freshmen of the Geography degree course and the instructor of their Geology course. Geography, Geology, Biology, Chemistry, Physics, and Mathematics were chosen as elements of the repertory grid. The elicitation of constructs took place at the beginning and at the end of the academic year. To see how much the students agreed with their instructor the EXCHANGE methodology was used: students were given a copy of the instructor's grids with all the ratings removed and were asked to fill in the ratings using his constructs. Cluster analysis was used to group the elements which appear to have common features in the eyes of people from whom the constructs have been elicited. Principal components analysis was used to determine the students' constructs with the most important epistemological value.

Five categories emerged based on students' and teachers' constructions. 1. Objects, areas and techniques of investigation. 2. Nature of science. 3. Application of science and its professional aspects. 4. Affective aspects. 5. Characteristics of the courses. At the end of the Geology course only the first category revealed an increase of the relevant constructs. Results of the EXCHANGE methodology indicate that, surprisingly, the agreement between the students and their instructor about the image of Geology and Geography has been reduced in some cases. Cluster analysis of the students' constructs elicited at the beginning of the Geology course shows a prevalent aggregation between Geology and Geography, and a minor grouping between Geology and Biology, while at the end of the course the association between Geology and Geography is slightly increased. Results of the principal component analysis indicate that, in the earliest elicitation, constructs belonging to the first two categories were the most important to discriminate the disciplines, with a particular role played by constructs connected to some extent with the mathematical aspects of science. Constructs related to the other classes showed a lesser influence. This situation is strongly enhanced in the final elicitation where the vast majority of constructs that have the status of differentiating the geosciences belong to the first two categories but, in this case, with a minor control exerted by the maths' constructs. Contrasting results seem to indicate that the change of students' constructed definition of geosciences must come from within the students' cognition. The awareness of students' beliefs should be the starting point to allow for this cognitive reconstruction.

Keywords: geological education, higher education, image of science, personal construct psychology, repertory grid

INTRODUCTION

Science educators consider students' perceptions of science a fruitful area of research for improving the teaching/learning process. The rationale to acknowledge the importance of this kind of studies includes:
1) the need for science education to improve scientific literacy and hence public understanding of science to appropriately prepare informed citizens who can fully participate in a modern democracy; such an understanding implies i) an appreciation of the purposes of science, ii) the comprehension of the nature and status of scientific knowledge, iii) the recognition of the social structure of the scientific endeavour (in other words, the epistemology and sociology of science).

2) the need to elicit the images of science that students are likely to hold when they enter the science classroom; such knowledge is the baseline from which it departs any teaching activity aimed at restructuring students' ideas.

Recently most of these issues have been thoroughly revised by some authors [1, 2, 3]. Additional references, lacking in their bibliographies, significantly implement the general picture [4-41]. In works concerning the image of science and scientists, investigations were carried out indirectly by means of analysis of textbooks, newspapers, lessons and television programs, or directly by interviewing people (laymen, students, teachers) and analysing written drafts and drawings produced by the interviewees. The techniques used to collect the data in the latter investigations vary widely. They range from a "nomothetic" approach in which people's perceptions compare with the "correct response", to an "ideographic" approach in which individuals express their ideas in their own terms. With the former approach multiple-choice, Likert-scale questionnaires were the usual survey instrument, with conventional or innovative format such as the "Views on Science-Technology-Society" (VOSTS) by Aikenhead and Ryan [6]. In turn, more or less structured interviews constitute the preferred tool to explore the images and collect the data in more naturalistic observation studies (*e.g.* Driver *et al.* [1]).

To inspect both students' and teachers' ideas about certain science education issues (such as contents, learning, teaching) and/or their attitudes towards the aspects of science, other authors [8, 42-53] used a different type of methodological approach, the repertory grid technique, whose theoretical framework resides in the Personal Construct Psychology (PCP) of George Kelly [54]. This theory is based on the assumption that individuals psychologically work in accord with their attempts to give their surrounding world a meaning. The central spirit of a Kellyan approach is to gain insight into aspects of the individual's construction of reality and to reveal her/his "unique psychological space" through the explicitation of personal constructs in terms that reflect attitudes, thoughts, and feelings in a personally valid way. Constructs are categories of mind used to identify things and each mind is populated by a collection of these constructs. What is distinctive about Kelly's theory is that he held that each person differentiates things differently, although using different or, at time, the same terms. Thus, when people use terms, these terms do not reflect, entirely, what people is thinking because they do not attribute to the words the same "alternatives". Repertory grid enables people to use their own language to convey meanings: thus it renders explicit what individuals hold tacitly, and enables researchers to explore the way people construe events with their constructive alternatives. The lack of space prevents all the aspects of Kelly's theoretical framework regarding both methodologies and techniques of grids' analysis from being fully described. In addition to the above mentioned papers, references of the details, as well as the various fields where PCP has been applied (psychology, psychotherapy, industry, education), can be found in other articles [42, 55-57].

Researchers choosing the repertory grid argue that this elicitation technique reveals strictly personal (cognitive, value-related, affective) dimensions and meanings which are true indicators of the uniqueness of the real dimensions of personal constructions and are free from external influences.

Besides, the repertory grid overcomes the difficulties, emphasised by some authors (*e.g.* Aikenhead and Ryan [6], Driver *et al.* [1]), inherent with the collection of data with "traditional" instruments of investigation which assume that students perceive and interpret the statements of the tests with the same meaning as given by the researchers. Problems of interpretation also exist with the clarification of responses to open items and multiple-choice questionnaires may force responders into pre-determined channels depending upon the cultural assumptions and purposes of the researchers. On the contrary, as pointed out by Lakin and Wellington [48], when using the repertory grid there are no previously established dimensions, except those of the system of constructs, used by the subject to give meaning to his/her own experience, to anticipate events or to build, subjectively and idiosyncratically, his/her surrounding reality.

In general, research on the image of science was mainly oriented towards general aspects rather than to single disciplines; for geology, in particular, there is a lack of specific works, apart from Bezzi [8] who used the repertory grid to elicit information about students' outlook on some geosciences and relevant instructors. This study demonstrated that this technique is useful to understand better the nature and range of students' ideas about the image of geosciences and confirmed Eichinger's findings [17] that teachers' personalities and teaching styles have great influence and are likely to be the most significant factors affecting student perception of science. The aims of the present investigation are: i) to explore the perceptions held by a university geology instructor and by some of his students about some science disciplines, and in particular about some geosciences, at the beginning of their degree course; ii) to verify the consensus existing between the students and their teacher regarding these perceptions before and after the teaching intervention; and iii) to investigate to what extent (if any) the teaching of geology can affect the students' images of these geosciences.

The subjects who took part in this research and who are referred to, in this paper, by pseudonyms, were initially nine University freshmen (four females and five males) of the Geography degree course, and the instructor of the Geology course attended by the students. The learners' number slightly decreased, because one of them, for unknown reasons, quitted the course towards the very end of the academic year. The students' high-school background was various, the majority of them coming from different types of vocational, technical schools and only three from liceo, a secondary school with speciality in classical and scientific studies. Despite this original difference, all the students experienced at school the disciplines (Mathematics, Physics, Chemistry, Biology, Geology and Geography) that were chosen as elements of the repertory grid. The choice for such a range of convenience (the Kellyan definition to indicate a specific group of elements to which any given construct only applies) resides in the opportunity to locate the geosciences in the more general framework of science in order to have coherent terms of comparisons and, at the same time, to draw a wider scientific background from the elicitees.

At high school, Geography contains both strictly geographic and geomorphologic topics, while often Geology includes most of the aspects of Earth Sciences, such as the Universe, Meteorology, Oceanography. The respective ratio among the above mentioned "basic" sciences, in terms of teaching hours, is roughly 5:3:1:1:1,5 with slight variations according to the different schools. This quantitative estimation does not reflect any qualitative evaluation about the impression that could affect the students, being the imprint based on a lot of undetermined variables such as the students' attitude towards science, the content achievement, the school success in some disciplines, the teachers' personality and teaching style, and so on. Nonetheless, it is worth to point out the quantitative predominance of Physics and especially Mathematics, because their extensive prevalence could represent a reasonable factor for their "imprinting" in young people's minds. Though students' perception of all these sciences could derive also from exposure to the wider culture prior to formal instruction, it is likely that the images of these very specific subjects ("school-science") basically derive from learning science at school [1].

	A		B	
N	emergent pole	contrast pole	emergent pole	contrast pole
1	laboratory	fieldwork	laboratory	fieldwork
2	axioms	observation data	axioms	observation data
3	applied	basic	conc.ed with real	conc.ed with abstract
4	synthesis	analysis	related to territory	related to objects
5	modifies reality	preserves reality	modifies reality	preserves reality
6	calculations	descriptions	calculations	descriptions
7	rigorous	approximate	rigorous	approximate
8	conc.ed with real	conc.ed with abstract	study the matter	study the mankind
9	formulas	concepts	inductive	deductive
10	inductive	deductive	formulas	concepts
11	analyse the effects	analyse the causes	scientific	humanistic
12	study the matter	study the territory	synthesis	analysis
13	objective	subjective	conc.ed with social	conc.ed with material
14	use of maps	use of substances	basic	applied
15	experimental	theoretical	use of maps	use of substances
16			experimental	theoretical

Table 1. List of the personal constructs elicited from the instructor at the beginning (A) and at the end (B) of the Geology course. N = number of constructs. conc.ed = concerned.

Since the main aim of the study was to monitor the effectiveness of a teaching intervention on the already existing views, the elicitation of constructs took place at two different stages, first at the beginning and then at the end of the academic year. The method devised by Kelly to gain insight into the cognitive structure of a person and therefore to elicit the personal constructs used to account for his/her world is the repertory grid. A particular set of elements in a rectangular matrix (the repertory grid) are compared to each other. The constructs referred to the elements are generated by having the subject sort out in terms meaningful to him- or herself how any two in a triad of elements (indicated by dots in the grid) are alike and how the third is different. These constructions are essentially bipolar in nature (simple examples of constructs are laboratory/fieldwork, inductive/deductive, basic/applied; a complete set of constructs is given in Table 1) and these terms describe the respective poles of the constructs' dimension. The elements in the triad that show the likeness are rated 1 and are described by the "emergent pole" of the construct; they are written on the left hand side of the grid. The other element is described by the "contrast pole" of the construct; it is rated 5 and written on the right hand side of the grid. The subject then decides whether the other elements in the grid are more like one or the other pole of the construct dimensions rating every element in the grid on a scale of 1 to 5. If the construct does not apply to some of the elements, then a 0 is entered. This procedure is repeated until the subject's repertory of constructs has been exhausted.

Because elements are rated on constructs, it is possible to apply to the grid statistical methods, such as cluster analysis, and principal component analysis, performed with a specific computer program (in this case RepGrid 2.1 of the Centre for Person-Computer Studies, Calgary, Alberta, Canada). The patterns resulting from the clustering emphasise the similarities that the elicitee attributes to both constructs and elements, reflecting coherent domains of meanings that this person uses to explain certain issues. Principal components analysis maps both elements and constructs in a two-dimensional space in which the two axes are the first two components. This spatial distribution can reveal specific relationships between element and constructs that are peculiar for each elicitee (a useful and simple introduction to the process of completing a grid and to these methods of analysis and interpretation can be found in Fetherston [58]). Another helpful technique for investigating how much an individual understands someone else and how much he/she agrees with that person is to assume his/her construct perspectives in order to see "through his/her eyes". The EXCHANGE methodology is an attempt to achieve this and the essential mechanism is very simple [59]. Students are given a copy of the instructor's grid with all the ratings removed and are asked to fill in the ratings using his constructs. When this procedure is completed, a Socio Difference Grid [60] is performed comparing the instructor's original grid with the students' EXCHANGEd grids.

Repertory grids thus represent the way the teacher or students construct those elements which are personally meaningful and significant and which provide the focus for subsequent elaboration and reflection. In this study this speculation has been carried on essentially analysing the types of constructs elicited and using the previously outlined techniques on data collected at the beginning of the academic year, prior to the teaching of the Geology course, and successively when students completed their attendance to the course. The span of time in which the collection of data occurred varies (particularly in the latter case) because the students returned the grids not at the same moment. This delay can only have a slight influence on the inferred considerations since the freshmen had no more exposure to further Geology teaching.

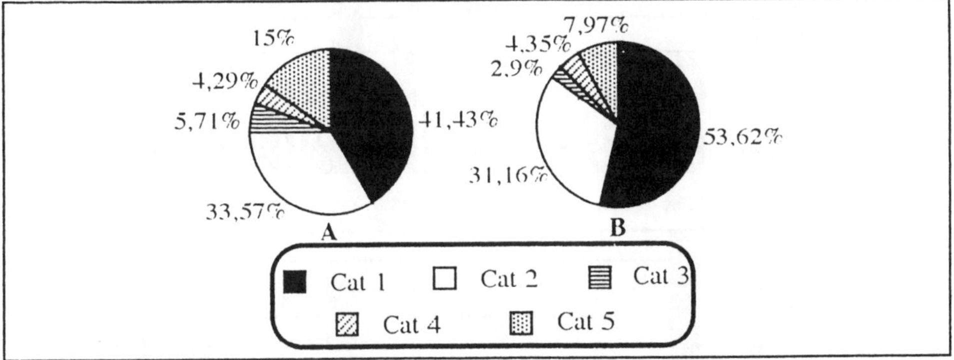

Figure 1. Analysis of categories of the elicited personal constructs concerning the science disciplines. A: elicitation at the beginning of the academic year (1994); B: elicitation at the end of the academic year (1995). Cat = category. Cat 1. Objects, areas and techniques of investigation. Cat 2. Nature of science. Cat 3. Application of science and its professional aspects. Cat 4. Affective aspects. Cat 5. Characteristics of the courses.

ANALYSIS OF THE GRIDS CONTENTS

In an initial analysis of the repertory grids the elicited constructs were grouped together into common categories based on the author's judgement of the meaning of the constructs. Some examples of the types of constructs that form these categories are: it studies the matter/it studies the territory; it uses/it does not use maps, charts (Cat. 1); objective/subjective; simple/composite science (Cat. 2); many/less working chances; it modifies/it preserves the natural environment (Cat. 3); I like/I dislike; difficult/easy to me (Cat. 4); elementary/high school; with lab/without lab (Cat. 5). Figure 1 shows that constructs related to the scientific essence of the disciplines (Cat. 1 and 2) by far exceed the other categories, with a significant increase in the first and second category at the end of the teaching episode, essentially at expense of professional aspects and characteristics of the courses.

However, in a Kellyan qualitative approach, which tends to point out the idiosyncratic system of the individual constructs, general considerations are undoubtedly less significant (also because they are intrinsically biased by the small number of subjects), while the personal "configurations" of people are much more important. Figure 2 shows the individual "profiles", as they resulted at the beginning (A) and at the end (B) of the academic year, and reveals the wide range of constructs through which people construe their idiosyncratic image of these disciplines. Originally, positions range from a more strictly "scientific" image (basically represented by categories 1 and 2) (like in Mel, Gian, and instructor) to a more balanced image (like in Max, Sim and Van) based on all the categories. At the end, some positions remain fundamentally unchanged (instructor, Mel and Max) while others display significant variations essentially marked by the increased number of constructs pertaining to the first category (notably Alex, and to less extent Sim and Van). Also the number of constructs can give some hints about students' background: the higher the number the wider the cultural horizon through which these disciplines are perceived. However, Ros' low number of constructs at the end of instruction represents more her unwillingness to continue the survey (as she frankly confessed) than a real cultural decline.

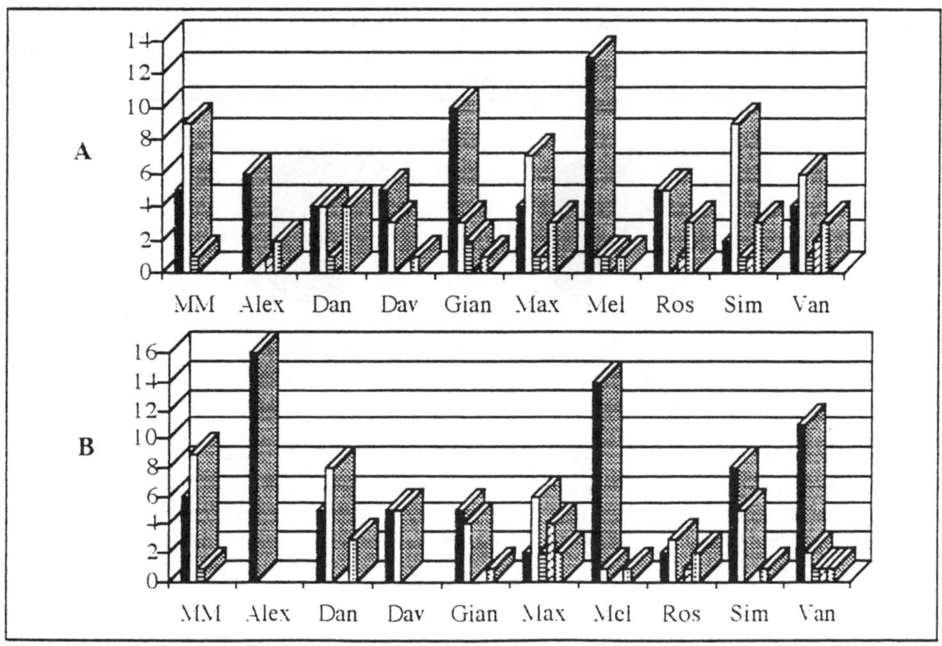

Figure 2. Analysis of the categories of the personal constructs, elicited in single individuals, in relation to all the disciplines. The symbols of the legend and of letters A and B are explained in Figure 1. y axis = number of constructs. x axis: MM = geology instructor; the others are students' pseudonyms.

Analysis of consensus and contrast between students and instructor

Some results of the elaboration of data after the EXCHANGE methodology and the performance of Socio Difference Grids (SDG) are presented in Figure 3. Each student was given the teachers' Repertory Grid, with his constructs left (Table 1) but the ratings removed, and he/she had to fill in the ratings. This procedure implies not only to understand the meaning of the teachers' constructs but also to use them accordingly to discriminate the elements. Therefore, it reveals the consensus (if any) existing between learner and instructor when the former tries to see the disciplines through the teachers' eyes. The SDG program matches the ratings for each construct and element giving the relevant percentage of "similarity match" that graphically and numerically indicates the degree of (dis)agreement between the two persons. Using data extracted from the SDGs, it is possible to have a more detailed insight into the situation of the geosciences. Figure 3 represents the consensus between students and their instructor measured with the similarity matching of elements on the EXCHANGEd grids performed prior to and after the geological instruction. The degree of agreement shows a general improvement which ranges from low (as it is the case for Ros about Geology and Sim about both geosciences) to very high increases, as shown by the other students, apart from Dan and Gian who represent a particular situation. It is quite surprising to realise that, after one year of geological instruction, these students have more difficulties in making sense of their instructor's constructs just about the geosciences.

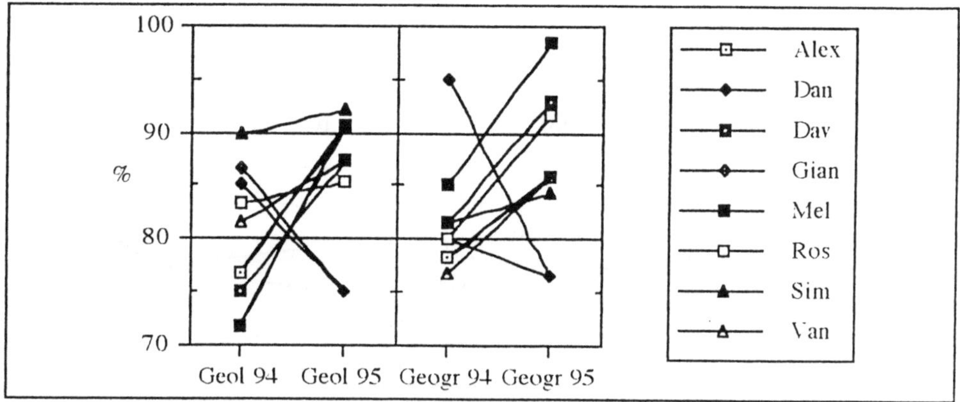

Figure 3. Representation of the consensus on geosciences between students and instructor measured with the percentage of similarity matching (vertical axis) of elements at the beginning (94) and at the end (95) of the academic year.

Cluster analysis of students' and instructor's grids.

Clustering algorithms such as FOCUS [60] enable to measure the distances between the sets of elements and constructs that are sorted and reorder the grid in such a way that similar elements and similar constructs are close together. The program portrays this similarity by arranging hierarchically groups of elements and constructs, starting with the most similar and adding the next most similar, and so on, until all the elements and constructs are clustered. Therefore the process groups elements and constructs which, thanks to a high similarity match between one and the next, appear to have common features in the eyes of people from whom the constructs have been elicited.

Figure 4 shows, as an example, some cluster layouts concerning the instructor (MM) and a student (Mel) restricted to the elements (disciplines) and based on constructs elicited at the end of the Geology course. The pattern of the instructor's clusters (on the left hand side in Figure 4) remained substantially unchanged, as it was expected since his constructs are essentially the same (Table 1A and 1B). The comparison between the two elicitations shows that only the images of geosciences, that at the beginning of the academic year were very similar (98.3%), seem to be more differentiated (84.4%). The percentage of similarity in both situations indicate that Physics and Chemistry continue to be matched at the same level (78.3%-79.7%) and are similarly linked to Biology (73.3%-71.9%). Mathematics is likewise attached to this cluster (56.7%-59.4%) and all are connected to geosciences always at a very low level (56.7%-53.1%). It is possible to interpret this construction as a group of "core" experimental sciences (Physics, Chemistry, and Biology) that are to some extent alike and that share some similarities with Mathematics; while geosciences appear to be construed fairly away from this assemblage. With the principal component analysis we shall see which are the instructor's and students' constructs with the major epistemological value responsible of their construing. Mel's initial construction showed a pattern given by two three-disciplines clusters grouping together Geol/Geogr/Biol (68.8%-60.9%) and Chem/Phys/Math (65.6%-60.9%).

The two clusters were connected at 59.4% and this connection allows to envisage a certain congruence among the images of these sciences as they are perceived by this student. In any case, it is worth to note that the clustering seems to respond to criteria such as "natural sciences" on one side and "laboratory or experimental sciences" on the other. As we can see from Figure 4, Mel's perception of the images of these sciences is extensively changed at the end of the course as it is illustrated by the occurrence of three two-disciplines clusters. The six elements are coupled as follows: Phys/Math and Geol/Geogr show the highest matching (76.6%), then Biol/Chem (71.9%) that are connected to the geosciences at 59.4%; the whole is joined at only 50.0%. The influence of the Geology course is mainly pointed out by the higher similarity of the geosciences.

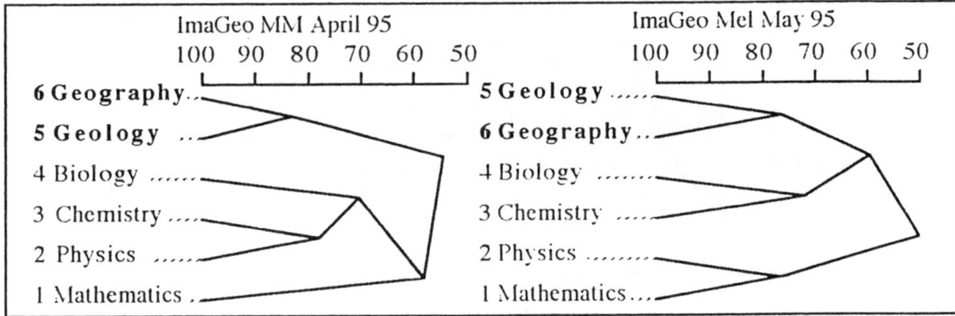

Figure 4. Clusters of the elements (disciplines) by means of the FOCUS program, on the basis of personal constructs elicited from each subject at the beginning and at the end of the academic year. The horizontal scale shows the percentage of the similarity match.

Principal component analysis of instructor's and students' grids

Principal component analysis [61] maps elements and constructs in a two dimensional space defined by two components. RepGrid 2.1 plots elements as cross and constructs as dots with their verbal labels: therefore in the computer layout it is easy to envisage particular relationships among them. Normally the two axes of the figure are the first two components: to be significant, they must explain a high percentage of variance, otherwise elements and constructs that may appear close on a two dimensional space could be actually far away if a third or higher component is considered (the first three components must account for over 80% of variation to give useful indications) [58]. The loadings on the components are responsible of the correlation of elements and constructs in the map: the higher the loading, the greater the significance of a certain construct to characterise an element. Elements and constructs plot within the four quadrants, defined by the two axes, according to their coordinates: therefore their positions in this two dimensional space are constrained by the mutual influence of the components [62].

Table 2 is based on the original grids of the students elicited at the beginning (A) and at the end (C) of the academic year. It contains all the construct poles with the highest correlation with the geosciences in the dimensional space defined by the three main axes (components). The table shows the broad range of constructs on which the real students' epistemology is founded and confirms the efficacy of a research tool that retrieves subjects' cognition in their own terminology leaving intact both intellectual and emotional aspects. Such a table could appear simple compared to a traditional, sophisticated questionnaire: but, normally, the latter offers only a nomothetic framework defined by researchers' background that can hardly imagine and therefore include all the possible idiosyncrasies related to actual students' cognitive and affective structure.

Stud.	Geology	Geography
Dan A	use of photographs no constant application to study composite discipline less mnemonic	territory aspects no laboratory taught to children
Dan C	not logic science oral exams fieldwork use of maps in high school	not based on laws and postulates recent science based on other disciplines
Gian A	visual studies concerned with planetary motions studies the planet no calculations related to underground	studies the changes of the planet not actually scientific no formulas easy to understand no laboratory
Gian C	territory very significant not experimental science study the rocks	empirical science no use of formulas
Mel A	not in industry sectors concerned with earth globe rock analysis important no use of maths methods	concerned with biogeography no laboratory not concerned with energy sources
Mel C	studies earth phenomena does not study the cell no use of calculations	studies crust deformations use of maps studies minerals
Sim A	composite discipline based on past events little maths conversational helps to know the world studies also biosphere	composite discipline based on past events little maths conversational minor influence on mankind
Sim C	research implies raw materials not particularly significant for progress	minor use of maths and physics on contingent situations
Van A	I like less hard to me no use of numbers	use maps less job chances elementary school
Van C	studies volcanoes studies the earth	not based on calculations studies natural phenomena does not use formulas I like concerned with daily life

Table 2. Constructs poles related to the geosciences with the greatest epistemological value (*i.e.* with the highest loadings on the first three components) resulting from the principal component analysis performed on students' original grids prior to (A) and after (C) the geological instruction.

DISCUSSION AND CONCLUSIONS

In accord with the findings of the Authors cited in the introduction, also the results of this paper seem to confirm that the repertory grid is a powerful heuristic tool for having access to a person's underlying construction system without splitting human functioning into intelligence, emotion, and motivation. Therefore, all the dimensions of single individuals emerge without any pre-determined constraint and their variations can be monitored over a period of time. The limited number of subjects prevents the making of more general statements about the image of (geo)sciences, but it is more significant to provide critical cases that delineate unique configurations of people's thought and feelings than to carry out assessments of a domain ontology prepared by "experts". Thus the images delineated by the elicitation of instructor's and students' constructs illuminate the actual realm of private personally viable meanings on which individuals build up their own reference network used to construe these sciences. It could be argued that (paradoxically) the more the constructions reflect the subjectivity of individuals, the more "objective" they are. Therefore, the overall picture is not an average expression resulting from the statistical analysis of definitions prearranged by others. On the contrary, it is the sum of single images that allows a comprehensive representation without the removal of singularities.

Besides, the repertory grid offers a number of different analysis possibilities: the EXCHANGE technique [59] is especially useful to reveal the degree of agreement and understanding existing between different people when they use the same verbal labels. This issue, in the educational context, has rather critical implications since teaching requires the skill of being able to enter sufficiently into another's system of personal meanings. The findings of this study quantitatively illustrate the need for a further joint work of negotiation between teacher and learners when the aim is to change a set of personal meanings until they come together. This specific mediation was beyond the scope of both the actual teaching practice and this paper, but it could be claimed as exemplar of how the structure of mind reflects the rules of thinking and feeling which are idiosyncratic for each individual and are the expression of his/her own accumulated cultural background and experience. In this study the EXCHANGE procedure used prior to and after the geological instruction has also allowed to verify the existing consensus on the images of the disciplines constructed with the use of the instructor's constructs and to monitor the relevant changes over time. In particular for the geosciences, it has been possible to ascertain the constructs responsible of the mismatch and to verify that, in this case, one year of geology teaching has sometimes increased the ambiguity between the learners and their instructor. Should these constructs represent aspects of science contents to be taught, then the good teacher must negotiate such misconstruing and mismatches to transform their personal meaning into an intrapersonal and public (scientifically correct) framework [42]. If the teaching aim is to produce any permanent conceptual change, then the identification of specific element and construct disagreement is the starting point for a process in which meaning is explored and shared. Only when knowledge has been thoroughly reconstructed from within student's cognition it will become personal, significant, and long-lasting [59].

The cluster analysis procedure (FOCUS; [60]) builds up a series of groups based on the strongest associations among constructs whose ratings lead to major groupings in elements. Both within and between subjects comparisons are possible, assuming identical element sets. This enables assessments of change over time, as well as the establishment of a construct set shared among a given group, or, as in this case, the comparison of element groupings produced by people with different levels of subject-matter expertise. The clusters are the resultant expression of the common features that the elicitees attributed to the disciplines when they completed the repertory grid. This hierarchical arrangement unfolds the similarities derived from people's constructions. This technique performed on the students' constructs elicited at the beginning of the Geology course shows that the aggregation between the geosciences as well as among all the natural sciences (opposed to the hard sciences) predominate (five out of eight); when Geology is not linked to Geography, then it gathers always with Biology. At the end of the course the majority of the subjects (five out of eight) fundamentally maintained their initial pattern, but the association between Geology and Geography is slightly increased (six out of eight). The cluster with all the natural sciences appears only three times at the expense of an increase in similarity with the whole pattern of the instructor's clusters that rises from one to three cases. The constructs elicited from teacher and students are generally verbally different, but this procedure allows one to realise that constructs, which may or may not use similar terminology, can be applied in a similar way to the different elements. Therefore, the resulting analogies after one year of geological instruction should purport the implicit influence (if any) to which learners have been subjected. When the set of constructs is the same (i.e. in the EXCHANGE processing), prior to geological instruction all the students see through the instructor's eyes the geosciences linked together, but, surprisingly, only seven out of eight do the same at the end of the Geology course, indicating a sort of reduction in comprehension of teachers' constructs concerning the geosciences already illustrated by Figure 3. On the contrary, after the geology teaching the similarity with the whole pattern of the instructor's clusters occurs one time more (four vs. three) indicating that half of the students reached a deeper general agreement with their teachers' construing.

A joint examination of both cluster and principal component analysis layouts can give an insight into the reasons of such (dis)agreement. This allows one to see if and how elements are differentiated by the individual. A detailed analysis of constructs allows the identification of the epistemological dimension on which Geology and Geography are paired. They are originally seen as subjective, approximate, and synthetic sciences based on concepts; both analyse the effects and use the description of observations carried out during fieldwork on the territory with the aid of maps. The geosciences are opposed to Physics and Chemistry considered objective, rigorous, and analytic sciences based on formulas. The former starts from axioms, analyses the causes and studies the matter, while the latter, epistemologically similar, is also construed as an experimental, inductive science that studies the matter in laboratory, with the use of substances and the aim to modify the reality. Biology is linked to these sciences because it shares much of Chemistry's construction and its experiments are pointed to real, applied objectives. As indicated also by its link in the clusters, Mathematics partakes of the epistemological dimensions of Physics and Chemistry, but it is more theoretical, basic, and interested to abstract. The picture remains virtually alike in the latest grid where the cartography of the territory seems to be the cause of displacement between Geology and Geography, confirmed by the lower similarity in their cluster.

In order to reveal students' epistemological dimensions, a careful scrutiny of the results obtained with the principal component analysis elucidates how, in the earliest elicitation, constructs belonging to the first two categories (Figure 1) are the most important to discriminate the disciplines, with a particular role played by constructs connected to some extent with the mathematical aspects of science. Constructs related to the other classes showed a lesser influence, though their number was relatively high. This situation is strongly enhanced in the final elicitation where the vast majority of constructs that have the status of differentiating the geosciences belong essentially to the first two categories but, in this case, with a minor control exerted by maths' constructs. The shifting towards a more "scientific" construction of the image of these sciences is not so positive as it could appear, firstly because it reflects an inadequate account of geosciences methodology, and, secondly, because it occurs at expense of a decreased attention towards the professional aspects. Due to the idiosyncratic nature of learners' constructions, it is difficult to summarise some general considerations about their epistemological dimensions, but it is possible to discern some features emerged from the findings. From an overall review of all data, some stereotyped images of science do appear with a characteristic antithesis between Physics (often joined to Chemistry and Mathematics) and the geosciences. As Frodeman [63] points out, Physics is traditionally considered as the paradigmatic science which exemplifies the true nature of "scientific method" with a certain, precise, and analytically derived knowledge of the world. Moreover, arbitrary classifications that rate sciences according to their mathematical sophistication put the experimental sciences at the top of this hierarchy [64]. In contrast, geosciences are seen as having many problems that undercut their claims to knowledge such as the incompleteness of data, the lack of experimental control, the huge span of time required for geological processes, the difficulties in making direct observations. Besides, historical narratives, typical of Earth Sciences, are often considered a vague form of knowledge lacking the logical rigour appropriate to the "hard" sciences. Sciences that do not meet the Physics standards have been hardly (but mistakenly) considered as such. This continued viewing geosciences from the perspective of Physics emerges throughout all the outcomes of this study, and in particular from data illustrated from the principal component analysis where many of the constructs matching with this commonplace image seem to be a persistent keystone of teachers' and students' epistemologies.

Further considerations must be claimed in relation to cultural and societal aspects inherent with geosciences, taking into account also the results illustrated in Figure 1 which show that aspects connected with the profession and the application of science decreased at the end of the academic year. This disturbing finding seems to confirm in some way Ledbetter's outcome [26] that Earth science students are less likely to see the connection between geology and the "real world". Though restricted to a few instances, some constructs do cause a certain concern: when geosciences are considered to have a minor influence on the mankind and, after one year of geological instruction, they are regarded as not particularly significant for the progress; when at the end of the course only one student (they were three at the beginning) gives to application of geosciences some epistemological value; when the concern to society does not emerge as a significant conceptual dimension, then it is legitimate to suspect that, at least in some cases, important educational aims have been missed. The understanding of the past, present, and future behaviour of the whole Earth system must be aimed to cope with the societal challenges (resources, hazards, environments, global change) [65]. If these issues do not emerge from people's cognitive framework, then this lack indicates that they are not perceived as being fundamental, suggesting that something has to be done to include, implicitly or explicitly, this matter into any Earth Science educational project. It is the responsibility of geologists to transform geoscience education in a process that must go beyond mere teaching and learning the facts, laws, and theories: it must involve understanding the nature of geoscience and its relationships with society [66].

More than coming to a real conclusion, I ought to say that all the data presented and discussed should constitute a starting point for further research since any interpretation must be confirmed with subjects from whom the constructs have been elicited [58]. Such confirmation has been accomplished only in a few cases (with the instructor and a couple of students) due to the difficulties in which educational research in the domain of geosciences can be carried out within Italian university. The lack of cultural tradition, personnel and technological facilities, joined to a limited financial support and a hostile academic environment, forcefully constrains the research to standards inappropriate to serious investigations. In any case, my claim is that this study adds some more data to the great picture drawn out by the many Authors quoted in the introduction about the images of (geo)sciences. The particular methodology applied discloses the learners' huge range of alternative ways of construing/making sense of the same elements and illuminate the rich diversity of intellectual and affective frameworks implied in the construction of the physical and conceptual world. These proved different "ways of seeing" reality confirm Kelly's philosophical position of constructive alternativism and reinforce the paradigm of a constructivist perspective of teaching and learning to which is committed the vast majority of science education researchers [57], despite some petty attempts [67] to deny with trivial arguments the important achievements of constructive epistemology and philosophy of science [1, 3, 68].

Acknowledgements

The Italian Ministry for University and Scientific Research provided financial support to this paper. The author thanks the Geology instructor, Marino Marini, and the students enrolled in Geology 1994/95 who agreed to be the focus of a research investigation.

REFERENCES

1. R. Driver, J. Leach, R. Millar and P. Scott. *Young people's images of science*. Open University Press, Buckingham, UK (1996).
2. M. Matthews. *Challenging NZ science education*. The Dunmore Press, Palmerston North, NZ (1995).
3. R. Porlan. *Constructivismo y Escuela. Hacia un modelo de ensenanza/aprendizaje basado en la investigacion*. D'ada Editora S. L., Sevilla (1993).
4. S. K. Abell and D. C. Smith. What is science?: preservice elementary teachers' conceptions of the nature of science, *International Journal of Science Education* **16:4**, 475-487 (1994).
5. J. A. Acevedo Diaz. Que piensan los estudiantes sobre la ciencia? Un enfoque CTS', *Ensenanza de las Ciencias, Nœmero extra* (IV Congreso Internacional sobre Investigacion en la Didactica de las Ciencias y de las Matematicas) 11-12 (1993).
6. G. S. Aikenhead and A. G. Ryan. The development of a new instrument: 'Views on Science-Technology-Society' (VOSTS), *Science Education* **76:5**, 477-491 (1992).
7. M. Alvarez, G. Soneira and I. Pizarro. Como percibe el alumnado algunas interacciones entre Ciencia-Tecnolog'a-Genero-Sociedad, *Ensenanza de las Ciencias, Nœmero extra* (IV Congreso Internacional sobre Investigacion en la Didactica de las Ciencias y de las Matematicas) 19-20 (1993).
8. A. Bezzi. *Geology: a science, a teacher, or a course? How students construct the image of geological disciplines and that of their teachers.* Proceedings of the Science Education Research in Europe Conference, Leeds, UK, April 1995 (in press).
9. A. Bezzi. *Lo institucional y lo personal en la Ciencia moderna: el papel de las epistemolog'as privadas en el proceso de ense–anza/aprendizaje*. Keynote paper presented at the XII Bienal-125 Aniversario Real Sociedad Espa–ola de Historia Natural, Madrid, Spain, March 11-15 (1996) (Abstract in Libro de Resœmenes, 43-44, 1996).
10. J. Carrascosa, I. Fernandez, D. Gil and A. Orozco. Analisis de algunas visiones deformadas sobra la naturaleza de la Ciencia y las caracter'sticas del trabajo cient'fico, *Ensenanza de las Ciencias, Nœmero extra* (IV Congreso Internacional sobre Investigacion en la Didactica de las Ciencias y de las Matematicas) 43-44 (1993).
11. V. Cavalli-Sforza, A. W. Weiner and A. M. Lesgold. Software support for students engaging in scientific activity and scientific controversy, *Science Education* **78:6**, 577-599 (1994).
12. W. W. Cobern. College students' conceptualizations of nature: an interpretive world view analysis, *Journal of Research in Science Teaching* **30:8**, 935-951 (1993).
13. M. Dunne and J. Johnston. An awareness of epistemological assumptions: the case of gender studies, *International Journal of Science Education* **14:5**, 515-526 (1992).
14. R. A. Duschl and D. H. Gitomer. Epistemological perspectives on conceptual change: implication for educational practice, *Journal of Research in Science Teaching* **28:9**, 839-858 (1991).
15. J. V. Ebenezer and U. Zoller. Grade 10 students' perceptions of and attitudes toward science teaching and school science, *Journal of Research in Science Teaching* **30:2**, 175-186 (1993).
16. K. M. Edmondson and J. D. Novak. The interplay of scientific epistemological views, learning strategies, and attitudes of college students, *Journal of Research in Science Teaching* **30:6**, 547-559 (1993).
17. J. Eichinger. College science majors' perceptions of secondary school science: an exploratory investigation, *Journal of Research in Science Teaching* **29:6**, 601-610 (1992).

18. J. J. Gallagher. Perspective and practicing secondary school science teachers' knowledge and beliefs about the philosophy of science, *Science Education* **75:1**, 121-133 (1991).
19. E. Guasch, J. De Manuel and R. Grau. La imagen de la ciencia en alumnos y profesores. La influencia de la ciencia escolar y de los medios de comunicacion, *Ensenanza de las Ciencias, Nœmero extra* (IV Congreso Internacional sobre Investigacion en la Didactica de las Ciencias y de las Matematicas) 77-78 (1993).
20. D. Hammer. Epistemological considerations in teaching introductory physics, *Science Education* **79:4**, 393-413 (1995).
21. M. Z. Hashweh. Effects of science teachers' epistemological beliefs in teaching, *Journal of Research in Science Teaching* **33:1**, 47-63 (1996).
22. D. Hodson. In search of a rationale for multicultural science education, *Science Education* **77:6**, 685-711 (1993).
23. S. Jarvela. The cognitive apprenticeship model in technologically rich learning environment: Interpreting the learning interaction, *Learning and Instruction* **5:3**, 237-258 (1995).
24. B. B. King. Beginning teachers' knowledge of an attitude toward history and philosophy of science, *Science Education* **75:1**, 135-141 (1991).
25. V. Koulaidis and J. Ogborn. Science teachers' philosophical assumptions: how well do we understand them? *International Journal of Science Education* **17:3**, 273-283 (1995).
26. C. E. Ledbetter. Qualitative comparison of students' constructions of science, *Science Education* **77:6**, 611-624 (1993).
27. N. G. Lederman. Sutching on the nature of scientific thought: are we anchoring curricula in Quicksand? *Science and Education* **4:4**, 371-377 (1995).
28. P. Lumb and P. Strube. Disturbing the boundaries: the science/literature membrane, *Science Education International* **4:1**, 5-8 (1993).
29. D. Maor and P. C. Taylor. Teacher epistemology and scientific inquiry in computerized classroom environments, *Journal of Research in Science Teaching* **32:10**, 839-854 (1995).
30. B. Martin and W. Brouwer. Exploring personal science, *Science Education* **77:4**, 441-459 (1993).
31. Y. J. Meichtry. The impact of science curricula on student views about the nature of science, *Journal of Research in Science Teaching* **30:5**, 429-443 (1993).
32. J. E. Nesbitt. A look at two minority families and their beliefs about who can become a scientist, *Science Education International* **3:4**, 23-25 (1992).
33. D. P. Newton and L. D. Newton. Young children's perceptions of science and the scientist, *International Journal of Science Education* **14:3**, 331-348 (1992).
34. M. Ogawa. Science education in a multiscience perspective, *Science Education* **79:5**, 583-593 (1995).
35. S. Ohlsson. Epistemic obstacles and the marriage of fantasy to rigor: a response to Sutching, *Science and Education* **4:4**, 379-389 (1995).
36. D. Pomeroy. Implications of teachers' beliefs about the nature of science: comparison of the beliefs of scientists, secondary science teachers, and elementary teachers, *Science Education* **77:3**, 261-278 (1993).
37. G. Rankin. A challenge to the theory view of students' understanding of natural phenomena, *Science Education* **79:6**, 693-700 (1995).
38. W. -M. Roth. In the name of constructivism: Science education research and the construction of local knowledge, *Journal of Research in Science Teaching* **30:7**, 799-803 (1993).
39. R. Ruggieri, C. Tarsitani and M. Vicentini. The images of science of teachers in Latin countries, *International Journal of Science Education* **15:4**, 383-393 (1993).
40. A. G. Ryan and G. S. Aikenhead. Students' preconceptions about epistemology of science, *Science Education* **76:6**, 559-580 (1992).
41. R. A. Schibeci. Public knowledge and perceptions of science and technology, *Bulletin of Science, Technology and Society* **10:2**, 86-92 (1990).

42. A. Bezzi. Use of repertory grids in facilitating knowledge construction and reconstruction in geology, *Journal of Research in Science Teaching* **33:2**, 179-204 (1996).
43. A. H. Corporaal. Repertory grid research into cognitions of prospective primary school teachers, *Teaching and Teacher Education* **7:4**, 315-329 (1991).
44. A. Descals and F. Rivas. *Aproximacion a la estructuracion cognitiva del estudiante en la situacion educativa universitaria*. Paper presented at II Congreso Internacional de Psicolog'a de la Educacion, Madrid, Spain, November 1995.
45. A. R. Fetherstonhaugh. Using the repertory grid to probe students' ideas about energy, *Research in Science and Technological Education* **12:2**, 117-127 (1994).
46. J. C. Happs and K. Stead. Using the repertory grid as a complementary probe in eliciting student understanding and attitudes toward science, *Research in Science and Technological Education* **7:2**, 207-220 (1989).
47. D. Kalekin-Fishman. *Constructing a concept of teaching: construing the role of teacher in a culturally plural society*. Paper presented at 11th International Congress on Personal Construct Psychology, Barcelona, Spain (1995).
48. S. Lakin and J. Wellington. Who will teach the "nature of science"?: teachers' views of science and their implications for science education', *International Journal of Science Education* **16:2**, 175-190 (1994).
49. M. L. Pope and P. Denicolo. The art and science of constructivist research in teacher thinking, *Teaching and Teacher Education* **9:5-6**, 529-544 (1993).
50. B. L. Shapiro. The use of personal construct theory and the repertory grid in the development of case reports of children's science learning. In: *Development and Dilemmas in Science Education.* P. J. Fensham (Ed). pp. 251-271. London, Falmer Press (1988).
51. E. L., Jr. Shaw. The influence of methods instruction on the beliefs of preservice elementary and secondary science teachers: preliminary comparative analyses, *School Science and Mathematics* **92:1**, 14-22 (1992).
52. J. Solas. Investigating teacher and student thinking about the process of teaching and learning using autobiography and repertory grid, *Review of Educational Research* **62:2**, 205-225 (1992).
53. K. Stead. Insights into students' outlooks on science with personal constructs, *Research in Science Education* **13**, 163-176 (1983).
54. G. A. Kelly. *The psychology of personal constructs*, Vols I and 2. Norton, New York (1955).
55. A. Bezzi. Aplicaciones de la tecnica de la psicolog'a de los constructos personales en el proceso de ensenanza/aprendizaje de las Ciencias De La Tierra. *Ensenanza de las Ciencias de la Tierra*, Volumen extra, 22-31 (1996).
56. B. R. Gaines and M. L. G. Shaw. Knowledge acquisition tools based on personal construct psychology, *Knowledge Engineering Review* **8:1**, 49-85 (1993).
57. M. L. Pope. *Constructivist educational research: a personal construct psychology perspective*. Keynote paper presented at the 11th International Congress of Personal Construct Psychology, Barcelona, Spain (1995).
58. A. R. Fetherston. *Using repertory grids in classrooms*. MASTEC Monograph Series n° 3. Edith Cowan University, Perth, WA (1995).
59. L. F. Thomas and S. Harri-Augstein. Self-organized learning. *Foundations of a conversational science for psychology*. Routledge and Kegan Paul, London (1985).
60. M. L. G. Shaw. *On Becoming A Personal Scientist. Interactive Computer Elicitation of Personal Models of the World*. Academic Press, London (1980).
61. P. Slater. *The measurement of intrapersonal space by grid technique - Vol. 2 - Dimensions of intrapersonal space*. John Wiley and Sons, London (1977).
62. M. L. Pope and R.T. Keen. *Personal Construct Psychology and Education*. Academic Press, London (1981).
63. R. Frodeman. Geological reasoning: Geology as an interpretive and historical science, *GSA Bulletin* **107:8**, 960-968 (1995).

64. W. Alvarez. The gentle art of scientific trespassing, *GSA Today* **1:1**, 29-31, 34 (1991).
65. National Research Council. *Solid-Earth Sciences and Society*, National Academy Press, Washington, DC (1993).
66. V. R. Baker. The geological approach to understanding the environment, *GSA Today* **6:3**, 41-43 (1996).
67. J. H. Shea. Constructivism in science education, *Journal of Geoscience Education* **4:3**, 242 (1996).
68. W. M. Roth. *Authentic school science. Knowing and learning in open-inquiry science laboratories*. Kluwer Academic Publishers Group, AH Dordrecht, The Netherlands (1995).

Proc. 30th Intern. Geol. Congr., Vol. 26, pp. 139-145
Wang *et al.* (Eds)

Earth Science Education and Engineering Education from the Standpoint of Practical Experiences

TAKESHI TANAKA
Japan Consulting Engineers Association, 4-1-20, Toranomon, Minato Ku, Tokyo, 105, Japan. Formerly Hokkaido University Sapporo, Japan

Abstract

Engineering geology is considered to be one of the fields of geology, but from its purpose, it may be called geotechniques and is one of the fields of engineering. Geological surveying is only a means of accomplishment for the purpose of construction. The importance of geological surveying is a matter of course in case of the necessity of large construction in the geologically ill conditions. The contents in the survey reports must be easily understood connecting with design and execution. and the knowledge of applied mechanics is indispensable. The gap of the knowledge between civil engineers and geologists is so large that its solution is urgently required. The best way of the solution of this gap is the refreshing education for engineers as member of society. In the course of the refreshing education, considerations for global environment must be taken into account both in earth science and in engineering education.

Keywords: earth science, engineering education, geotechniques, applied mechanics, refreshing education

INTRODUCTION

The writer received and earth science education in the university and was engaged in the geological survey for mineral and geothermal resources prospecting and construction engineering in a mining company. After practical experiences in the company for thirty two years the writer became a teacher in engineering education to the students of Department of Mineral Resources Development Engineering, Faculty of Engineering, Hokkaido University, and gave lectures on hydrogeology, geothermal and underground environmental engineering. The writer interested to express opinions on what earth science education and engineering education should be from a standpoint of practical experiences.

ENGINEERING GEOLOGY IS ONE OF THE FIELDS OF ENGINEERING

Engineering mentioned here corresponds to geosystem engineering including resources development engineering, construction engineering (civil engineering, architecture), environmental engineering and energy engineering.

The Association of Engineering Geologists defines engineering geology as "... the application of geologic data, techniques and principles to the study of naturally occurring rock and soil materials or subsurface fluids. The purpose is to assure that geologic factors affecting the planning, design, construction, operation, and maintenance of engineering structures and the development of groundwater resources are recognized, adequately interpreted, and presented for use in engineering practice" [1].

Engineering geology is considered to be one of the fields of geology. But from its purpose, it may be called geotechnique, and is one of the fields of engineering. For example, the purpose of dam survey is to determine the dam site, type (gravity, arch, fill type, *etc.*) and scale (height, length, *etc.*). The purpose of tunnel and road survey is to determine the optimum route, and that of landslide survey is to determine the countermeasure method of landslide prevention(Figure 1-4), Geological survey is only a means of accomplishment of these purposes.

The importance of geological surveying is a matter of course in case of the necessity of large construction in the geologically ill conditions. Those who read the reports written by geologists are civil engineers in many cases. The contents in the reports must be easily understood and closely connected with design and execution. Among the reports by geologists some are reports similar to a Degree Thesis in which difficult rock formation names and technical terms are frequently used. To such a report, it may be difficult to ask for understanding of civil engineers. If possible, the usage of special rock or formation names and technical terms had better be avoided.

Description of rocks in the reports must be stressed on the dynamical data, that is, compressive strength or tensile strength and Poisson's ration rather than the macroscopic description and the chemical analysis, which are however necessary for weathered or altered rocks.

FOR ENGINEERING GEOLOGY, THE KNOWLEDGE OF PHYSICS AND MATHEMATICS IS INDISPENSABLE

As mentioned above, engineering geology is one of the fields of engineering. Therefore the knowledge of applied mechanics (rock mechanics, soil mechanics, hydraulics, *etc.*) and physics (material mechanics, solid mechanics, elastic theory, plastic theory, rheology, thermodynamics, statistical mechanics, fluid mechanics, *etc.*) is indispensable (Table 1.)

Objects	Applied Mechanics	Fundamental Mechanics, Physics
Rock	Rock mechanics	Material mechanics, Solid mechanics, Elastic theory, (Plastic theory), Rheology
Soil	Soil mechanics	Plastic theory, (Elastic theory), Rheology, (Thermodynamics), (Statistical mechanics)
Ground water	Hydraulics	Fluid mechanics
Landslide	Soil mechanics Hydraulics	Plastic theory, Rheology, Fluid mechanics

Table 1. Aspects of applied mechanics, fundamental mechanics and physics for some geological objects [2].

Complicated formulae sometimes appear in rock mechanics, soil mechanics, *etc.* Among these, there are empirical formulae derived from fundamental mechanics (Newtonian mechanics), the confirmation of which is sometimes impossible. In such cases, the solution can often be gained by recognizing the concerned formulae being right, by knowing well their application limit, and by avoiding any mistake of dimension and unit. The typical formulae which brought about a revolution to physics, are unexpectedly simple as exemplified in the following.

Figure 1. Kurobe dam, the highest dam (H=186m) in Japan, Hokuriku district.

Figure 2. Hokuriku express way along Japan Sea, Hokuriku district.

Figure 3. Kamui tunnel, Hokkaido district.

Figure 4. Fukuchi landslide, occurred 1976, 9, 13, Kinki district.

$F=ma$ F: force, m: mass, a: acceleration (Newton's formula)
$S=k\ln W$ S: entropy, k: Boltzmann' constant, W: volume of phase space (Boltzmann formula)
$E=h\nu$ E: energy, h: Planck constant, ν: frequency (Planck formula)
$E=mc^2$ E: energy, m: mass, c: light velocity (Einstein formula)

The knowledge of mathematics to understand these mechanics is quite simple. The knowledge of partial calculus and double integral calculus plus that of mathematics studied in high school are sufficient. Higher mathematics such as topology and group theory are not necessary for the present. Recently the computing disposal of enormous data is so usual that familiar computer using of simulation is desirable.

NECESSITY OF REFRESHING EDUCATION

Civil engineers can learn these applied mechanics and physics in the course of engineering education, but geologists can rarely learn these subjects in the course of earth science education. On the contrary, civil engineers can rarely learn earth science subjects such as geomorphology. structural geology, petrology, mineralogy, and a part of geophysics in connection with geophysical survey. Now, the gap of knowledge between civil engineers and geologists is so large that its solution is urgently required.

The students of Hokkaido University are fortunate as they can learn the earth science subjects in the course of education in mineral resources exploration engineering as shown in Table 2.

Table 2. Curriculum of the course of education in mineral resources exploration engineering Hokkaido University.

UNDERGRADUATE CURRICULUM
Required Subjects
Applied Geology and Laboratory
Rock Mechanics and Laboratory
Prospecting
Mineral Resources Engineering
Underground Environmental Engineering
Natural Resources
Powder Technology and Laboratory
Field Practice
Graduation Thesis
Geology
Mineralogy and Petrology and Laboratory
Solid Mechanics and Exercise
Applied Fluid Mechanics and Exercise
Surveying and Practice
Measurement Engineering and Laboratory
Materials Science
Physical Chemistry
Engineering Mathematics I and Exercise
Numerical Computing Technology
Exercise on Computer Calculations
General Mechanical Engineering I and II
General Electrical Engineering I, II and Laboratory
General Electronic Engineering

Selective Subjects
Hydrogeology
Geothermal and Hot Spring Engineering
Drilling Technology
Mineral Processing
Explosives Theory and Practice
Mine Surveying and Practice
Geochemistry
Soil Mechanics
Concrete Engineering
Analytical Chemistry I
Mechanics
Introduction to Modern Physics
Engineering Mathematics II
Industrial Economics
Biology

GRADUATE CURRICULUM
Master's Courses:
Advanced Applied Geology
Advanced Mineral Resources Engineering
Advanced Mineral Processing
Advanced Geological Measurement
Advanced Rock Mechanics
Advanced Underground Transfer Phenomena
Advanced Physical Chemistry in the Earth
Advanced Economic Geology
Advanced Geothermal Engineering
Advanced Civil Engineering Geology
Advanced Rock Drilling and Well Logging
Advanced Blasting Engineering
Advanced Applied Computational Mechanics
Advanced Geo-Resources Engineering and Geotechniques Seminar
Advanced Geo-Resources Engineering and Geotechniques Laboratory

Doctor's Courses:
Studies on Special Subject I and II

The best way of the solution of this gap is the refreshing education for engineers in industries (corporate engineers) and it has become increasingly important in Japan [2].

Industries have faced with rapid change in business and social structure caused by technological evolution. Recently, these exist urgent requirements for the industries to contribute for the harmonious coexistence in society and to preserve ecological environments.

For these purposes, every company is making effort to train engineers having wide and profound knowledge in technology and understanding various points of views. The government and industries are making combined studies of the way of cooperation between universities and industries for continuing engineer education [3].

The results of the survey by the Japanese Society of Engineering Education (JSSE) indicate that

Japanese industry well have an increasing need of refreshing education for engineers and of corresponding demands to universities or other organizations for such programs [4].

For the present, many universities in Japan are giving refreshing education courses for bachelors, masters and doctors. In this case, it is not only necessary for the engineers in industries to exert self motivation and encouragement, but the generous support of the companies where they work is also important.

In the course of refreshing education, considerations for global environment must be taken into account both in earth science and engineering education.

Acknowledgements

I thank all members of the Department of Mineral Resources Development Engineering, Faculty of Engineering, Hokkaido University, Sapporo, Japan, where I has been engaged during 1990-1996 as a professor of geotechniques and safety engineering.

REFERENCES

1. P. H. Rahn. *Engineering geology: an environmental approach.* Elsevier, New York (1986)
2. T. Tanaka. *Geotechniques for practical engineers: a guide to its survey.* Riko Tosho, Tokyo (1984) (In Japanese).
3. Y. Gamo and W. Shimada. Refreshing Education for industry's engineers towards next era, *Journal of JSEE* **40:4**, 124-131 (1992) (in Japanese).
4. Y. Shimizu and M. Nakayama. Needs and issues of refresher education for working engineers. *Journal of JSEE*, **40:4**, 116-123 (1992) (in Japanese).

HISTORY OF GEOLOGY
PERSONS AND INSTITUTES

Proc. 30th Intern. Geol. Congr., Vol. 26, pp. 147-156
Wang *et al.* (Eds)

Professor Amadeus William Grabau (1870-1946) — Respected Teacher and Beloved Friend of Chinese Geologists

WANG HONGZHEN
China University of Geosciences, Beijing 100083, P. R. China

Abstract

Professor Amadeus W. Grabau ranked with the greatest geologists in the early half of this century. His academic career was wide and profound, and his scientific thought was creative and comprehensive. His scientific activities include two periods. The first period was from about 1890 to 1920, when he had built up his reputation in the foremost rank in palaeontology and stratigraphy in North America. The second period began with his emigration to China in 1920 to his death in 1946. In the later part of his life, Grabau had made great contributions to the geological research and geological education in China. The Geological Society of China, of which he was a founding member, dedicated to him two volumes of its bulletin, the Anniversary Volume of his 60th birthday (Vol. 10, 1930), and the Memorial Volume (Vol. 27, 1947). He was buried by his will in the compound of the Geological Building of the Peking University in 1946, and his grave was removed to the new campus of the University in 1982. Grabau's academic achievements are a treasure and seem to have acquired an ever increasing esteem with time. As a respected teacher and beloved friend, Grabau will live forever in the hearts of Chinese geologists.

Keywords: Grabau, respected teacher, beloved friend, Chinese geologists

INTRODUCTION

Professor Amadeus William Grabau (1870-1946), the American palaeontologist and stratigrapher who enjoyed a global fame in the first half of this century, had devoted the later part of his life to the geological research and geological education of China. The year 1996 was the semicentennial of his death in Beijing after a long illness and hard life under Japanese occupation. The Palaeontological Society of China held a memorial meeting in May 1996 at its 18th Annual Conference in Peking University, and planned to dedicate a special issue of *Acta Palaeontologica Sinica* of 1997 in memory of him. It was further deemed appropriate to pay more tribute to him at the occasion of the 30th IGC Beijing in 1996. Accordingly a seven-person group for this purpose was set up in the Spring of 1996, representing the Peking University, the Palaeontological Society of China, the Geological Society of China and the 30th International Geological Congress, in addition to Professor Gerald M. Friedman of the City University of New York, U. S. A., and the author representing the China University of Geosciences and also as a student of Professor Grabau. Consequently a joint meeting was held on August 14, 1996, which consisted of two parts. The first part, the Grabau Semicentennial Memorial Meeting was co-chaired by Professor Zhang Miman (Chang Meeman), then President of the Palaeontological Society of China and of the International Palaeontologists' Association, and Professor Friedman, and the second part, the IGC Session 22-2 was co-chaired by Professor Friedman and the author. At the meeting Professor Zhang gave first a short address, praising Professor Grabau as a great scholar, a great teacher and a great friend. Professor Friedman gave a very vivid and interesting talk about the early career and life of Professor Grabau. Then several papers in regard to Professor Grabau were presented.

The present paper includes two parts. First, Grabau as a respected teacher and beloved friend of Chinese geologists. Second, Grabau's main scientific achievements as a great scholar of global calibre.

A RESPECTED TEACHER AND A BELOVED FRIEND

Through the invitation of the late Doctor V. K. Ting, Grabau arrived at Beijing in 1920 to serve as Professor of Palaeontology of the National University of Peking, and at the same time Chief Palaeontologist of the Geological Survey of China. Since then he had been teaching in the University and carrying out research in the Survey until 1937, when the Sino-Japanese war broke out. After 1937 Grabau lived in Beiping and continued his active scientific research until 1941 when he was actually imprisoned by the Japanese invaders. When the Japanese surrendered, he was already very ill physically and mentally. He died on March 20, 1946 in Beiping. In 1930 he made a will to bequeath his private library to the Geological Society of China. In 1946, he was buried, by his will, in the compound of the Geological Building of the University. His grave was removed, through the joint effort of the Geological Society and the University, to the new campus of the University by the Unnamed Lake.

For many years, the well known founders of geological undertaking in China, V. K. Ting (Ding Wenjiang), W. H. Wong (Weng Wenhao), H. T. Chang (Zhang Hongzhao) and J. S. Lee (Li Siguang) had been Grabau's close friends. Most of the first generation of geologists trained in China, especially the palaeontologists and stratigraphers, were his pupils, or were benefited through his supervision and advice. He had lectured on many courses in the Geology Department of Peking University during the period 1920-1937, and the author was very fortunate indeed to be in the last class having heard him lecturing on historical geology. His success as a teacher was complete, and his inspiring enthusiasm and genial personality were immediately overwhelming and attractive. By the middle of the thirties many of his pupils and young colleagues in the University and in the Geological Survey had accomplished a number of monograph volumes of the *Palaeontologia Sinica*, which he helped V. K. Ting to set up in the early twenties. Grabau shared a genuine happiness with the success of his pupils. He organized a special congratulation party at the publication of the first volume of *Palaeontologia Sinica* written by the Chinese author Y. C. Sun, his first assistant and young colleague in China. He was always ready to give lectures or wrote articles for the Geological Society of the University of Peking, which was organized by the students and had kept active for many years.

The Geological Society of China, in the organization of which Grabau had taken an active part, dedicated two volumes of its bulletin to him. The *Grabau Anniversary Volume* (Vol. 10, 1930) in congratulation of his sixtieth Birthday, contains a portrait of him made by Sven Hedin (Fig. 1), a biographic note written by V. K. Ting [78]), and a reminiscence paper by C. Y. Wang [79]. Both H. D. Thomas and H. W. Shimer wrote in his memory after his death [80, 81]. The *Grabau Memorial Volume* (Vol. 27, 1947), contains a complete list of his works compiled by Sun [82], and an antithetic couplet composed by H. T. Chang in his memory. J. M. Weller in a memorial written in 1950 [83] described Grabau as one of the most remarkable geologists of modern times. There has been a new appreciation of Grabau's accomplishment since the seventies in North America as well as in China [84-86]. His name was included in the *Encyclopaedia Britannica* of 1980 [87] and in the *Encyclopaedia Sinica* of 1993 [88]. Recently G. M. Friedman even praised him as the father of modern Sedimentology [89].

Grabau bore a deep love for the geological undertaking and geological education in China, as well as a great sympathy for the country. The author clearly remembers he was sympathetic with the students' movement against the Japanese aggression in the mid-thirties. During the war years he more than once refused to do teaching work under Japanese occupation. The *Geological Volume* (1948) of the Fiftieth Anniversary Papers of the University also paid special tribute to him, in which the author mentioned him in the Foreword [90] as one of the most loyal to the geological cause in China and to the University. Indeed, he fully deserved the title of a respected teacher and beloved friend in the geological circle of China, and at the same time a true friend of the country. Figure 2 is a group picture taken at Grabau's home in 1933. In the back rows are the first generation and younger geologists trained in China, the youngest graduating in 1930.

A GREAT SCHOLAR OF GLOBAL CALIBRE

Grabau's scientific career may be divided into two periods. The first period in North America (1890-1920), when he had already won a world fame as a pioneer in the frontiers of sedimentary geology and palaeontology. The second period in China (1920-1944) may be even more important, as his main theories in geology had evolved and developed in this period.

Grabau's scientific contributions lasted a period of more than a half century (1890-1944) and covered wide fields in geology. His publications numbered more than 290 amounting to no less than 18900 pages of over nine million words. This enormous productiveness is well coordinated with a generalizing comprehensiveness and scientific accuracy. In the list of major works of Grabau at the end of this paper, the author has attempted to select his important publications which include papers initiating new geological concepts and presenting representative examples, as well as a number of big monographs and books, many of which are monumental and epoch-making.

The First Period (1890-1920)

In the first half of his scientific life, Grabau began to publish papers in 1890. His early research in geology was about the physiography and glaciation topography along the Genesee River near Boston [1, 9], and was concentrated in the Devonian and also the Lower Palaeozoic palaeontology and stratigraphy of eastern New York [2-6, 10-13, 17-19, 23]. In respect of palaeontology he studied brachiopods, corals and mollusks [7, 14, 16], among which the phylogeny of *Fusus* formed the main theme of his doctor thesis at Harvard University. He showed also interests in bionomy and sedimentation [8, 15]. Grabau's major summarizing work in the first decade of this century consists in the two big volumes of *North American Index Fossils* (with H. W. Shimer), which contains a very useful systematic stratigraphy [21]. An article on "*palaeontology*" (with H. F. Osborn) in the *Encyclopaedia Britanica* [22] was also important.

A masterwork of Grabau in the second decade of this century was *Principles of Stratigraphy* (1185 pages) [25], which laid the foundation of modern stratigraphy. Unlike most of his American contemporaries, he was keen in following the advanced ideas of the European masters. He admired Johannes Walther, and attempted to treat sedimentation in the context of geology as a whole. This book was an integrated treatise of nearly all the branches of geology of the time. In the same period Grabau discussed the various kinds of sediments and salt deposits [24, 26-29], which led to his summarizing work on non-metallic mineral deposits [31]. Another important paper worth noting is his review of Darwinism [30]. In 1920-1922 appeared Grabau's two big volumes of *Text-book of Geology* [32, 33], which marked the acme of his career of teaching. The second volume for Historical Geology was the first American text-book treating the geological history in a global scale.

The Second Period (1920-1944)

The second half of Grabau's scientific career spanned from 1920 to 1944 after his settlement in China. The first decade of his stay in Beijing had been very productive both in palaeontology and stratigraphy.

Figure 1. Professor Amadeus W. Grabau in 1930 (by Sven Hedin).

Figure 2. A group picture taken in Grabau's home in 1933.
Front row, left to right: H. T. Chang, V. K. Ting, A. W. Grabau, W. H. Wong, Teilhard de Chardin.

By 1930 he had published four volumes of *Palaeontologia Sinica* [34, 35, 40, 43] and separate papers dealing with corals, brachiopods and mollusks [42, 48, 51]. His discussion on graptolites, fossil man [45, 46, 49] and on science in general [47] fully testifies his wide interest and general concern for science in China. In stratigraphy his paper on the Sinian System [36] was on a global scale and his two volumes of Stratigraphy of China [38, 44], in addition to his excellent summary of the Cenozoic of Asia and his serial palaeogeographical maps of Asia [41, 39], were probably the most comprehensive and complete treatise of the geological history of the whole Asia that ever appeared.

In the later part of Grabau's life after 1930, his achievements were more concentrated in establishing his global theories, the pulsation theory and the polar control theory. Of these theories U. B. Marvin have given excellent summaries and comments [86, 91]. Always with an aim to interpret the earth as a whole, Grabau naturally took an interest in geotectonics. As early as 1919, he published an abstract of a paper on migration of geosynclines. In his book '*The Permian of Mongolia*" in 1931 [52], he had extended his views in this respect. His earliest articles about the pulsation theory appeared in 1933 [56, 58]. Based on this concept, Grabau reclassified the Palaeozoic into fourteen systems [65] and published in 1936-1938 four big volumes on the Lower Palaeozoic pulsation systems [67, 68, 70, 72], amounting to 3223 pages in total. These led to the publication of his most important synthetic and summarizing work, *The Rhythm of the Ages* in 1940 [75]. During these years he had also explained his viewpoints in stratigraphy in several papers [69, 73],

The polar control theory of Grabau, first published in 1937 [70] and subsequently reviewed in 1939 [74], represents a mobilist and catastrophic doctrine, in which he called in a by-passing star to pull all the sialic crustal masses on the earth into an Antarctic Pangaea, which was finally disrupted, the parts drifting apart from each other in the Mesozoic. Thus Grabau was an advocate of continental drift and believed in the presence of Pangaea. This was very outstanding indeed, if we think of the dominant geoscience thinking of uniformitarianism and fixism at that time in North America and in the world as a whole. It may be noted that, when the Wegener hypothesis was in its lowest point in the thirties, Grabau was amongst the very few geologists who stood boldly by this brilliant idea, which was to prove its merit only some thirty years later in the geoscience revolution in the sixties. It may also be pointed out that Grabau's interpretation of the pericontinental geosynclines and of the pushing of palaeocontinental fronts onto the adjacent oceanic floor bears to a certain extent a similarity to the modern idea of subduction and obduction in continental margins. As adequately indicated by U. B. Marvin [91], although many aspects of Grabau's global theories are unacceptable, some of them are closer in spirit to those we favour today than were the theories held by most of his contemporaries.

In his scientific research after 1930, Grabau retained his interest in palaeontology. Apart from his three volumes of *Palaeontologica Sininca* mainly on Devonian and Permian brachiopods [53, 57, 64], his studies of gastropoda consisting of five parts written from 1902 to 1928 were collected into one volume and reprinted in 1936 [62], and his studies of Brachiopoda in the early thirties were also reprinted in one volume in 1936 [63]. In addition, he presented with V. K. Ting two papers on the classification of the Carboniferous and Permian at the 16th International Geological Congress in Washington [59, 60]. He wrote also about salt deposits [37, 55], fossil man [61, 76], and sedimentation, orogeny and regional geological history [50, 54, 66, 71]. The last book by Grabau was written under extremely adverse conditions in 1944, and was published pothumously through the effort of Professor V. C. Juan of Taiwan University in 1961 [77, 92].

The academic achievements and scientific thoughts of Grabau are a great treasure store, from which enlightening inspiration and precious experience may be drawn constantly. The memory of this great scientist, respected teacher and beloved friend will live forever in the hearts of Chinese geologists.

REFERENCES: THE MAJOR WORKS OF A. W. GRABAU (1894-1944) AND PAPERS IN MEMORY OF HIM

1. The glacial channel of the Genesee River, *Proc. Boston Sec. Nat. Hist.* **26**, 259-369 (1894).
2. Palaeontology. Eastern Massachussets. In: *Amer. Assoc. Adv. Sci. Anniversary Meeting. Guide to the localities illustrating the geology, marine zoology and botany of the vicinity of Boston*. A. W. Grabau and J. E. Woodman (Eds). pp. 37-62. Salem. Mass. (1898).
3. Marine invertebrates (of Massachusetts). In: *Amer. Assoc. Adv. Sci. Anniversary Meeting. Guide to the localities illustrating the geology, marine zoology and botany of the vicinity of Boston*. A. W. Grabau and J. E. Woodman (Eds). pp. 67-96. Salem. Mass. (1898).
4. Geology and palaeontology of Eighteen Mile Creek and the lake shore sections of Erie County, New York, Part I, *Buffalo Soc. Nat. Hist. Bull.* **6**, 1-91 (1898).
5. The palaeontology of Eighteen Mile Creek and the lake shore of Erie County, N. Y., *Buffalo Soc. Nat. Hist. Bull.* **6**, 93-403 (1899).
6. The faunas of the Hamilton Group of Eighteen Mile Creek and vicinity in western New York, *New York Geol. Surv. 16th Ann. Rept.* 227-340 (1899).
7. Moniloporidae, new family of Palaeozoic corals, *Boston Soc. Nat. Hist. Bull.* **28**, 409-424 (1899).
8. The relation of marine bionomy to stratigraphy, *Buffalo Soc. Nat. Sci. Bull.* **4**, 319-367 (1899).
9. Lake Bouve, an extinct glacial lake in the southern part of the Boston Basin, *Boston Soc. Nat. Hist., Occ.* Papers IV, **3**, 564-600 (1900).
10. Palaeontology of the Cambrian terranes of the Boston Basin, *Boston Soc. Nat. Hist. Occ.* Papers IV, **3**, 601-694 (1900).
11. Guide to the geology and palaeontology of Niagara Falls and vicinity, *Buffalo Soc. Nat. Sci. Bull.* **7**, 1-280 (1901).
12. Stratigraphy of the Traverse Group of Michigan, *Mich. Geol. Surv. Ann. Rpt. for 1901* 163-210, (1902).
13. Stratigraphy of the Becraft Mt., Columbia County, New York, *N. Y. State Museum Bull.* **69**, 1030-1074, (1903).
14. Palaeozoic coral reefs, *Geol. Soc. Amer. Bull.* **14**, 337-352 (1903).
15. On the classification of the sedimentary rocks, *Amer. Geol.* **33**, 228-247 (1904).
16. Phylogeny of *Fusus* and its allies, *Smith. Misc. Coll.* **44**, 1-157 (1904).
17. Guide to the geology and palaeontology of the Schoharie Valley in eastern New York, *N. Y. State Museum Bull.* **9**, 277-386 (1906).
18. Physical and faunal evolution of North America during Ordovicic, Siluric and early Devonic time, *Jour. Geol.* **17:3**, 209-252, (1909).
19. New upper Siluric fauna from southern Michigan (with H. W. Sherzer), *Geol. Soc. Amer. Bull.* **19**, 540-563 (1909).
20. *North American Index Fossils. Invertebrates* (with H. W. Shimer). **I**. A. G. Seiler, New York (1909).
21. *North American Index Fossils* (with Shimer). **II**. A. G. Seiler, New York (1910).
22. Palaeontology (with H. F. Osborn). *Encyclopaedia Britanica*, 11th ed., 579-592 (1911).
23. Ueber die Einteilung des nordamerikanischen Silurs, *Int. Geol. Congr. XI, Stockholm 1910, Compte Rendu* 979-995 (1913).
24. Continental formations in the North American Palaeozoic, *Int. Geol. Congr. XI, Stockholm 1910, Compte Rendu* 997-1003 (1913).
25. *Principles of Stratigraphy*. A. G. Seiler, New York (1910).
26. The origin of salt deposits with special reference to the Siluric salt deposits of North America, Min. Meteor, *Soc. Amer. Bull.* **6:2**, 33-44 (1913).
27. Early Palaeozoic delta deposits of North America, *Geol. Soc. Amer. Bull.* **29:3**, 399-528
28. Problems of the interpretation of sedimentary rocks, *Geol. Soc. Amer. Bull.* **28**, 735-744 (1917).
29. Significance of the Sherbourne sandstone in upper Devonic stratigraphy, *Geol. Soc. Amer. Bull.* **30**,

423-470 (1919).
30. Sixty years of Darwinism, *Natural History* **20:1**, 58-72 (1920).
31. *Geology of the Non-Metallic Mineral Deposits Other Than Silicates. Vol. 1, Principles of Salt Deposition.* MacGraw Hill Book Co., N. Y. (1920).
32. *Text-book of Geology, Vol. I.* D. C. Heath and Co., New York (1920).
33. *Text-book of Geology, Vol. II.* D. C. Heath and Co., New York (1922).
34. Ordovician Fossils of North China, *Palaeontologia Sinica, Ser. B* **I**, 1-127 (1922).
35. Palaeozoic Corals of China, Pt. I. The Tetraseptata, *Palaeontologia Sinica, Ser. B* **II:1**, 1-76 (1922).
36. The Sinian System, *Bull. Geol. Soc. China* **1**, 48-88 (1922).
37. Geological conditions bearing upon potash prospecting in China, *The China Institute of Mining and Metallurgy Bull.* **3**, 1-28.
38. *Stratigraphy of China, Pt. I.* Geological Survey of China (1924).
39. *Palaeogeographic maps of Asia. 36 maps and explanations.* Geological Survey of China (1925).
40. Silurian Faunas of Eastern Yunnan, *Palaeontologia Sinica, Ser. B* **3:2**, 1-100, (1926).
41. A summary of the Cenozoic and Psychozoic deposits, with special reference to Asia, *Bull. Geol. Soc. China* **6**, 23-3, 151-264 (1927).
42. *Shells of Peitaiho (Grabau and King)*, 2nd edition. Peking Soc. Nat. Hist. (1928).
43. Second Contribution to the Knowledge of the Streptelasmoid Corals of China and Adjacent Tedrritories, *Palaeontologia Sinica, Ser. B* **2:2**, 1-175 (1928).
44. *Stratigraphy of China, Pt. II.* Geological Survey of China (1928).
45. *Origin, distribution and preservation of the graptolites.* Memoir of the Research Institute of Geology, Academia Sinica (1929).
46. *Geological studies in China: notable information, "Peking man" and other early men.* Peking Leader Press (1928).
47. The outlook for science in China. Address at the Annual Dinner of the Science Society of China, Aug. 25, 1929, *Peking Leader Reprint* **49**, 1-15 (1929).
48. Terms of the shell elements in the Holochoarites, *Bull. Geol. Soc. China* **8:2**, 115-123 (1929).
49. Asia and the evolution of man, *The China Journal* **12:3**, 152-163 (1930).
50. An outline of the geological history of North China, *Bull. Peking Soc. Nat. Hist.* **5:1**, 1-13 (1930-31).
51. Corals of the Upper Silurian Spirifer tingi beds of Kweichow, *Bull. Geol. Soc. China* **9:3**, 223-240 (1930).
52. The Permian of Mongolia — A Report on the Permian Fauna of the Jisu Honguer Limestone of Mongolia And Its Relations to the Permian of Other Parts of the World, *Nat. Hist. of Central Asia* **4**, 1-665 (1931).
53. Devonian Brachiopoda of China, Pt. I. Devonian Brachiopoda from Yunnan And Other Districts in South China, *Palaeontologia Sinica, Ser. B* **3:3**, (1931) (pls. I-LIV issued 1933).
54. Palaeozoic centers of faunal evolution and dispersal, *Bull. Geol. Soc. China* **11:3**, 227-239 (1931).
55. The availability of the bar theory in the elucidation of ancient potash deposits, *Bull. Geol. Soc. Nat. Univ. Peking* **4**, 1-26 (1932).
56. *The rhythm of the ages--The pulsation theory, a new aspect of earth history.* The Peking Chronicle, May (1933).
57. Early Permian Fossils of China, Pt. I: Brachiopods, Pelecypods, and Gastropods of the Lower Permian Beds of Kweichow, *Palaeontologia Sinica, Ser. B* **8:3**, 1-214 (1934).
58. Oscillation or pulsation? *Rpt. XVI Int. Geol. Congr. Washington, 1933* **1**, 539-553 (1936).
59. The Carboniferous of China and its bearing on the classification of the Mississipian and Pennsylvanian (with V. K. Ting), *Rpt. XVI Int. Geol. Conger., Washington 1933* **1**, 555-571 (1936).
60. The Permian of China and its bearing on Permian classification (with V. K. Ting), *Rpt. XVI Int. Geol. Conger., Washington 1933* **1**, 663-677, (1936).
61. Tibet and the origin of man. *Sven Hedin 70th Anniversary Publ., Geografisca Annular* 317-325 (1935).

62. *Studies of Gastropoda* (reprint in one volume, pts. 1-5, 1902-1928), pp. 159 (1935).
63. *Studies of Brachiopoda* (reprint in one volume, pts. 1-4, 1931-1932), pp. 117 (1935).
64. Early Permian Fossils of China, Pt. II, Fauna of the Maping Limestone of Kwangsi and Kweichow, *Palaeontologia Sinica, Ser. B* **8:4**, 1-411 (1936).
65. Revised classification of the Palaeozoic systems in the light of he pulsation theory, *Bull. Geol. Soc. China* **15:1**, 23-51 (1936).
66. The great Huangho plain of China, *Jour. Assoc. Chinese and American Engineers* **17:5**, 247-266 (1936).
67. *Palaeozoic Formation in the Light of the Pulsation Theory, Vol. I, Taconian and Cambrian Pulsation Systems, 2nd edition.* The University Press, Nat. Univ. Peking (1936).
68. *Palaeozoic Formations in the Light of the Pulsation Theory, Vol. II, The Cambrovisian Pulsation System, Part I, Caledonian and St. Lawrence Geosynclines.* The University Press, Nat. Univ. Peking (1936).
69. Fundamental concepts in geology and their bearing on Chinese stratigraphy, *Bull. Geol. Soc. China* **16**, 127-176 (1936-1937).
70. The polar control theory of earth development, *Jour. Assoc. Chinese and American Engineers* **18**, 202-223 (1937).
71. The evaluation and dating of Palaeozoic orogenies, *Jour. Assoc. Chinese and American Engineers* **18**, 371-379 (1937).
72. *Palaeozoic Formations in the Light of the Pulsation Theory, Vol. 4, Ordovician Pulstion System, Pt. 1.*
73. The significance of the interpulsation periods in Chinese stratigraphy, *Bull. Geol. Soc. China* **18:2**, 115-120 (1938).
74. Present status of the Polar control theory of earth development, *Bull. Geol. Soc. China* **19:2**, 189-205 (1939).
75. *The Rhythm of the Ages: Earth History in the Light of the Pulsation and Polar Control Theories.* Vetch, Beiping (1940).
76. Tibet, the cradle of the human race, *Collec. Common Sinodalis* **15**, 62-78 (1943).
77. *The World We Live in — A New Interpretation of Earth History.* Taipei, Taiwan, China (1961)
78. V. K. Ting. Biographical Note (with scientific career and honours, and selected bibliography), *Bull. Geol. Soc. China* **10** (*Grabau Anniversary Volume*), iii-xviii (1931).
79. C. Y. Wang. Dr. Amadeus W. Grabau, a reminiscence, *Bull. Geol. Soc. China* **10** (*Grabau Anniversary Volume*), xi-xx (1931).
80. H. D. Thomas. Obituary of Prof. Amadeus W. Grabau, *Nature* **158:4003**, 89-90 (1946).
81. H. W. Shimer. Memorial to Amadeus William Grabau, *Proceedings Volume of Geological Society of America, Annual Report for 1946* 155-166 (1947).
82. Y. C. Sun. Biographical note of A. W. Grabau (with scientific career and honours, and publications), *Bull. Geol. Soc. China* **27** (*Grabau Memorial Volume*), 1-26 (1947).
83. J. M. Weller. Review of the A. W. Grabau Memorial Volume (Geological Society of China Bulletin, vol. 27, 398 pp.), *Journal of Geology* **58**, 598 (1950).
84. M. E. Johnsson. A. W. Grabau and the fruition of a new life in China, *Jour. Geol. Education* **33:2**, 106-111 (1985)
85. Wang Hongzhen. Professor A. W. Grabau — a respected teacher and beloved friend in the geological vircle of China. In: *The Early History of Geological Undertaking in China.* Wang Hongzhen *et al.* (Eds). pp. 81-93. Peking University Press (1990).
86. U. B. Marvin. Amadeus W. Grabau's global theories in the light of current models. In: *Interchange of Geoscience Ideas between the East and the West.* Wang Hongzhen *et al.* (Eds). pp. 55-71. China University of Geosciences Press (1991).
87. The New Encyclopaedia Britannica, Micropaedia, 15th ed. **IV**, 658 (1980).
88. Wang Hongzhen. A. W. Grabau. *Encyclopedia Sinica, Geology Volume*, 209 (1993) (in Chinese).
89. G. M. Friedman. In Memory of Professor Amadeus William Grabau (1870-1946) on the Semi-

Centennial of his Death. This volume. 157-164.

90. Wang Hongzhen. Foreword to the Geological Volume, *Fiftieth Anniversary Papers of the National Unoiversity of Peiking* (1948).
91. U. B. Marvin. The global theories of Amadeus W. Grabau (1870-1946): a retrospective view. This volume. 165-175.
92. V. C. Juan. Introduction to Amadeus W. Grabau: "*The World We Live in — A New Interpretation of Earth History*". vii-x (1961) (Taipei, Taiwan, China).

Proc. 30th Intern. Geol. Cong., Vol. 26, pp. 157-164
Wang *et al.* (Eds)

In Memory of Professor Amadeus William Grabau (1870-1946) on the Semi-Centennial of his Death

GERALD M. FRIEDMAN
Department of Geology, Brooklyn College and Graduate School of the City University of New York, Brooklyn, NY11210, and Northeastern Science Foundation affiliated with Brooklyn College, Rensselaer Center of Applied Geology, 15 Third Street, P.O. Box 746, Troy, NY 12181-0746, U.S.A.

Abstract

A. W. Grabau (1870-1946), one of the world's greatest sedimentary geologists divided his working life between the United States (1890-1920) and China (1920-1946). My early acquaintance with the publications of Grabau was through one of my fellow students, Mrs. Mary Welleck Garretson (1896-1971), who had been a student of Grabau at Columbia University. Grabau served as Professor of Geology and Mineralogy at Rensselaer Polytechnic Institute before he joined the faculty of Columbia University. Grabau may truly be considered the Father of Modern Sedimentology. This paper focuses on (1) Mary Garretson's recollections of Grabau; (2) surviving correspondence in the Archives of the Geological Survey of China, which include handwritten letters in German script addressed by Johannes Walther (1860-1937) to Grabau, which reveal an intimate personal friendship; and (3) glimpses of Grabau's personal life. Grabau's tomb is located in a pleasant memorial park on the campus of the new site of Peking University, where a white marble headstone with English and Chinese inscriptions adorns his grave.

Keywords: Grabau, United States, China, Sedimentology, Garretson, Rensselaer, Columbia University

INTRODUCTION

Amadeus W. Grabau (1870-1946), one of the world's great sedimentary geologists, served as professor of Geology at Rensselaer Polytechnic Institute (RPI) (Troy, NewYork) at the turn of the century. The contributions spanning his career include 6 books, 6 monographs and 300 papers. He may be considered the father of both modern stratigraphy and sedimentology, and he was a leading palaeontologist of his time. From my papers *Geology at Rensselaer: A Historical Perspective* [1, 2] and *Geology at Rensselaer Polytechnic Institute: an American Epitome* [3] I quote the following: "The vacancy created after James Hall's death and the ensuing unavailability of John M. Clark because of his full-time commitments with the New York State Geological Survey opened the opportunity for another giant to enter the halls of Rensselaer: Amadeus W. Grabau". Like his predecessors Grabau had close working relationships with the New York State Geological Survey. With the support and cooperation of the Buffalo Society of Natural Sciences and the New York State Geological Survey, Grabau prepared a *Guide to the Geology and Paleontology of Niagara Falls and Vicinity* (New York State Museum Bulletin 45, 1901), probably one of the best prepared and most professional of the New York State Museum Bulletins [4]. His title and address in this publication are listed as Professor of Geology at Rensselaer Polytechnic Institute, although in the archives of RPI he is listed as Professor of Geology and Mineralogy. In the preface to New York State Museum Bulletin 45 John M. Clark introduced Grabau. Grabau may truly be considered the

Father of Modern Sedimentology.

Figure 1. Portrait of Amadeus W. Grabau on his tombstone on the campus of Peking University (author's photograph).

To backtrack and digress, one of the most effective pioneers in making the doctrine of actualism useful as a stratigraphic tool for a better understanding of the rock record was the German geologist Johannes Walther (1860-1937) [see 5, p. 9-10; 6]. His writings present some of the first real data for use in the interpretation of sedimentary strata in the bedrock. Some of Walther's observations form the cornerstone of modern stratigraphy. He explained that lithologies whose antecedent sediments formed beside one another in space, such as point-bars sands beside overbank muds and next to marshes, lie on top of one another in vertical sequence. Geologists neglected Walther's prolific writings; but Grabau picked them up. Grabau's textbook *Principles of Stratigraphy* [7], a classic far ahead of its time, followed in the footsteps set by Walther. In fact Grabau dedicated his book to Walther. As his writings attest; the pioneer sedimentologist W. H. Twenhofel, continued the tradition of Walther and Grabau. By their philosophy, Twenhofel's influential books, *Treatise on Sedimentation* (1926, 1932), and *Principles of Sedimentation* (1939, 1950) [8-11] assured the continued influence of Grabau. Unfortunately for Rensselaer, Grabau, and American geology, Grabau later transferred to Columbia University, where he became a victim of political infighting in the Department of Geology, which led to his emigration (some even say expulsion) from the United States. In retrospect Grabau probably treasured his association with Rensselaer. As an example, in his *Textbook of Geology* (1921) [12] he makes sure that from the title page readers realize that he was formerly Professor of Mineralogy and Geology in the Rensselaer Polytechnic Institute (note here that in the Rensselaer archives his title is reversed as Professor of Geology and Mineralogy). Among Grabau's other books should be mentioned *Geology of the Non-Metallic Mineral Deposits* (1920), *The Rhythm of the Ages–Earth History in the Light of the Pulsation and Polar Control Theory* (1940), and *The World We Live In* (1961).

His working life was divided between the United States (1890-1920) and China (1920-1946). Once in China, despite the physical handicap of arthritis, his contributions were enormous: his pulsation theory, rhythm of the ages, fundamental concepts in geology and their bearing on Chinese stratigraphy, migration of geosynclines, Permian of Mongolia, the world we live in, among others. He made outstanding contributions to the stratigraphy and palaeontology of China and to the training of Chinese geologists in both the Geological Survey of China and at Peking University. Grabau became the springboard for the renaissance of Chinese geology. As Charles Berkey (1867-1955) of Columbia University stated at the time of his death he was probably the greatest in geology that this country (USA) had ever produced. Waldemar Lindgren (1860-1939), distinguished mineralogist and economic geologist and professor at the Massachusetts Institute of Technology, in presenting the medal of the National Academy of Sciences to Grabau on April 27, 1937 stated "The Committee on the Mary Clark Thompson Fund, meeting in 1936, decided unanimously to award the medal provided for in this Fund for most important services to Geology and Palaeontology to Amadeus William Grabau, Professor of Palaeontology in the National University of China and Chief Palaeontologist to the Chinese Geological Survey. Among the recipients of this medal since 1924, chiefly palaeontologists, are C. D. Walcott, E. de Margerie, J. M. Clarke, J. Perrin Smith, W. B. Scott, E. O. Ulrich, David White, F. A. Bather, and in 1934, Charles Schuchert. The Committee noted the distinguished services of Professor Grabau in general and stratigraphic geology, in the science of non-metallic mineral deposits, and particularly in palaeontology. His palaeontological researches include the Palaeozoic of New York and Michigan, and during the last seventeen years the Palaeontology of China. The results of this work are contained in a splendid series of monographs on Chinese Palaeozoic and also Mesozoic fossils, the last volume of which was published like the rest by the Academia Sinica, in Peking".

REMINISCENCES

When I was a graduate student at Columbia University in the late 1940s one of my fellow students was Mrs. Mary Welleck Garretson (1896-1971). Mary had been a student of Grabau at Columbia University. My early acquaintance with the publications of Grabau was through Mary Garretson. Mary was a hard-working student and much older than the rest of the graduate student population. She was determined in her quest for a Ph. D. degree. Unfortunately in the early 1950s, when she presented herself for the Oral Ph. D Examination, Grabau's concepts were not in style. As Walter H. Bucher (1888-1965) explained to me: instead of new and up to date publications Mary harked back to dated concepts of Grabau's, so she failed her Ph. D. examination and never graduated with this degree. While a student at Columbia University (1946-1951), Mary taught at Hunter College in New York City and thereafter was a consulting geologist in Haiti and the United States. My last visit with Mary was in 1969 at the annual convention of the Northeastern Section of the Geological Society of America, in Albany, New York, where she expressed to me her disappointment with her failing efforts at Columbia University. As Behre (1975) [13] put it while at Barnard, she (Mary Garretson) came under the stimulating and enthusiastic teaching of Professor Amadeus W. Grabau, who gave courses at Barnard and Columbia. Later she assisted him in research and publications.

A special tribute must be given to the outstanding aid, previously mentioned, that Mrs. Garretson gave to Professor Grabau. In the introduction to his Textbook of Geology, Professor Grabau said of her: "My former student, Miss Mary Welleck, A. M., has been my assistant throughout the arrangement of this text for the press, and has been of the greatest service in securing illustrations". She was for many years not only his admiring student and understudy but might almost have been referred to as his right-hand organizer. Grabau went to Peking as chief palaeontologist to the Chinese Geological Survey, and Mary graciously represented him in the United States until his death. Robert Rakes Shrock (1904-1993) of the Massachusetts Institute of Technology was a lifelong admirer of Grabau. At a national geology convention he offered to turn over his Grabau file to me for further work and ultimate publication. When I was ready for this transfer Shrock had just died, and the Institute Archivist Helen W. Samuels made copies of the correspondence in the Grabau file available to me. The written material included an oral history which Mary Garretson recorded on October 24,1968. A few remarks will be culled from Mary Garretson's testimony. She noted that Grabau worked at all hours, day and night. He had quantities of coffee, strong coffee that kept him awake. Sometimes he would work all night. He ate if his housekeeper put something before him. He dressed in old slacks except when he was at the university, where he was very properly and formally dressed. Grabau had arthritis, and when he was in China he was wheelchair bound. Mary Garretson met Grabau for the first time at Columbia University's Journal Club, a regularly scheduled seminar for students and faculty. This first meeting dated back to 1917 when Grabau was already afflicted. By about 1920 he was on crutches. Mary served as Grabau's assistant from 1918 to 1919.

The questions that always arise are the reasons that forced Grabau to resign from the Columbia University faculty. I have heard all kinds of rumors from various sources. Marshall Kay (1904-1975), at the time chairman of Columbia University's Department of Geology, confided to me that he destroyed the correspondence between Grabau and Columbia University, because its release would not have been a credit to the university. The interest in Grabau's resignation has spread widely; in fact, publication of my abstract on Grabau in the abstract volume of the 30th International Geological Congress in Beijing, China [14] intrigued geologist Shen-su Sun from the Australian Geological Survey Organization to write me (September 12, 1996): "Professor Marshall Kay once told to me something about the reasons for Grabau leaving Columbia (University). I am interested to know whether Professor Kay's explanation was right".

One of the reasons generally given why Columbia University forced Grabau to resign was his heritage. Although born in Wisconsin he was of German parentage, in fact his knowledge of the German language was so excellent that I suggest his parents must have spoken German at home. His grandfather Johann Andreas August Grabau (1804-1879) was a Lutheran minister who fled Germany on November 8, 1837. Likewise his maternal grandfather, Henry von Rohr fled Germany at about the same time [15]. Grabau's father likewise was a Lutheran minister. This German Lutheran background was his undoing. Grabau's forced resignation in 1919 followed the end of World War I. According to Mary Garretson a colleague wanted Grabau's position and approached Columbia University president Nicholas Murray Butler (1862-1947) and accused Grabau that he was teaching a pro-German viewpoint in his lectures. However, this course of lectures happened to be a class in palaeontology in which Mary was one of two students, and Grabau never mentioned the war. According to Mary Garretson this colleague told Nicholas Butler that this Germanism was going on and Nicholas Murray Butler was looking for an excuse to get rid of professors.

The German geologist Johannes Walther (1860-1937), another giant of geology, was a friend of Grabau (Friedman, 1986). Grabau's great classic *Principles of Stratigraphy* (1913) was dedicated to Johannes Walther, a leader in the field of knowledge herein explained. In his preface Grabau cites (p. VIII) "it was this time also that we in America first became acquainted with those monumental contributions to Lithogenesis and Biogenesis, that had been and were being made by the then Haeckel Professor of Geology and Palaeontology at Jena (Germany), Dr. Johannes Walther, now Professor of Geology and Palaeontology at Halle". Surviving correspondence in the archives of the Geological Survey of China include handwritten letters in German script addressed by Walther to Grabau which reveal an intimate personal relationship. From this correspondence it emerges that Grabau and his wife Mary Antin (1881-1949) had visited Walther and his wife in the fall of 1910 at a time when Grabau was working on his *Principles of Stratigraphy*. Walther's note, dated September 12, 1910, says briefly (my translation) "we are happy that you will come to us. The travel plans will depend on the weather. With cordial greetings from (my) wife to (your) wife, your J Walther".

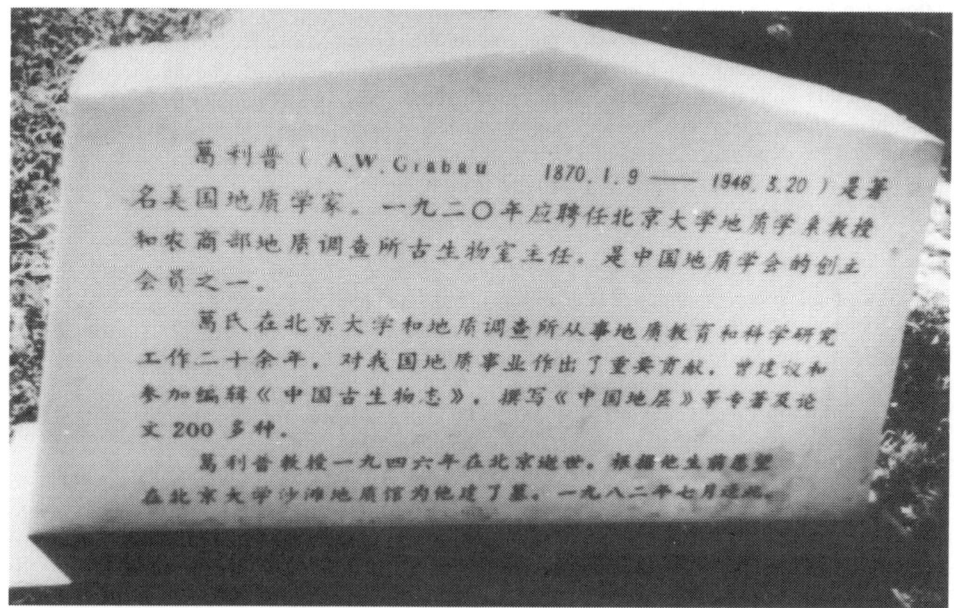

Figures 2 and 3. Tombstone and memorial park of Amadeus W. Grabau on the campus of Peking University (author's photograph).

Apparently fieldwork was combined with this visit. This observation is confirmed in another letter dated January 8, 1941, written by Erik Nystrom of Stockholm, Sweden, which stated that in 1910 he (Grabau) spent one year on holiday in Europe, studying personally the most significant geologic sections. On May 11, 1911, Walther wrote a letter (my translation from German) which says: "dear friend and colleague" and then explains that one of Walther's favorite students whom Grabau had planned to take to Columbia University will not come. Walther noted that he probably cannot recommend this student anyway since he has much yet to learn. In this same letter Walther refers to his fieldwork during the preceding semester. Walther had been to Egypt, Libya, Nubia, Sudan, and Ethiopia and he deplored the end of his travels, and assuming once again academic duties. The letter ends "with cordial greetings to your wife from us and be yourself greeted in friendship from your devoted J Walther".

There is a nine year gap in the letters of Walther to Grabau. The next letter is dated October 11, 1920. Since the last correspondence World War I (1914-18) had intervened and Grabau had left the United States for China. Walther addressed this letter, written on October 11, 1920, "dear old friend". He had heard from Rudolf Ruedemann (1864-1956), state palaeontologist of New York, another German-American, disturbing news about Grabau. He wrote (my translation from German): "several months ago I received word. . . about you which saddened and pained me, especially since I did not have precise information about your permanent location". According to Mary Garretson Grabau simply dropped out of the geological world when he went to China. In 1927, Walther was in the United States as a visiting professor at Johns Hopkins University and wrote to Grabau in China on March 21 (my translation from German): "My dear Grabau it is very painful to me that I am in your homeland, without finding you here. At the beginning of April I will present five lectures at Columbia (University)", where, I am sure, Walther heard much of the gossip about Grabau's resignation from the University. "Recently someone (whose name I cannot decipher in Walther's letter) reported to me about your life and related many things which filled me with sorrow about you. How much I would like to help you! If I could". A letter of November 27, 1929 is more cheerful (my translation from German): "Dear friend! I have received with pleasure your two packages of books, firstly as a sign that you are alive and exude admirable creativity and then because of the book's most interesting contents. Your studies of Peking man generate much publicity in our daily newspapers. In old friendship. Your Johannes Walther".

Interestingly, although Grabau was versed in English, French, and German, he never learned Chinese. Grabau's wife Mary Antin was of Russian-Jewish origin. She came to the United States at the age of thirteen as the daughter of Israel and Ester Weltmann and received her education in the public schools and girl's Latin school in Boston. She transferred to Columbia University's Teachers College and Barnard College of Columbia University in 1901, after meeting Grabau. They married in 1901, but she never received a Columbia degree. Mary wrote three books *From Polska to Boston* (1899), *They Who Came To The Gate* (1904), and *The Promised Land* (1912). She wrote her first book in Yiddish at the age of 11. Her own translation into English appeared two years later, with a foreword by Israel Zangwill (1864-1926), English Jewry's greatest writer of the turn of the century. The last of these three books was especially popular. Ursula Marvin (personal communication, 1996) told me that in her native Vermont this book was widely read. Grabau helped her in writing it, but [as Mary Garretson put it] he was a little bit miffed when at first (United States) President Theodore Roosevelt called him Mr. Antin. His wife and daughter Josephine left him when he went to China [16-18]. They remained in the United States, but he was at least briefly reunited with Mary Antin at the time of the International Geological Congress in Washington in 1933 [15]. According to May [18] Mary Antin was the daughter of Russian Jews who left their homeland to escape religious persecution, Antin achieved fame with the 1912 publication of The Promised Land, her autobiography chronicling Jewish life in Czarist Russia as well as describing life for a Jewish American in the slums of Boston. The book, regarded as one of the most popular immigrant autobiographies ever written, was ultimately issued in thirty-four editions and sold nearly eighty-five thousand copies. Its publication earned Antin credibility as a commentator on Jewish immigrant life, and as a result she spent several years lecturing throughout the United States on immigration and patriotism. After 1920, when Grabau left for China, Mary Antin's popularity as a writer and lecturer waned. Her brief period of fame was followed by years of hardship, when she supported herself by doing social work [17]. She contributed to periodicals such as Atlantic Monthly, Outlook, and Common Ground. After a long illness, she died in 1949 in Suffern, New York [16, 18].

After Grabau attended the International Geological Congress in the United States in 1933, on his way back to China, the Japanese would not let him get off the boat because they felt he had made slighting remarks about them. From 1941 to 1945 the Japanese interned him in a camp in the old British Embassy [15]. He was not allowed to receive geological publications; he just deteriorated and after liberation he died in 1946. Still, despite this handicap he was writing a manuscript, known as The World We Live In, which has been published posthumously. He was buried in the compound of the Geology Department of Peking University. In the early 1980s his grave was moved to a new location in a pleasant memorial park of the campus of the new site of Peking University (Figs. 2 and 3). A white marble headstone adorns his grave with English and Chinese inscriptions in memory of his major contributions to the geology of China. A large block of karstified bedrock in this park provides a geological theme.

Acknowledgements

I wish to acknowledge my indebtedness to the Geological Survey of China in whose archives I discovered handwritten letters in German script which Johannes Walther addressed to A. W. Grabau, and which reveal an intimate personal friendship. My thanks are extended to Professor Y. Wang for his help in making these letters available to me. Nostalgically I recall my visits with Professor Robert Shrock at meetings of the Geological Society of America in which we discussed the life cycle of A. W. Grabau. As pointed out in the text of this paper he urged me to take over the correspondence and manuscripts relating to Grabau, and after his death Archivist H. W. Samuels of the Massachusetts Institute of Technology provided me with copies of Shrock's Grabau file.

My thoughts go back to Mary Garretson. She was a special fellow-student at Columbia University, much older than her peers in the graduate program and Grabau's student and last assistant in the United States. She acquainted me for the first time with the contributions of Grabau. Thanks are extended to Professor Ye Lianjun who took me to see Grabau's tomb on the campus of Peking University in 1984.

REFERENCES

1. G. M. Friedman. Geology at Rensselaer: Address of the Retiring President of the New York State Geological Association. In: *Guidebook. Joint Annual Meeting of New York State Geological Association 51st Annual Meeting and New England Intercollegiate Geological Conference, 71st Annual Meeting*. G. M. Friedman (Ed). pp. 1-19 (1979a).
2. G. M. Friedman. Geology at Rensselaer: A historical perspective, *The Compass of Sigma–Gamma–Epsilon* **57**, 1-15 (1979b).
3. G. M. Friedman. Geology at Rensselaer Polytechnic Institute: An American Epitome, *Northeastern Geology* **3**, 18-28 (1981).

4. A. W. Grabau. Guide to the geology and paleontology of Niagara Falls and vicinity, *New York State Museum Bulletin* **45**, 1-284 (1901).
5. G. M. Friedman and J. E. Sanders. *Principles of Sedimentology*. John Wiley & Sons, NewYork, Chichester, Brisbane, Toronto, and Singapore (1978).
6. A. W. Grabau. *Principles of Stratigraphy*. New York, A. G. Seiler & Co. (1913).
7. G. M. Friedman. Glimpses of pioneer sedimentologists Johannes Walther and A. W. Grabau Correspondence 1911-1936. *12th International Sedimentological Congress, 24th–30th of August 1986 Canberra, Australia,* 111 (1986).
8. W. H. Twenhofel and Collaborators (The University of Wisconsin, Department of Geology and Geography, Madison, Wisconsin). *Treatise on Sedimentation.* Baltimore, The Williams & Wilkins Company (1926).
9. W. H. Twenhofel. *Treatise on sedimentation.* Baltimore, Williams & Wilkins Co. (1932).
10. W. H. Twenhofel. *Principles of sedimentation.* New York and London, McGraw-Hill Book Company Inc. (1939).
11. W. H. Twenhofel. *Principles of sedimentation.* New York-Toronto-London, McGraw-Hill Book Company (1950).
12. A. W. Grabau. *A Textbook of Geology*. Boston, D. C. Health & Co. (1921).
13. C. H. Behre. Memorial of Mary Welleck Garretson 1896-1971, *Geological Society of America Memorials* **4**, 72-73 (1975).
14. G. M. Friedman. From Rensselaer to China: Amadeus W. Grabau on the Semicentennial of his Death (Abstract). *30th International Geological Congress Abstracts, Beijing, China, 4-14 August 1996,* **3**, 528 (1996).
15. Gert Michel and Wang Genyuan. Amadeus William Grabau (1870-1946) Geologe in China, *Zeitschrift der deutschen geologischen Gesellschaft* **142**, 1-12 (1991).
16. De Burgh Marshall *et al.* . Mary Antin. In: *Concise Dictionary of American Biography*, 3rd edition. p. 26. Charles Scribner's Sons, New York (1960).
17. Encyclopaedia Judaica. Mary Antin. *Encyclopaedia Judaica, Jerusalem* **3**, 67 (1972).
18. Hal May (Ed). Antin, Mary 1881-1949, Contemporary authors **118**, 22. Gale Research Co. , Michigan (1986).

Proc. 30th Intern. Geol. Congr., Vol. 26, pp. 165-175
Wang *et al.* (Eds)

The Global Theories of Amadeus W. Grabau (1870-1946): A Retrospective View

URSULA B. MARVIN
Harvard-Smithsonian Center for Astrophysics, Cambridge, Massachusetts

Abstract

While working in China during the latter part of his career, Amadeus W. Grabau formulated his Pulsation Theory, based on his encyclopedic knowledge of Palaeozoic stratigraphic sequences in North America, Europe and Asia, that the Earth's history has been dominated by transgressions and regressions of sea level due to rhythmic pulsations of the ocean floors. He also formulated his Polar Control Theory in which he tentatively called upon the gravitational force of a stellar body that approached the Earth's south pole and caused the sialic crust to rip open and amass into a single, huge southern supercontinent, Pangaea. As he worked out the Palaeozoic history of Pangaea, Grabau traced an apparent migration of the south polar icecap from a position centered on present day Egypt in Cambrian time to Antarctica in the Jurassic. He concluded that Pangaea had drifted northward while remaining intact until the Jurassic, when it began to split into drifting continental fragments. Thus he became an advocate of both climatic polar wandering and continental drift. For the advent of life on the Earth he called upon a seeding of our planet by impacts of extraterrestrial bodies containing living organisms. Now that the theory of plate tectonics validates horizontal motions of crustal plates, and meteorites from Mars have been found to carry possible evidence of past life, we will reexamine Grabau's global theories in the light of our present knowledge. We will find that although many aspects of Grabau's global theories are unacceptable, some of them are closer in spirit to those we favor today than were the theories held by most of his contemporaries.

Keywords: Grabau, A. W., pulsation theory, polar control theory, continental drift, Pangaea, plate tectonics

INTRODUCTION

Born in Wisconsin in 1870, as the third of ten children to parents of German heritage, Grabau became interested in natural history during his early years in public and parochial schools. In 1896 he received his B. S. degree in palaeontology at the Massachusetts Institute of Technology. In 1900, Grabau earned his Ph.D. degree in geology from Harvard University with a thesis titled: *Phylogeny of Gastropoda. I. The Fusidae and their Allies*. By then, he already was teaching palaeontology and stratigraphy at the Rensselaer Polytechnic Institute in Troy, New York, where he remained for two more years. In 1903, Grabau moved to New York City where he accepted a position as Professor of Palaeontology at Columbia University.

Principles of Statigraphy: 1913

In 1913, Grabau published his early masterwork, *Principles of Stratigraphy* [1], a book of 1,185 pages in which he ventured: "To bring together those facts and principles which lie at the foundation of all our attempts to interpret the history of the earth from records left in the rocks." Grabau focused primarily on sedimentation, a subject that did not exist as a science when he began his book. To place sedimentation in the context of geology as a whole, he included sections on structural geology, igneous and metamorphic petrology, atmospheric and oceanic sciences, and the dynamic and chemical interaction of the hydrosphere and lithosphere. In presenting current depositional environments as guides to past sedimentary and faunal assemblages he laid the groundwork for the discipline of palaeoecology.

Grabau dedicated his book to the German geologist Johannes Walther (1860-1937), whose ideas he helped to introduce to his English-speaking readers. As early as 1893-1894 Walther had emphasized the importance of lateral changes of facies–both lithofacies and biofacies–within sedimentary formations [2]. He had pointed out the importance of primary processes such as changes in sea level that may cause coeval facies, deposited during a single epoch of sedimentation, to be superimposed with no intervening interval of erosion or non-deposition. The facies principle gained some acceptance in Europe, and also in America until it met with powerful opposition from a brilliant and highly influential stratigrapher, Edward Oscar Ulrich (1857-1944), of the United States Geological Survey.

Ulrich argued that our knowledge of geologic time intervals was too rough to allow for distinguishing coeval facies within stratigraphic formations. He believed that each sedimentary series consists of beds deposited by ephermeral epicontinental seas that periodically advanced and retreated from basin to basin in response to rhythmic diastrophic motions of the lands. Local hiatuses and unconformities served to separate strata of constant lithology and fauna. With his extensive knowledge of the Phanerozoic deposits of North America and his unmatched powers of persuasion, Ulrich dominated the field. His monograph: *Revision of the Palaeozoic Systems*, published in 1911 [3], led to a remapping of stratigraphic sections in the mid-continent during the next two decades. In retrospect it seems clear that Ulrich hindered progress in American stratigraphy for the first 30 years of this century. The history of science records all too many examples of such obstruction by well-established authorities.

At Columbia, Grabau was recognized for his excellence in teaching and research and he began to gain influence among American stratigraphers. However, with the onset of World War I, he began to meet with hostility from the University faculty and administrators because of his Germanic heritage and his declared admiration for German science. Feelings intensified with America's entry into the war and created an atmosphere that made it difficult for Grabau to continue his work. At the same time he became estranged from his wife, Mary Antin, a Polish immigrant and author of *The Promised Land* [4], who passionately and publicly espoused the allied cause.

GRABAU IN CHINA, 1920-1946

In 1919, Grabau, left Columbia University and accepted an invitation to serve jointly as Professor of Palaeontology at the National University at Peking and Chief Palaeontologist of the Geological Survey of China. The move provided him with opportunities to introduce his views on stratigraphy to a whole new group of students and to learn about an important area that was largely unknown to him. Thus, at the age of 50 Grabau, who already was partially crippled, began a new and enormously productive career in China.

Grabau directed his students in extensive field investigations in China. From their maps and his encyclopedic knowledge of the literature he worked out detailed correlations between the stratigraphic deposits of Asia with those of North America and Europe. These efforts only strengthened his faith in the facies principle and led him to formulate his pulsation theory to explain what he saw as a rhythmic sequence of marine transgressions and regressions that were synchronous worldwide.

Grabau's Pulsation Theory

In 1933, Grabau presented the earliest version of his pulsation theory at the XVIth International Geological Congress that met in Washington, D.C. He referred to it as a working hypothesis and invited his fellow stratigraphers to test it. Grabau designated each global marine transgression and its subsequent regression as a pulsation period, and identified the sediments of each pulsation period by its characteristic fauna. During transgressions, Grabau envisioned a luxuriant mixing and diversification of fauna as sea waters flood littoral lowlands and all the world's geosynclines. During regressions, the draining away of waters from all but shallow epicontinental seas or landlocked basins would raise stress levels and lead to species competition and extinctions. In Grabau's system, each pulsation is followed by a hiatus–an interpulsation period when ocean waters would occupy only the pelagic and abyssal basins and the continents would undergo severe erosion, rapid sedimentation, volcanism, and tectonic movements.

In his IGC talk of 1933 (published in 1936) [5], Grabau displayed a diagram showing eleven Palaeozoic pulsation periods separated by ten hiatuses. He noted that his hiatuses correlated remarkably well with the periodic tectonic disturbances described in 1928 by Professor Hans Stille, of the University of Berlin [6]. He borrowed Stille's names, such as "Sicilian folding" and "Taconic folding," for the ten hiatuses on his diagram. When Grabau concluded his presentation, Dr. Stille stood up in the audience and expressed his strong support for the pulsation theory. Grabau was gratified, but he differed with Stille's belief that worldwide disturbances originate in the continents rather than in periodic pulses of the ocean floors.

Grabau discussed his pulsation theory and all its ramifications for sedimentary deposits of the Palaeozoic Era in his highly technical 4-volume *Palaeozoic Formations in the Light of the Pulsation Theory*, published between 1933 and 1938 [7]. In 1940 he presented his full view of the Earth's history in *The Rhythm of the Ages, Earth History in the Light of the Pulsation and Polar Control Theories* [8]. He regarded this book as his most significant contribution to science and, inasmuch as a reconciliation had taken place by then, he dedicated it to his wife and their daughter and two granddaughters.

Grabau (1940:24) [8] described his pulsation theory as follows:

"So regular was the encroachment of the sea upon the land, and again its retreat, that it is possible to record a sequence of transgressions and regressions which correspond, with certain modifications, to the independently established successions of geological periods, and for the establishment of which periods in the first place, the changes in the marine faunas were almost the only reliable criteria."

To account for the periodic pulsations of the ocean floors, Grabau adopted the theory of John Joly [9] that radiogenic heat periodically builds up to melting temperatures in the world-encircling basaltic substratum; the heat is then lost by conductivity through the ocean floors. Grabau (1940:15) [8] wrote:

"Periodic increase of heat is followed by loss of heat through conductivity through the cold ocean floors. Expansion is followed by contraction, and the ocean rises and falls with the changes in its floor; transgression is followed by regression, and a pulsation period is completed."

Grabau tentatively adopted Joly's estimates that the accumulation of heat would require at least 30 million years, and the subsequent cooling would require another 30 million years; an additional 30 million years would elapse before the build up of sufficient heat to initiate a new episode of heating.

Thus, Grabau assigned 60 million years to each pulsation period and 30 million years to each hiatus (interpulsation period). He did not dwell on the logical conclusion that these figures would sum to a minimum time span of 1,260 million years for the 14 pulsations and 14 hiatuses he then was postulating for the Palaeozoic Era. Inasmuch as Arthur Holmes, in 1937, had listed several independent lines of evidence that converged on about two billion years as the age of the Earth, Grabau's time scale for the Palaeozoic Era was too long, by far [10]. Today, when the Earth is dated as 4.6 billion years old, only about 325 million years are assigned to the Palaeozoic Era.

Recognition of Grabau in America

Grabau did not remain in America after the IGC of 1933, and so he was not present to make his voice heard when the first strong challenges were brought against some of E. O. Ulrich's stratigraphic interpretations. During the 1930s, the facies principle began to regain a following in America, but with no reference to the work of Grabau. In the upper Mississippi Valley, for example, American stratigraphers reexamined the Croixan Series in which Ulrich had mapped ten formations with a total thickness of 656 feet. By applying the facies principle, they reduced the number of formations to three and the total thickness to 460 feet [11]. Similar work was performed along the west flank of the Appalachian mountain chain and in the Ozark Mts. of the mid-continent. In 1948, a ground-breaking conference titled *Sedimentary Facies in Geologic History* was convened at the annual meeting of the Geological Society of America. Six major papers and the remarks of 15 members of the audience were published in the proceedings [12] without a single citation of any work by Grabau.

But recognition of Grabau's accomplishments soon was forthcoming. In 1949, Francis J. Pettijohn, at the University of Chicago, published his influential textbook, *Sedimentary Rocks*, in which he introduced thousands of American readers to Grabau's "descripto-genetic" classification of sedimentary rocks, calling it complete, logical and, in many ways, very modern — one of the two best yet devised [13]. Years earlier, as a young faculty member, Pettijohn had been alerted to the importance of Grabau's ideas by his older colleague, J. Harlan Bretz (Siever, personal communication, 1991). Pettijohn pointed to the value of some of Grabau's terms that such as "exogenetic" and "endogenetic" to indicate, respectively, sediments derived from outside areas and those from source materials at their immediate locality. However, for his own book Pettijohn adopted the much more recent classification, devised in the 1940s by Paul D. Krynine at Pennsylvania State University, who had factored in the tectonic settings in which sedimentary deposits accumulate. Two years later, in *Stratigraphy and Sedimentation*, W. C. Krumbein and L. L. Sloss described Grabau's *Principles of Stratigraphy* as the classic English-language text and reference on the entire field of stratigraphy [14]. Thus, shortly after his death in 1946, leading American stratigraphers were beginning to appreciate the importance of Grabau's contributions.

Alternative Theories of Global Periodicity

The concept that the Earth's history is marked by periodic episodes of diastrophism, linked with worldwide fluctuations of sea level, was by no means new when Grabau proposed his pulsation theory. Thomas C. Chamberlin, at the University of Chicago, had introduced such a concept as early as 1898, and, in 1909 he declared that episodes of continental diastrophism provide the basis for global correlations of stratigraphic systems [15]. Others besides Chamberlin and Stille, who had envisioned global periodicities, included Eric Haarmann in Germany and R. W. van Bemmelen in the Netherlands.

In 1930, Haarmann published his oscillation theory in which he ascribed all structural deformations to vertical forces acting in the upper mantle [16]. He described giant "geotumors" rising out of the primeval ocean in response to the gravitational pull of a cosmic force, of unspecified nature, that approached the Earth. As the geotumors eroded, sediments were deposited in adjacent geodepressions. Haarmann postulated that shifts in the position of the cosmic force would set up oscillatory up-and-down movements in the crust. When new geotumors arose out of geodepressions, the sediments would glide downslope and come to rest in intricate folds. Subsequently, secondary phases of tectogenesis would raise the folded sediments into mountain ranges such as the Alps, revealing their spectacular nappes.

In 1931, Van Bemmelen announced his undation theory in which uplifts and depressions were caused by subcrustal flow generated by the differentiation of a molten basaltic substratum into rising granitic and sinking ultrabasic materials [17]. He believed that this process provided sufficient energy to produce magmatic cycles linked with orogenies. In place of Haarmann's vertically-directed oscillations, van Bemmelen pictured wave-like motions, or undations, in the crust, ranging from mega-undations of the geoid, 10,000 km in wave length, to local 1-km undations resulting from disturbances of the upper tectonosphere.

In Grabau's paper *Oscillation or Pulsation*, of 1936, he contrasted his own theory with what he called Eric Haarmann's "much more orthodox" doctrine of oscillations [5]. To Grabau, the oscillation theory appeared orthodox because, following Chamberlin, Stille, and others, Haarmann assigned the primary role to continental tectogenesis, leaving marine transgressions and regressions as secondary phenomena. Grabau did not join in the widespread criticism of Haarmann's invocation of a cosmic force; he himself would call upon one a few years later.

Grabau's *Rhythm of the Ages* appeared in China, and unfortunately was little read elsewhere, at a time when interest still was rising in the idea of global rhythms [8]. A prime example is found in the works of J. H. F. Umbgrove, in the Netherlands, who published *On Rhythms in the History of the Earth* [18], *The Pulse of the Earth* [19], in which he spoke (p. xxi) of "...oscillations of sea level, the pulsation of folding and mountain building, the periodicity of the ice-ages, the rhythmical cadence of Life..." In 1950 he published the more complicated *Symphony of the Earth* [20]. Umbgrove called upon convection currents in the mantle to explain the dominant periodicities, but he regarded as still mysterious the causes of smaller ones.

In the 1920s to 1950s many geologists believed in periodicities in the Earths history; and most of them also believed strongly in a fixed distribution of continents and ocean basins. Chamberlin, Stille, Haarmann, van Bemmelen, and Umbgrove ruled out all hypotheses of continental drift–Umbgrove specifically ruled out Grabau's version. Grabau worked outside this tradition. He was one of the few American scientists of professorial rank to support the idea of continental drift.

Continental Drift and the History of Pangaea

Grabau examined and rejected three principal theories that were formulated to explain the strange, asymmetrical distribution of the Earth's continents and ocean basins.

1. The continents and ocean basins are permanent features that have been rooted in place since the planet cooled from a molten globe. This theory, proposed in 1847 by James D. Dana at Yale [21] and by Elie de Beaumont in Europe in 1852 [22], enjoyed widespread support until the advent of plate tectonics.

2a. Cooling and contraction of the Earth's interior sets up compressional stresses that result in crustal shortening — folding, faulting, mountain-building–along zones of weakness between rigid crustal blocks. Successive orogenies have welded new orogenic belts to the margins of the early shield areas, thus contributing to the continual growth of continents at the expense of ocean basins. This theory was popular until after the discovery of radioactivity.

2b. In a reversed version of the contraction theory, proposed by Eduard Suess [23], tangential stress causes the rigid, sialic crust to split open and great blocks of it to founder into the depths, creating the ocean basins. Suess saw the growth of the ocean basins at the expense of the continents as an ongoing process that will continue until all parts of the crust fit onto a smaller globe and the Earth, once again, will be engulfed by a universal ocean.

3. The Moon was ripped out of the Earth's crust and mantle during the Mesozoic Era, leaving behind the vast scar of the Pacific Ocean Basin; the large remnant of sialic crust split into fragments that drifted toward the cavity–thus opening the Atlantic Ocean. This dramatic theory, put forward in 1881 by Osmond Fisher in England [24], had immense popular appeal and, for a time, was attractive to geologists as it accounted for three puzzling features: the Moon with its density of about 3.34 g/cm^3 matching that of the Earth's mantle, the Pacific Basin with its (unhealed) ring of fire, and the jig-saw fit of continents on either side of the Atlantic Ocean. However, the geophysicists, Harold Jeffreys [25] and Beno Gutenberg [26], showed it to be dynamically impossible for the Moon to escape from the Earth once the crust and mantle were solid. By that criterion, an escape in the Mesozoic, or even in the Precambrian Era, was out of the question.

After considering these ideas, with their many variations, Grabau concluded that the best explanation of the distribution of continents and ocean basins lay with some version of continental drift. He called continental drift "Wegener's hypothesis," although he also gave credit for it to Frank B. Taylor [27] in America and Alexander Du Toit [28] in South Africa. Grabau cited the 3rd edition of Wegener's book that appeared in 1922 [29]. This was the first of Wegener's writings to become widely known, although he published his first articles on continental displacement as early as 1912. Wegener introduced the first non-catastrophic hypothesis of continental drift. He depicted sialic continental blocks breaking away from a supercontinent, Pangaea, and moving through the ocean-floor sima under the impetus of tidal and rotational forces that were intrinsic to the Earth. At least two serious failings plagued Wegener's concept: the mechanisms he invoked were too weak by orders of magnitude (not until the 4th edition of his book in 1929 [30], did he ascribe the horizontal motion of continents to convection currents in the mantle), and he envisioned the breakup of Pangaea as a single relatively late event in the Earth's history, initiated in the Jurassic.

Grabau was more interested in Pangaea itself. To him, a single large landmass, laced with geosynclines, provided a more satisfactory framework for explaining the apparent global effects of pulsating sea levels throughout the Palaeozoic than did continental areas scattered through the ocean basins. Grabau created a detailed Palaeozoic history for Pangaea by plotting the stratigraphy of each period on a map he constructed showing all the continents grouped into that great landmass. He also extended his history of Pangaea backward in time by transferring to his map the patterns of Archaean trend lines that Rudolph Reudemann had plotted on a Mercator projection of the ancient shield areas of the world [31]. Grabau [8] cut a piece of green baise cloth to fit over a 14-inch globe and pulled it into one hemisphere. The frontispiece of *The Rhythm of the Ages* is a photograph of three stages in that operation, with the final one showing the cloth compressed into subparallel creases that, in his view, made a fair match with the trend lines on his map of Archaean Pangaea. Grabau concluded that Pangaea was the first landmass to rise out of the primaeval ocean.

The Polar Control Theory of the Origin of Pangaea

To account for the origin of Pangaea, Grabau put forward his polar control theory that called upon an extraterrestrial force. He was convinced that no force internal to the Earth could rip apart the primeval sialic crust and compress it into a thick mass of crushed and contorted gneisses occupying only one hemisphere. The globe, he wrote, could no more shrug half of its crust to one side than a man could wriggle out of his skin or, for that matter, out of his closely buttoned coat without using his hands or outside aid. As a working hypothesis, Grabau proposed that a stellar body approached the young planet and hovered over the south pole of its rotation axis. Under the gravitational pull of the body, the sial that covered the north pole rifted apart and was pulled into the southern hemisphere. As Grabau described the process (1940:19) [8]:

"...dry land appeared in obediance to the command, the gravitative urge, of some celestial visitor. This may have been the newly captured moon, or more probably some star...which at that early date approached the earth from a south polar direction and by its attraction caused the thin crust of light rock-material to gather together into a hemisphere of intensely folded and hence much strengthened sial-rock...All through the Archaeozoic and the Palaeozoic Eras, the earth's surface was divided into a southern Pangaea and a northern Panthalassa..."

Grabau's Pangaea did not remain fixed in place, however. In the back of his book, he included twelve colored maps showing the locations of the geosynclines and epicontinental seas of Pangaea for each successive period of the Palaeozoic Era. He also plotted the changing locations of the south polar icecap, which he showed as centered on present day Egypt on his map of Taconic (earliest Cambrian) Pangaea. In each later period the icecap progressed southward through Africa until it finally came to rest on what is now Antarctica. Grabau deduced this wandering of climatic zones from locations of glacial tillites and other types of deposits. He concluded that, while the rotation axis remained fixed within the body of the Earth, the buoyant sialic supercontinent periodically shifted northward following a zig-zag path that carried parts of it through the polar and the equatorial zones. A comparable theory of climatic polar wandering had been developed by Köppen and Wegener in their book *The Climate of the Geological Ages* [32]. However, those authors assumed that the body of the Earth itself has shifted with respect to its axis of rotation.

To account for the motions of Pangaea, Grabau pictured the sialic continent moving as a unit over a hot, weak substratum. He argued that such a large, irregular landmass would tend to be unstable on a rotating planet and that it might swivel back and forth around sites where intrusions of sima temporarily would serve as anchors. He believed that, eventually, rotational instability would splinter the supercontinent into fragments that would drift apart. His maps show the first rifting of Pangaea taking place in the "Comanchaean" (Jurassic) Period. He then traces the subsequent drift of the continental fragments on nine maps of Tertiary time to the present day. Grabau pictured the leading edges of the moving sialic fragments forming geosynclinal depressions where they pushed against the resistant sima, in which they were partially buried, and the trailing edges undergoing tensional drag accompanied by rifting and volcanism. This conception, with its faint intimations of sea floor-spreading and subduction, suggests that had he lived into the 1960s Grabau almost certainly would have been an active contributor to the development of plate tectonics theory.

The Advent of Life on Earth

Grabau called upon an extraterrestrial source to account for the advent of life on Earth. He speculated that the great primitive shoreless sea of the Earth was lifeless — unoccupied by even the germs of organisms — but that living organisms probably entered the waters before dry land appeared. Grabau (1940:17) [8] then offered his readers a choice of two unsatisfactory alternatives:

"Where they [the organisms] came from is an unsolved problem, but that they reached the earth from some extra-telluric source must be assumed, unless we are prepared to believe in spontaneous generation."

Spontaneous generation of living organisms from inanimate matter (in the sense that grubs were once thought to spring from offal) was, and still is, a totally unacceptable idea, but Grabau's alternative explanation simply transferred the problem of the origin of life to some other body out in space. There he left his argument, but we will return to it later.

Assessments of Global Theories in the 1950s and 1980s
A few years after Grabau died, Professor L. U. de Sitter [33] of the University of Leiden, published the following assessments of five theories of global tectonics that had been debated up to that time.

1. The classical contraction theory: largely abandoned after the discovery of radioactivity. In any case, it did not account for cyclic diastrophism.

2. John Joly's theory of expansion of the Earth from radioactive heating: abandoned as improbable. [Author's note: de Sitter's evaluation was unfair; Joly proposed a theory of recurrent thermal cycles with alternating expansion and contraction of the subcrustal layer. Expansion of the Earth itself was proposed in 1935 by the South African astronomer, J. K. E. Halm [34], but de Sitter did not refer to Halm.]

3. Wegener's theory of continental drift: abandoned for lack of an adequate driving force and its application to only a single, late episode in history; it did not account for periodic diastrophism, and, in de Sitter's view, the idea of polar wandering was suspect from the beginning.

4. Haarmann's and van Bemmelen's oscillation and undation theories: Haarmann's theory was spoiled by the call for a cosmic force and too much gravity sliding; van Bemmelen's showed some promise.

5. Vening Meinesz's convection-current theory: adjoining cells rotating in opposite directions are very unlikely to occur in a rotating Earth; convection may possibly help to account for major features (continents-ocean basins) but not for individual mountain ranges and other secondary features.

It is notable that de Sitter, who was an avowed fixist, expressed no strong support for any of these theories. Clearly, no satisfactory theory of the Earth's tectonic and stratigraphic history existed in the mid-20th century. At about that time, however, the great vogue for periodicity in episodes of global diastrophism and eustatic changes of sea level began to fall into decline and geologists started thinking of earth history as governed by tectonic processes that operate continuously.

About thirty years after de Sitter expressed his views, a new assessment was offered by Homer E. Le Grand, of the University of Melbourne, Australia. Le Grand [35] stated that during the 1930s theories of permanence vied for dominance with those of contraction; the theory of global expansion was accepted by a respectable minority, and Wegener's hypothesis of continental drift occupied the (lunatic) fringe. Le Grand then pointed out that by 1988 when he was writing his book, Wegener's mobilist hypothesis had evolved into plate tectonics, which enjoyed universal acceptance; both permanence and contraction were long dead, and, in Le Grand's view, only the expanding earth theory occupied a respectable minority position.

Grabau's Theories in the Light of Plate Tectonics and Planetary Science
Plate tectonics, a mobilist theory *par excellence*, was formulated in the late 1960s and, as geoscientists everywhere saw how well it helped to solve their problems, it quickly became the great unifying theory of modern geology. The same period also witnessed the beginning of the U. S. A. Apollo missions and the U.S.S.R. Luna missions to the Moon, and exploratory missions to other planets. From the images and samples that were returned, geologists are beginning to comprehend the enormously important role that collisions with meteorite and comet fragments have played in the Earth's history. How may we view Grabau's theories in the light of plate tectonics and our new

knowledge of the Earth's interaction with bodies from space?

In place of Grabau's Pangaea, plate tectonics envisions scattered continental fragments being rafted about on recycling oceanic plates from Precambrian time to the present. Efforts to reconstruct the shapes and paths of these fragments on the basis of palaeomagnetic measurements, palaeoclimatic data, fossil distributions, and tectonic features reveal that the coalescing of continental fragments contributed to an immense and growing continent in the southern hemisphere by the Cambrian Period. Thus, despite numerous errors, both in broad conceptions and in details, Grabau approached the truth when he postulated the existence of a Palaeozoic supercontinent. He was only partly correct in envisioning the vast Pacific basin as a remnant of Panthalassa. Much of the Pacific always has remained an ocean basin but continental fragments have drifted through parts of it and its floors have constantly been renewed by growth at the ridges and loss by subduction at the trenches.

Polar wandering now is fully accepted, although today the term is applied primarily to apparent motions of the Earth's magnetic poles, coupled with pole reversals. Grabau's view that climatic polar wandering has resulted from motions of continental crust with respect to the Earth's rotation axis is accepted in principle, as is an alternative mechanism — the "tumbling" of the globe around its rotation axis.

Grabau's main triumphs were his applications of the facies principle — especially its extension to China — and his comprehension of the key role played by the ocean floors in generating marine transgressions and regressions. According to the plate tectonics model, as oceanic ridges swell, rift, and spread, the waters rise and flood over continental areas. As ridges cool and subside, the basins deepen and waters withdraw from the land. However, these episodes never are as rhythmic and finely-tuned or as global in consequence as Grabau supposed. Additionally, Grabau, along with all other geologists today, would have to reinterpret the deposits he defined as geosynclinal in origin; geosynclines were a major casualty of plate tectonics theory.

Thermal instability in the mantle is seen as the driving force of plate tectonics, most likely without any orderly, large-scale patterns of convection cells. A fully satisfactory mechanism for causing horizontal crustal motion, the lack of which was a primary objection to Wegener's hypothesis, still has not been worked out. Today, however, the body of evidence favoring plate tectonics is too powerful to be denied for lack of a specific causal mechanism.

Grabau's (and Haarmann's) postulates that lands first arose out of the ocean due to the gravitational pull of a stellar body that approached the Earth is totally unacceptable. Such ideas indicate that many geologists before mid-century had no conception of celestial dynamics. A stellar body that approached the Earth closely enough to exert an effective gravitational pull on the crust would wreak havoc with the entire solar system. Planets and satellites would be perturbed into new orbits, and we would not survive to puzzle over the resulting geological problems.

Grabau's dictum that we must assume that life was carried to Earth by an extraterrestrial body does not bear serious scrutiny. The warm, wet planet Earth offers the most favorable environment we know of for enabling self-replicating organisms to spring from the chance combination of the requisite components — carbon, hydrogen, oxygen, nitrogen–most likely after millions of years of random mixing and combining of molecules. Amino acids and other building blocks of life occur in carbonaceous meteorites and also have been detected in galactic clouds and in the spectra of comets. To date, the hydrocarbon molecules that have been examined in carbonaceous meteorites are of non-biologic origin. However, recent reports by McKay et. al. [36], indicate that meteorites from Mars may contain evidence of past life. If this proves to be true, the transfer of life from planet to planet becomes a possibility, but the supposed unicellular organisms from Mars are so much smaller and appear to be so much more primitive than those of Earth that they seem an unlikely source for life on our planet. When offered Grabau's choice between "spontaneous generation"(in the old sense) and the introduction of living matter from space, we should prefer spontaneous generation by molecular combinations within the Earth's Precambrian waters.

In conclusion, although present models differ in many particulars from those proposed by Grabau, we find that his basic concepts relating to stratigraphic correlations, to thermal changes in the ocean floors, to moving continents and polar wandering were closer in spirit to those of today than were the ideas championed by most geologists of his time. It seems certain that had he lived to see them Grabau would have welcomed the advent of plate tectonics and the recent advances in the planetary sciences.

In a memorial written in 1950, Professor J. Marvin Weller [37], palaeontologist and stratigrapher at the University of Chicago, described Amadeus William Grabau as:

"...one of the most remarkable geologists of modern times...In influence, fame and the respect that he won from all, he can only be compared with some of the ancient Chinese heroes."

REFERENCES

1. A. W. Grabau. *Principles of Stratigraphy*. A. G. Seiler, New York (1913).
2. J. Walther. *Introduction to Geology as a Historical Research*. Gustav Fischer Verlag, Jena, 3 vols (1893-1894).
3. E. O. Ulrich. Revision of the Paleozoic Systems. *Geological Society of America Bulletin* **22**, 281-680 (1911).
4. Antin, Mary. *The Promised Land*. Houghton Mifflin Co., New York (1912).
5. A. W. Grabau. Oscillation or Pulsation. *Report of the XVIth International Geological Congress* **1**, 539-553. Washington, D. C. (1936) .
6. H. W. Stille. *Introduction to the Phases of Paleozoic Mountain Building*. Deutsch. Geol. Gesell. Zeitschr. **80:1** (1928).
7. A. W. Grabau. *Palaeozoic Formations in the Light of the Pulsation Theory*. 4 vols. Henri Vetch, Peking (1933-1938).
8. A. W. Grabau. *The Rhythm of the Ages, Earth History in the Light of the Pulsation and Polar Control Theories*. Henri Vetch, Peking (1940).
9. J. Joly. *Radioactivity and The Surface-History of the Earth*. Halley Lecture, Clarendon Press, Oxford (1924).
10. Arthur, Holmes. *Age of the Earth*. Thomas Nelson & Sons, Ltd., London (1937).
11. G. E. O. Merk. Ulrich's Impact on American Stratigraphy. In: *Geologists and Ideas: A History of North American Geology, Centennial Special Volume 1*. E. T. Drake and W. M. Jordan (Eds). pp. 169-187. The Geological Society of America, Inc., Boulder, Colorado (1985).
12. C. R. Longwell. Sedimentary Facies in Geologic History. *Geological Society of America Memoir* **39**, 1-171 (1949).
13. F. S. Pettijohn. *Sedimentary Rocks*. Harper and Brothers, New York (1949).
14. W. C. Krumbein and L. L. Sloss. *Stratigraphy and Sedimentation*. W. H. Freeman and Co. San Francisco(1951).
15. T. C. Chamberlin. Diastrophism as the Ultimate Basis for Correlation. *Journal of Geology* **17**, 685-693 (1909).
16. E. Haarmann. *The Oscillation Theory, a Clarification of the Crustal Motions of the Earth and Moon*. Ferdinand Enke Verlag, Stuttgart (1930).
17. R. W. Van Bemmelen. The Undation Theory. *Natuurk. Tidschr. Nederl. Indie* **92:1**, 85-242;

94:2, 373-402 (1932).

18. J. H. F. Umbgrove. On Rhythms in the History of the Earth. *Geological Magazine* **LXXVI:897**, 116-129 (1939).
19. J. H. F. Umbgrove. *The Pulse of the Earth.* Martinus Nijhoff, The Hague (1942).
20. J. H. F. Umbgrove. *Symphony of the Earth.* Martinus Nijhoff, The Hague (1950).
21. James, D. Dana. On the Volcanoes of the Moon. *American Journal of Science* **2**, 335-355 (1846).
22. Elie de Beaumont. *Notice sur les symtèmes de montagnes.* 3 vols. P. Bertrand, Paris (1852).
23. E. Suess. *The Face of the Earth.* Translated by H. B. C. Sollas under direction of W. J. Sollas **4**, Clarendon Press, Oxford (1909).
24. O. Fisher. *Physics of the Earth's Crust.* Macmillan and Co., London (1881).
25. H. Jeffreys. *The Earth: Its Origin, History, and Physical Constitution.* Cambridge University Press (1924).
26. B. Gutenberg. Structure of the Earth's Crust and the Spreading of the Continents. *Bulletin of the Geological Society of America* **47**, 1587-1610 (1936).
27. F. B. Taylor. Bearing of the Tertiary Mountain Belts on the Origin of the Earth's Plan. *Geological Society of America, Bulletin* **21**, 179-226 (1910).
28. A. L. Du Toit. *Our Wandering Continents.* Oliver and Boyd Ltd., Edinburgh (1937).
29. A. Wegener. *The Origin of Continents and Oceans.* Methuen and Co. Ltd., London, 3rd ed., (1922).
30. A. Wegener. *The Origin of Continents and Oceans.* 4th edition, revised. Frederich Vieweg & Sohn, Braunschweig (1929).
31. R. Ruedemann. On Some Fundamentals of Pre-Cambrian Paleogeography, *Proceedings of the National Academy of Science* **5**, Washington (1919).
32. W. von. Koeppen and A. Wegener. *The Climate of the Geological Ages.* Verlag von Gebrueder Borntraeger, Berlin (1924).
33. L. U. de Sitter. *Structural Geology.* McGraw-Hill Book Co., N.Y. (1956).
34. J. K. E. Halm. An Astronomical Aspect of the Evolution of the Earth. Presidential Address. *Astronomical Society of South Africa* **4**, 1-28 (1935).
35. H. E. Le Grand. *Drifting Continents and Shifting Theories.* Cambridge University Press, Cambridge (1988).
36. D. S. McKay, E. K. Jr. Gibson, K. L. Thomas-Keprta, H. Vali, C. S. Romanek, S. J. Clemett, X. D. F. Chillier, C. R. Maechling and R. N. Zare. Search for Past Life on Mars; Possible Relic Biogenic Activity in Martian Meteorite ALH84001. *Science* **273**, 924-930 (1996).
37. J. M. Weller. Review of the Amadeus William Grabau Memorial Volume (*Geological Society of China Bulletin* **27**, 1947), *Journal of Geology* **58**, 598 (1950).

Proc. 30th Intern. Geol. Congr., Vol. 26, pp. 177-186
Wang *et al.* (Eds)

Robert Logan Jack, Geologist in China

D. F. BRANAGAN
Department of Geology and Geophysics, FO5, University of Sydney, 2006, Australia

Abstract

Robert Logan Jack (1845-1921), born in Scotland, worked on the Geological Survey of Scotland between 1867 and 1876. In 1876 he was appointed Geological Surveyor for North Queensland, Australia, becoming Government Geologist for all of Queensland in 1879. Until 1899 he carried out important work in many parts of Queensland, mainly concerned with economic mineral deposits, particularly coal, gold and underground water. In 1899 he resigned to work for an English mining company examining the metalliferous deposits of Sichuan, China. Jack arrived in China when there was considerable internal disturbance, but travelled by boat from Shanghai to Chongqing and thence to Chengdu which he made his headquarters. From Chengdu Jack, his son R. Lockhart Jack, J. F. Morris, interpreter Chung Chui Lin and party made several expeditions to examine mineral properties, including a very arduous one north to Wong-Shan Kan Mines, a round trip of 1200kms. They then travelled southwest to examine the Maha mines near Mien-ning, although travel was again difficult. When the Boxer Rebellion conditions made return to Chengdu unwise, Jack and party set out to travel to Burma, a journey of some 1500 kms through some very difficult and little known country. Jack's geological work in China, like his Australian work, was marked by careful observation and analysis and was in part incorporated.

Keywords: Jack, China geology, Queensland Geology, metalliferous mines, Burma

R. L. JACK'S EARLY YEARS

Robert Logan Jack (1845-1921) (Fig. 1) was born in Irvine, Ayrshire, Scotland on 16 September 1845, the son of Robert Jack, a cabinet-maker, and his wife Margaret (nee Logan). After study at the University of Edinburgh (but not graduating) he worked on the Geological Survey of Scotland between 1867 and 1876, where he came under the influence of Sir Archibald Geikie, B. N. Peach and J. Horne (with whom he published, in 1877 a significant paper on glaciology), and, according to Oldroyd [1], influenced Geikie in return, in relation to the Southern Uplands of Scotland. Both Peach and Horne spoke highly of Jack's work in mapping the Silurian sequence in southern Scotland. However much of Jack's work in Scotland involved mapping of coal deposits.

APPOINTMENT TO NORTH QUEENSLAND

In March 1876 Jack applied for the position of Geological Surveyor for North Queensland, Australia, proferring excellent references from Geikie, Peach, Horne and others in the Scottish establishment. He was appointed, but requested time and funds to study topographic mapping (in Scotland or England), because there were almost no topographic maps available in Queensland. Although the Colonial Government of Queensland refused an allowance to cover the cost of training and equipping himself, Jack was permitted to take time at his own cost to become efficient in such work. He finally took up his duty in Townsville, North Queensland on 10 July 1877, bringing with him his new wife Janet Simpson Love. Two years later he became Government Geologist for all of Queensland when the renowned explorer, A. C. Gregory, retired [2, 3, 4].

Figure 1. Robert Logan Jack in his early Queensland days.

With his new appointment Jack remained in Townsville, although the seat of Colonial government was in Brisbane, 1200km to the south. However he began to cover the vast extent of the Colony, including the far interior, traversing some of the least explored country in Australia. Until he resigned in 1899 he carried out important work in many parts of Queensland, mainly concerned with economic mineral deposits, particularly coal, gold and underground water, travelling mainly by horse. His search for gold-bearing country took him into Cape York Peninsula in the wet season of 1879-1880. It was a virtually unexplored country for Europeans at that time. The native Aboriginal people were quite hostile and his camp was attacked at night and Jack was speared through the neck. Despite the wound he continued his survey to its completion.

Jack first recognised a structure favourable for artesian water between Townsville and Cloncurry in 1881. This began his investigation and utilisation of what became known as the Great Artesian Basin, Australia's largest region of underground water.

CO-OPERATION WITH ROBERT ETHERIDGE, JUNIOR

From his arrival in Australia in 1877 Jack sent fossil specimens to his former colleague at the Geological Survey of Scotland, Robert Etheridge Jnr. (1846-1920), who was then working at the British Museum, and Etheridge published on this material. The two collaborated at a distance to publish a very valuable catalogue of works, papers, reports and maps on Australian geology in 1881[5]. Nothing comparable to this reference list appeared until T. G. Vallance's history of Australian geology paper of 1975 [6].

Etheridge, who had worked in the Geological Survey of Victoria between 1866-68, returned to Australia in 1887 to become Palaeontologist to the Geological Survey of NSW and Australian Museum, Sydney. He and Jack collaborated again to publish in1892 an important summary of the Geology and Queensland and New Guinea [7].

QUEENSLAND ASSISTANTS

W. H. Rands was appointed Jack's Assistant in 1883, and while Jack was based in Townsville Rands worked from Maryborough, central Queensland. A. Gibb Maitland joined Jack in Townsville in 1887. All three geologists moved to a central office in Brisbane, the colony's capital in 1892, bringing together an important collection of rocks and minerals, which Jack formed into a mining museum (in opposition to the Queensland Museum), visited by more than 9 000 people each year. S. B. J. Skertchley, who had earlier worked in China, was another geologist employed in 1895 , W. E. Cameron and B. Dunstan (1897) [8, 9].

During the 22 years of Jack's work in Queensland 144 reports and maps were published, including maps of the whole colony, an area of some 1 750 000 km^2, the first in 1886 and a final version in 1899.

In 1891 he was sent by the Government of Queensland to examine the newly-discovered huge orebody at Broken Hill in New South Wales. His report was the most complete geological assessment of the occurrence published by any geologist visiting the site prior to detailed work begun by the Geological Survey of New South Wales in 1892 [10].

During his work in Queensland Jack used every opportunity to publicise the mineral wealth of the colony using displays of maps, minerals, rocks, fossils and photographs. In the last six years of his time in Queensland Jack was increasingly involved in administrative matters, but he succeeded in getting a surveyor and a draftsman appointed which helped in the preparation of geological maps and reports for publication.

THE 1899 LONDON EXHIBITION

Because of Jack's success at preparing material for display at the Queensland International Exhibition he was appointed a Commissioner for Queensland at the Mining, Metallurgical and Machinery Exhibition, held in conjunction with the Greater Britain Exhibition at Earl's Court in London in May 1899 and he left Queensland with the minerals and other displays early in 1899. He was honoured that same year by the award of an honorary LL. D degree by the University of Glasgow.

When in England in 1899 he was offered the opportunity to work for W. Pritchard Morgan M. P. , Director of The English Pioneer Company, a mining company which proposed examining the metalliferous deposits of Sichuan, China, perhaps at the invitation of the Chinese Government. At this time, the regulations for mining in China were undergoing considerable change. In 1898, by Imperial edict, a central bureau was established to control the mines and railways of the country, and to encourage enterprise. Ironically Chao-Shu-Chiam, one of the most anti-foreign member of Tsung-Li-Yam was placed in charge of the bureau. Furthermore new rules were set in place requiring strictly defined localities, that concessions would not be granted for all mines in a province or district, Chinese investment must be not less than 50%, and that control of all enterprises must remain in Chinese hands [11].

JACK IN CHINA: PREPARATIONS IN SHANGHAI

Jack spent a busy few weeks in Shanghai, preparing equipment, meeting representatives of mining companies and government officials, and learning of the intrigues between companies, specially the competing English, French and German groups interested particularly in profitable metal mines, but with an eye to the possibilities of discoveries of coal, oil and gas. He also had to prepare a system of codes for sending important telegraphic messages. Jack accepted this interesting challenge, resigning his Government post. He arrived in Shanghai early in 1900 with two younger men, his son, Robert Lockhart Jack and his friend J. Fossbrook Morris, both recent engineering graduates of the University of Sydney. At this time there was considerable internal disturbance within China with rival Chinese and foreign groups trying to gain political power. However Jack was able to carry out some geological work, although travel was difficult [12, 13].

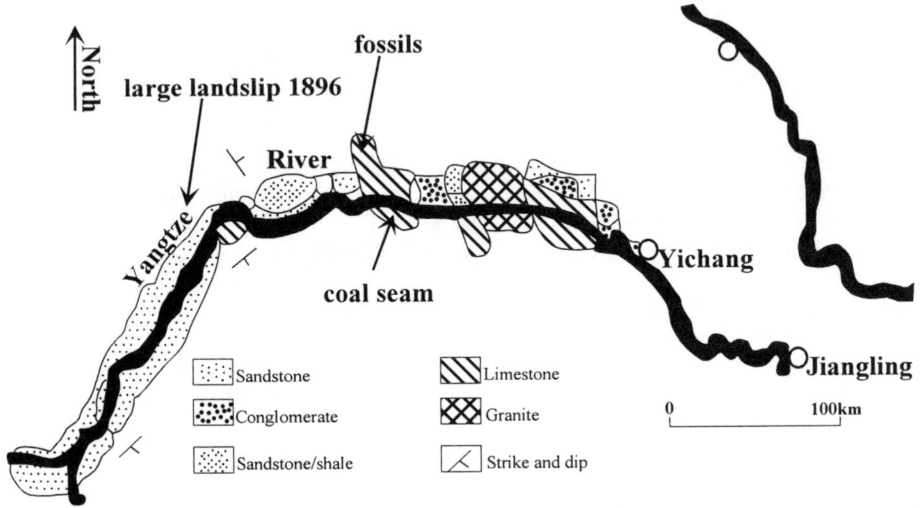

Figure 2. Geology of the Yangtze River recorded by R. L. Jack, 1900, summarised by D. F. Branagan from Jack's notes and drawn on Jack's base map.

Although Jack spent less than a year in China, it was an eventful period, which can be divided into six episodes: i) preparations in Shanghai; ii) journey up the Yangtze River to Chengdu; iii) negotiations in Chengdu and a visit to the relatively close Tung-Ling Tse mine; iv) a round trip of 1200kms north to Wong-Shan Kan Mines and return via Sung-Pan-Ting; v) Chengdu to the Maha mines in southwest Sichuan; vi) Maha to Burma (some 1500kms).

JOURNEY UP THE YANGTZE RIVER TO CHENGDU

Assisted by a young interpreter, Chung Chui Lin, Jack and his party set out in February 1900 by the only reasonable route to western China, the Yangtze River. The journey of 1750kms took them nearly ten weeks. The quick journey to Wuhan by steamer was followed by an extraordinarily difficult journey through the three gorges to Chongqing, the low water level and strong rapids being matched by the sleets of winter and the growing problems of social disturbance.

Nevertheless Jack managed to make significant geological observations along the river, taking every opportunity to record information on alluvial gold workings (first seen at the mouth of the Tai-Ping river 170 km below Yichang, noting that gravels were skimmed each year, no attempt being made to reach the bottom surface), dips of strata, the presence of faults, the ancient granite outcrops in the first of the three gorges. He collect specimens as had been done thirty years earlier by Raphael Pumpelly [14], but of whose visit Jack was apparently unaware, attributing the charting of the river and the first "scientifically correct narrative" to the Jesuit Chevalier. Jack used Chevalier's map and "spent time converting it into a geological map" (Fig. 2), which unfortunately has not been located.

Figure 3. R. L. Jack and party with Chinese officials, Chengdu (Jack is in the middle of the second back row; J. Fossbrook Morris, R. Lockhart Jack, and interpreter Chung Chui Lin are at the back).

In essence Jack recognised the old granite core overlain by fossiliferous limestones which comprise much of the three gorges, and the sandstones and shales of the eastern edge of the great Sichuan Basin,with its thin coal seams. He also noted the site of a large landslide which had occurred in 1896, 250km above Yichang.

Admiral Fu Ting Shing of the Yangtze Lifeboat Service, which acted as escort for the journey to Chongqing, showed him many specimens of ore minerals (such as copper pyrites, galena, fluorspar and black slate with pryites crystals) from sites mostly less than 150km from Yichang, giving Jack an inkling of the possibilities for mineral exploitation. Despite poor roads the final part of this preliminary journey from Chongqing to Chengdu was achieved with little difficulty across the broad Sichuan Basin.

NEGOTIATIONS IN CHENGDU AND A NEARBY VISIT

Jack and his party were well received in Chengdu by Government officials (Fig. 3) and he was provided with support and protection for his various journeys. In April he took his first expedition into the older rocks on the edge of the Sichuan Basin, about 90kms north-north-west of Chengdu to look for indications of gold and copper, and, in particular, to visit the Tung-Ling-Tse copper mines. Although the short expedition was a good "settling down" experience for the group Jack was disappointed in the mineral occurrences and did not feel they warranted further attention by his company (Fig. 4).

EXPERIENCE IN THE HIGH COUNTRY

The next expedition was more challenging, and, in fact, proved almost fatal. It involved a 500km trek north from Chengdu to examine the Wong-Shan Kan Mines. The party left Chengdu on 25 April observing the regional geology en route, sampling alluvial gold sites and noting possibly valuable unworked alluvial occurrences. There were other aspects of interest, such as the large round granite boulders in the Nea Waho River, (a stream that had been diverted into a canal now flowing at a lower level), very thin coal seams which were mined for local use, although larger seams were also worked. Jack was careful to note the various high level terraces in some valleys, one with a clay surface which was extensively excavated for tile-making. The terraces were backed by red foothills with snow covered mountains to the west. The expedition was climbing some 600m above Chengdu when Jack first observed soft grey sandstones associated with interbedded limestones and conglomerates. He observed a high range to the north with successive scarps indicating a set of thick beds dipping easterly.

North of An Hsien shafts were being sunk 30m on an old 50m high terrace on a ridge of limestone and conglomerate, the payable bottom material being 15m above the present river. Workmen were carrying casks of gravel nearly .5km from the terrace to the river for washing. Although attracted by quartz boulders the party was disappointed that only pyrites was obtained from crushing and washing this material. 178km from Chengdu, limestone (containing encrinites) with subordinate carbonaceous shalewas replaced by steeply-dipping greywacke and slate. They dollied quartz veins taken from the slates in search of gold, but also without success.

When the older rocks were reached there was greater concentration on mineral potential, R. Lockhart Jack and Morris taking a side trip up the west branch of the Mao Chow River through limestone into steeply-dipping clay slates with numerous quartz veins in the bedding planes, and undertaking assaying in a field laboratory they carried. The party was climbing steadily, being now more than 1700m above sea-level. At Ma Wang Go Jack examined an adit where a 15cm thick poor coal seam dipping 700 had been opened up.

At Kiangyou Hsien Jack observed flat-lying sandstone and shale beds underlying dipping limestone, cut by the Fou river. The limestone proved to be quite fossiliferous, containing bivalves, gastropods and corals. Near the remarkable wrought iron bridge at Pay Chow Hai river gravels occurred 130m above the river. Near Lungan Jack saw for the first time in China dykes (silica rich) intersecting the slates.

They reached the Wong-Shan Kan Mines on 16 May and spent some days mapping, sampling and assaying the pyritous talc schists, but again with disappointing results, obtaining only two colours. Jack then proposed to return to Chengdu *via* Sung-Pan-Ting, as there was a relatively easy route from there down the valley of the Minjiang River. The only problem was the need to traverse a 4000m high pass, still well-covered by late winter snows.

This proved an extraordinarily difficult task as the road, such as it was, was well-nigh impassible and few of the party had any experience of high altitude travel. In the event almost all the party, except Morris, were affected, although Morris's pony collapsed and died. One Chinese soldier accompanying them died several days after crossing the pass. Unbeknown to Jack this high country experience was to prove of great value to the party several months later.

Returning towards Chengdu Jack was particularly struck by the effectiveness of the ancient engineering works constructed about 250 B. C. at Dujiangyan by Li Ping and used to divert much of the flow of the Minjiang River to irrigate the Sichuan Plain [15]. A modified version of this ancient scheme continues to operate very effectively today.

Jack reported his disappointment at the mineralisation to Commissioner Li in Chengdu, saying the Sichuan mines were poor and that the company was wasting time unless it was shown better prospects. If there were better mines he also had to be certain that they were not held in reserve for the government, and would be available for the company. At this time Jack was informed that all mines of great value were in the vicinity of Yachan [Yuchan] Ningyan, and realised that he must move quickly to examine these, or risk the possibility of missing out on obtaining a concession.

He was given to understand that the government in Peking wished to encourage copper mining, even close to present government mines, and that gold mining would also be permitted, but not in close proximity to government workings. Jack also tried to learn something of the various business intrigues to obtain concessions for mining petroleum and other assets, and considered a consolidation between Chinese and British interests against French companies, although he was advised, soon after, to co-operate with one French group in particular local dealings.

CHENGDU TO THE MAHA MINES IN SOUTHWEST SICHUAN

Preparations began in earnest in early June for the 560kms journey into southwest Sichuan to examine the Maha Mine. Tension had built up in the region with confusing reports coming by telegraph and letter about conflicts in various parts of China. Jack was advised by company representatives in Shanghai, and by local officials, to abandon his journey, but he reasoned that the distant parts of Sichuan were as safe as anywhere else and decided to go ahead. Nevertheless he was glad of official military support and letters of safe custody, and eventually a party numbering nearly 200 set off from Chengdu on 16 June. By this time telegraphic contact with Shanghai had been cut off. Jack was later upset that the Chinese members of his party were aware of the extent of the troubles much earlier but did not inform him, his first real information being a delayed telegram he received on 26 June.

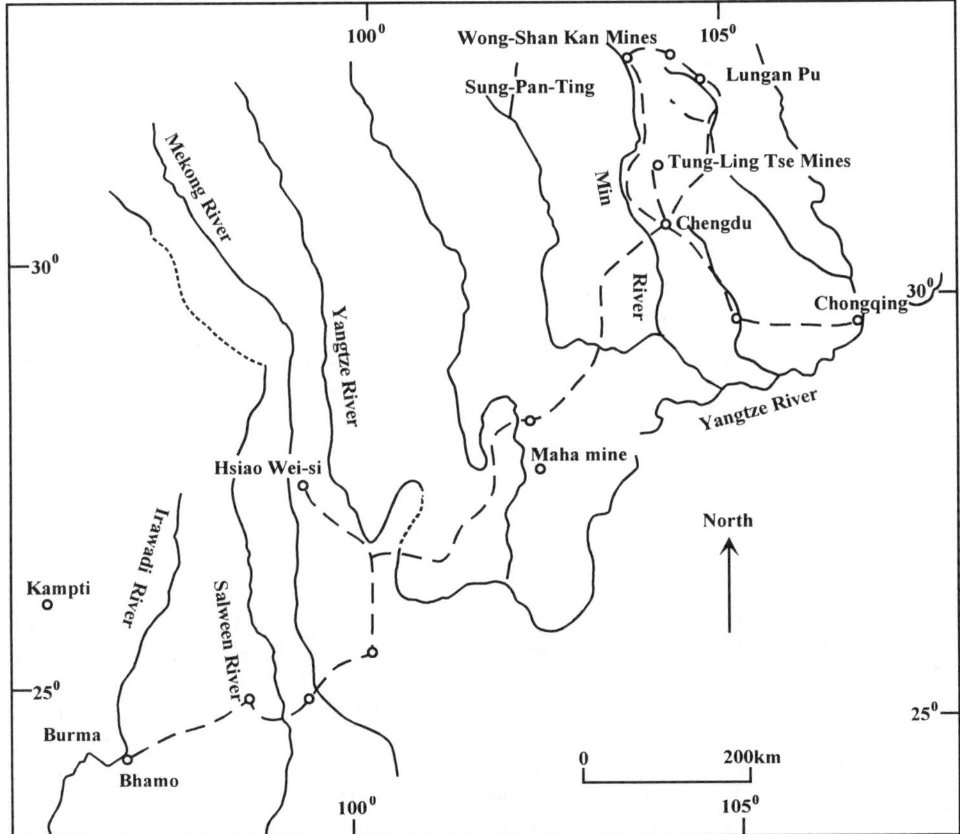

Figure 4. Routes of R. L. Jack's party in Sichuan, 1900.

The journey took almost a month over a series of mountain passes to Maha, reaching there on 13 July. Jack recorded the changes in geology, limestone in abundance underlain by quartzites, argentiferous pyrites and copper pyrites associated with the limestone, thin coals, yellow and grey sandstones and shales, but with no easily discernible regional structure, but he made some educated guesses about the ages of the various units. En route he also studied the many constructions using cyclopean masonry.

At Mien Shaw, after traversing a 3300m pass, Jack examined the iron works at a high-grade haematite site, the operation of two foundries being powered by water. The ore was brought 15km. Granite here was cut by numerous basic dykes. Jack was intrigued by the ancient techniques still used in iron smelting and founding. At Lu-ku he also visited three French priests, the last Europeans the party was to see for some months. At Sha-ba Mines he reduced the luggage to cross the 4000m pass to Maha, noting the transition from gneiss to granite to the top of the pass, then schists and slates. On 13 July they camped at the stamp mills- of which there were 80 in operation.

MAHA MINES - SUCCESS

Jack and his party set to work to assess the potential of the area, a large operation employing 15 000 workers. Jack mapped the underground workings which followed the reefs of iron and copper pyrites and galena, which were seldom less than 2m wide, while Robert and Morris sampled the rocks and soils of the district and prepared a trial crushing for analysis. Other nearby prospects were noted for visiting and sampling. Jack was astonished to find the mining machinery in use at Maha had been carried manually over the passes, some pieces weighing up to 500lbs (230kg) having apparently been carried by single labourers.

Jack was impressed with the mineralisation at Maha, where hard-rock mining had only been active for about twenty years. He immediately entered into negotiations with the local manager, Californian-trained Tong Sing Kow, to take over the mine and surrounding prospective area on behalf of his company, also sending off runners in the hope of getting messages to his principals in Shanghai and London.

Finally, left to make his own decisions, on 26 July he signed an agreement with Tong, written in English, to acquire the mine and adjacent properties. He must have been delighted when, on 29 July, samples from the nearby Ko Co Co copper mine yielded "incredible assay" results. The first crushing from the large Maha sample also produced, on the same day, a bar of gold. On the other hand alluvial pannings were disappointing.

GO TO BURMA!

Although he had not any reply to his earlier messages, on 29 July he did get a telegram, sent on 19th and brought by runner some 150kms from the nearest operating telegraph station, advising him not to attempt to return to Chengdu, but to take his party out of China by the most expeditious means. Preparations took a few days. There was a need to prepare reports on the various mineral occurrences and it was not till 7 August that Jack signed a second contract written in Chinese. The party was to be split up, many Chinese deciding to return north, and luggage was again reduced and packed for what was to be a very arduous journey of some 1500km. The previous year Captain Wingate of the 14th Bengal Lancers had made a 5 and 1/2 months journey from Shanghai to Bhamo with Chinese servants travelling through Yunnan Province, but Jack was probably not aware of this [16].

It turned out to be a journey in two parts. When Jack decided to leave China he certainly did not want to travel south via Kunming in Yunnan Province, where he had heard there was fighting and danger for foreigners. He therefore thought it would be best via Upper Burma (Kampti) or even Tibet. With this in mind the party set off from Maha on 10 August intending to go to Kampti, west of the Irrawadi River. The beginning did not augur well, as it took half a day to cross the river and less than 4kms was covered that day. The next few weeks involved travel over many mountain passes over 2700m in elevation, some high dry valleys in limestone country, some gold sites, brine wells, meeting bandits and people afflicted by goitre (evidence of iodine deficiency, although Jack attributed it to salt or lime). At the time there was still some uncertainty about the exact course of the Upper Yangtze River and Jack sketched what turned out to be a relatively accurate path taken by the river in its upper course.

A journey of 640kms took them to Hsaio Wei Si on the Mekong River. Here they were advised not to continue north into Tibet, because of the weather conditions and the virtual non-existence of a road. There was nothing for it but to turn around and retrace their steps south some 120 kms before branching off to Burma. From Wei Si to Bhamo (Sin Kai) was 930kms, climbing first from the Mekong valley to the Salwen and then over another range to the drainage of the Irrawadi River, which they accomplished between 10 August and14 September. Even on this part of the journey Jack was recording geological information, noting basalts of two distinct ages, lignites, a"tin mine" which proved to be a narrow antimony-lead vein and recording reports of good gold mines "two stages" from their track. The journey from Maha, a distance of 873 miles by Jack's calculations, had taken 73 days.

Despite his experiences Jack returned briefly to Shanghai from Burma, presumably to verify the validity of his agreements with the owners of the Maha mines and the Government, but there are few details readily available on this matter.

JACK'S LATER YEARS

Jack returned to England in 1901 and worked as a private mining geologist consultant until 1904, when he moved to Western Australia where he undertook mining consultant work during the next five years. The Government of Western Australia hired him to act as a Commissioner to report on the potential development of the Collie Coalfield and later to report on ventilation, sanitation and health problems in the state's underground mines. In particular he made an important study of the artesian water potential of the Kimberley region of northwest Western Australia.

About 1909 he moved to Sydney, where he undertook some consulting in NSW and Queensland, but worked mainly on his historical research, publishing a very important book on the history of exploration of northern Australia. He died on 6 November 1921 and is buried in Sydney.

CONCLUSION

R. Logan Jack was a very important pioneer in the story of geology in Australia, but also made a significant contribution in Scotland. His brief visit to China and publication on his travels helped to make the inland of China better known to Europeans. Jack's work was incorporated by J. W. and C. J. Gregory [17] in their important review.

Although undoubtedly greatly assisted by his interpreter, Chung Chui Lin, Jack's leadership, coolness and ability under duress ensured a successful prospecting expedition in a region of China where little geological investigation had previously been undertaken.

REFERENCES

1. D. R. Oldroyd. *The Highlands Controversy*. University of Chicago Press (1990).
2. D. Hill. Robert Logan Jack: A Memorial Address, *Proceedings of the Royal Society of Queensland* **58:7**, 113-124 (1947).
3. D. Hill. Jack, Robert Logan (1845-1921). In: *Australian Dictionary of Biography*, **4**. Douglas Pike (Ed). pp. 466. Melbourne University Press (1972).
4. R. D. Oldham. Robert Logan Jack, *Proceedings of the Geological Society of London* **78**, xlix (1922).
5. R. Etheridge and R. L. Jack. *Catalogue of Works, Papers, Reports and Maps on the Geology....of the Australian continent and Tasmania*. Stanford, London (1881).
6. T. G. Vallance. Presidential address: Origins of Australian geology, *Proceedings of the Linnean Society of New South Wales* **100:1**, 13-43 (1975).
7. A copy of *Geology of Queensland and New Guinea* by R. L. Jack & R. Etheridge Jnr was presented to the Geology Department of James Cook University, Townsville, North Queensland by Mr. R. J. Allen, Chief Government Geologist to mark the occasion of the award of the Inaugural Robert Logan Jack Scholarship in Geology, April 6, 1984.
8. H. G. S. Cribb. History of the Geological Survey of Queensland. In: *History and role of the Government Geological Surveys in Australia*. R. K. Johns (Ed). pp. 46-55. Government Printer, Adelaide (1976).
9. E. N. Marks. Skertchly, Sydney Barber Josiah (1850-1926). In: *Australian Dictionary of Biography*, **11**. G. Serle (Ed). pp. 621-2. Melbourne University Press (1988).
10. D. F. Branagan. The first ten years of Broken Hill Geology. In: *Mineral Search in the South-West Pacific Region*. 8th Edgeworth David Day symposium. D. F. Branagan and R. Facer (Eds). pp. 65-82. Edgeworth David Society, Department of Geology and Geophysics, University of Sydney (1995).
11. (London) *Times*, 23 July, 7 & 15 August, 4 & 1899.
12. R. Logan Jack. *Notebooks*. Oxley Library, Brisbane (1921).
13. R. Logan Jack. The Back Blocks of China, *A narrative of experiences among the Chinese, Sifans, Lolos, Tibetans, Shans, and Kachins, between Shanghai and the Irrawadi*. pp. xxii, +269. Edward Arnold, London (1904).
14. D. R. Oldroyd. On Being the First Western Geologist in China: The Work of Raphael Pumpelly (1837-1923), *Annals of Science* **53**, 107-136 (1996).
15. Zhenghai Song. Water Conservancy Projects and Knowledge of Hydrology. In: *Ancient China's Technology and Science* (Compiled by Institute of the History of Natural Sciences, Chinese Academy of Sciences). pp. 239-249. Foreign Language Press, Beijing (1983).
16. (London) *Times*, 27April, 5 & 29 April, 7 1899.
17. J. W. Gregory and C. J. Gregory. Geology and Physical Geography of Chinese Tibet, *Philosophical Transactions of the Royal Society of London*, Series B, **213**, 171-298 (1924).

Proc. 30th Intern. Geol. Congr., Vol. 26, pp. 187-196
Wang *et al.* (Eds)

The Changing Mission of the Geological Survey of Japan in a Changing World

HIROKAZU HASE AND KISABURO KODAMA
Geological Survey of Japan, 1-1-3 Higashi, Tsukuba, Ibaraki 305, Japan

Abstract

The Geological Survey of Japan (GSJ), since its birth in 1882, has striven to support the national growth in industry. At present its role has broadened to multidisciplinary geoscientific fields. Recently, we have been assessing GSJ's activities for the future, which includes the selection of important research subjects. There are a number of subjects which are difficult to achieve solely and the subject on the contribution from earth sciences to sustainable development in the east Asian region is widely supported by researchers. Cooperation with other geological organizations around the world is essential.

The result of our review was evaluated by specialists from outside GSJ representing various parts of the geological community. We have to face problems and weaknesses common to many of world's geological survey organizations. and some of the items pointed out by the evaluation committee seem to be applicable to geological surveys worldwide. However, we conclude that our near future is rather favorable, taking into account of particular domestic conditions. We should be able to contribute significantly to global geological problems.

Keywords: Geological Survey of Japan (GSJ), activities, evaluation, subjects, research cooperation

INTRODUCTION

The Geological Survey of Japan (GSJ) experienced a relatively calm period in the 1980s as a whole, supported by the economic boost of the country. But at the beginning of the 1990s, the overheated Japanese economy, nicknamed the "bubble economy", deflated abruptly and the country fell into a stagnant economic condition which is still continuing. Above all, since the late 1980s the world is changing rapidly and the uncertain future is throwing us into a confusing situation.

In particular, being informed that many of the world's geological surveys are facing difficulties, we are obliged to face the fact that GSJ will experience the same situation sooner or later. This is urging us on to examine how and what to do with our future.

A Symposium entitled "The role of geological survey organizations in market economics" was held on the occasion of the 30th International Geological Congress (IGC) Meeting at Beijing, China, and many scientists representing world geological survey organizations gathered together to introduce their situations and to examine future trends. We were given the opportunity to introduce what we had done in our recent review and the evaluation of GSJ from outside GSJ.

BRIEF OUTLINE OF GEOLOGICAL SURVEY OF JAPAN

GSJ was founded in 1882, fifteen years after the Meiji Reform in our modern history. The concept of the establishment was incorporated from European countries and the United States at that time, so that the principal function of the organization was similar to western geological surveys as is described for the British Geological Survey [1].

Figure 1. Geological Survey building at Tsukuba. They include main building, geological museum, laboratories, and so on. The white building at the top left of the view is the newest building completed in 1995, and is called the Center for Geosphere Information.

GSJ has changed its location several times and moved to the present place, Tsukuba, about 50 km north northeast of Tokyo, in 1979. Tsukuba is a newly constructed "science city" where one third of the government institutes are based. GSJ occupies a part of a complex of research institutes (Fig. 1). We have two other local offices, in Osaka, southwestern Japan and Sapporo, Hokkaido.

At present, GSJ is under the control of the Agency of Industrial Science and Technology (AIST) of the Ministry of International Trade and Industry (MITI). Our total budget comes from the government with most of it through AIST.

This situation is rather exceptional since most worldwide geological surveys, even in socialist countries, are shifting towards the market economy.
The budget situation of GSJ for fiscal year (FY) 1996 is given in Figure 2. The Japanese fiscal year starts from April and ends at the end of the following March.

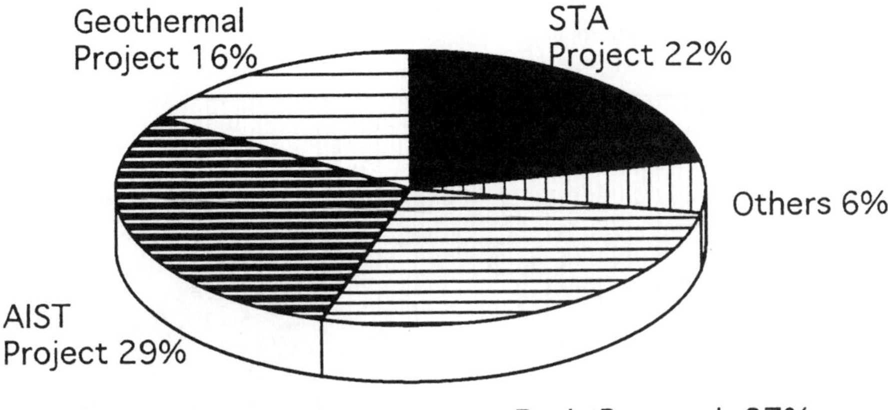

Figure 2. Total budget and sources for GSJ in FY 1996 (data; Research Planning Office, June 1996). All of the funds come from the government, mostly through AIST. Today, the source is diversifying and the budget from the Science and Technology Agency (STA) is increasing. The figure does not include a part of budget of demand to STA.

PRESENT MISSION

Along with our history, GSJ has shifted some precious functions to other semi-governmental organizations which have a market economy function. This shift might resemble the privatization movement as is discussed today. The following organizations have been founded as semi-government corporations with related functions to GSJ:

- Metal Mining Agency of Japan (MMAJ), established in 1963
- Japan National Oil Corporation (JNOC), established in 1972
- Geothermal Survey Department, New Energy and Industrial Technology Development Organization (NEDO), established in 1980

The cooperative style with these organizations is at present based on joint cooperative project implementation sending two researchers from (NEDO), project support sending a researcher to specific geophysical fields (JNOC), and numerous and frequent technical advice through committee activities (all).

In addition to these shifts, GSJ has been extending its functions into multidisciplinary geoscientific fields accompanied by continuous downsizing of the staff number. The scale of GSJ in terms of staff number was roughly constant until World War II. Then it expanded after the War, absorbing many scientists and technicians who had been in similar specialties. Downsizing followed after that and in 1968, the government adopted an overall policy for shrinking government which is running as two to five year programs that will continue to at least the year 2000 (Fig. 3).

The present staff includes ten female researchers and two foreign researchers. Recently the number of visiting scientists reached five to ten percent of the total number of researchers. Research areas range from traditional geological mapping to marine geology, environmental geology, resource geology, information geology and so on. There is increasing linkage of these fields through international cooperation. GSJ has made great efforts in international cooperation compared with other government research organizations.

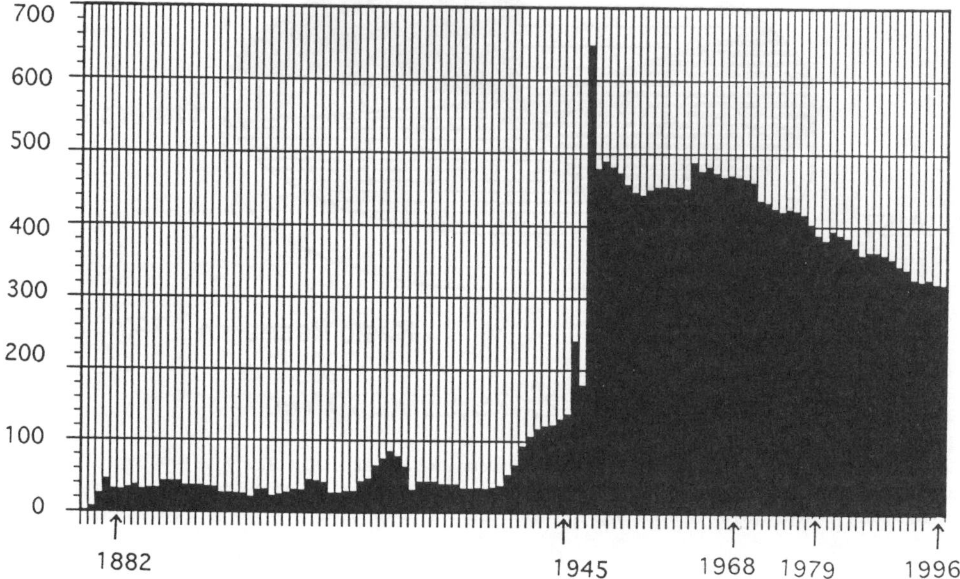

Figure 3. Change of the total number of staff of GSJ since its beginning.

Japan is increasing its dependency on other countries for natural resources. For this and other reasons we realize that we must contribute significantly to research efforts to mitigate geological problems, both in other countries and on a global scale. In January 1995, a big earthquake severely hit the Kobe and Awaji island area, southwestern Japan, and the active faults system in Awaji island was reactivated. It claimed more than 5000 human lives and was a strong reminder of the need for the publics awareness of active faults.

The Japanese Government has long been budgeting for researches into earthquake prediction. GSJ is one of the national organizations which shares the task mostly of "earthquake geology", or more specifically long term risk assessment of active faults systems and research on earthquake prediction by observation wells.

The year 1995 became a special year for us because we completed a detailed active fault survey mostly within and in the vicinity of the above mentioned area by using a part of an extra government budget. GSJ was allocated about 8 billion yen for the survey and more than sixty scientists were involved. The extra budget is roughly equivalent to three years of our ordinal budget scale.

REVIEW OF ACTIVITY

The downsizing of GSJ has been done in the manner of cutting the least number of researchers, which has meant that most of the cuts were made in research supporting personnel such as operators, technicians and administrators. This has accelerated the tendency of GSJ to become more of a research oriented organization. But as we have experienced in the active fault survey, the role to respond to government and public needs should be kept within GSJ.

Researchers are doing multidisciplinary works and it is rather difficult to define an individual's specific field. We categorize the specialty of our staff on the basis of their present work as shown in Table 1. Roughly speaking, half of our staff are geologists and a quarter are geophysicists, and the rest are geochemists.

Geology	(122)	
	Stratigraphy & Structure Geology	23
	Petrology & Mineralogy	17
	Resource Geology	36
	Marine Geology	21
	Quaternary Geology & Earthquake Geology	17
	Volcanology & Volcanic Rocks	8
Geophysics	(58)	
	Gravity & Magnetics	8
	Rock Physics	6
	Seismology & Seismic Exploration	20
	Electricity & Electromagnetics	6
	Geothermics	7
	Remote Sensing	7
	Information & Computer Science	4
Geochemistry	(47)	
	Analytical Chemistry	6
	Organic Geochemistry	7
	Isotope Geochemistry	10
	Geochemistry of Hydrothermal Systems	16
	Marine & Environmental Geology	8
Total		227

Table 1. Specialties of researchers in GSJ on the basis of their present research work. Total data is as of April 1, 1995 (after Matsuhisa).

Our former Director-General, Takeo Sato, realized that many of the world's geological surveys were changing their missions and roles. He called for all our staff to review our activities to establish clearly our future directions and scope of research. The work started at the beginning of April 1994. Two groups representing resource geology and environmental geology were formed to review the present situation and future direction, respectively. The activity of resource geology has long been the major part of our mission; however, in the case of mineral resources, we have to face the fact that almost all of our domestic mines are closing, mainly due to economic reasons. On the other hand, research opportunities in environmental geology are expanding these days, although we have difficulty in obtaining budgets from AIST, whose main policy is the promotion of industrial science and technology.

The resource geology group reviewed the situation from this view-point as, 1) a member of the national research institutes under MITI, 2) a member of the academic community, and 3) as a member of society. The environmental geology group placed emphasis on the selection of important research subjects for the future.

The two groups summarized their results in July, 1994. It was integrated into the following restructuring discussions of GSJ involving board members. Timely or not, the before mentioned big earthquake took place in the midst of the review, reminding us of another function which geological survey organizations ought to have.

One of the major results obtained through the review is the selection of important research subjects. The following subjects have been chosen as such important research subjects:

· Unconventional energy source (sources of natural gas-hydrates)
A cooperative project involving other national research institutes and private companies is starting. GSJ shares the role of making clear physico-chemical conditions of the formation and dissolution of gas-hydrate and resource assessment. The acceleration of research is proposed as another new project.

· Assessment of deep geothermal resources
This is the largest research project running at GSJ and it is implemented cooperatively with NEDO. The main research objective is shifting from resource assessment to fracture-controlled reservoir modelling, exploration, and evaluation.

· Global environmental research
Several research projects such as research on global geochemical cycles in the ocean, coral reefs as sinks of atmospheric carbon dioxide and so on, are being conducted. The subject should be covered systematically and cooperative research is inevitable for an effective outcome.

· Rational land-use of urban are and the assessment of human induced hazards
The main research objective at the moment is the detailed understanding of geological processes in Quaternary sedimentary basins where much of our population is concentrated. The importance of a geo-technological approach is recognized and new researchers are to be recruited.

· Earth science aspects of nuclear energy, particularly waste storage
The subject is increasing in importance as the first nuclear power plants approach the end of their lives and are decommissioned.

· Long- and short-term earthquake prediction
Within GSJ's role share, progress of research is expected in accordance with the recent active fault survey as well as the increase in number and improvement of the observation wells system.

· Prediction of volcanic eruptions
To date, scientists in GSJ have played an important role in the monitoring of volcanic activity such as the Oshima and Unzen volcanoes.

· Contributions from earth sciences to sustainable development in the East Asia region
Continuous effort is being made on a bilateral and multilateral cooperation basis, which is widely supported by the staff.

· Geological and geoscientific maps and other publications
We have covered about 70% of our land as 1:50 000 scale geological maps.
Geological maps and other publications are mailed to world-wide geological surveys and other official organizations in over 160 countries. This is the fundamental continuing activity of GSJ.

· Digitalization of geological information

Foreseeing the "network society", digital geological information is thought crucial and the effort to digitize massive earth data is to be continued.

The development of new concepts, methods and tools in frontier fields of earth sciences is strongly expected of GSJ. Such research work as the study of crustal fluids, magma systems and the processes related to subduction, geophysical technology as three-dimensional modeling and advanced application of remote sensing and so on, may be a part of them.

Discussions and analyses made by GSJ staff members have been invaluable and we have reconfirmed the strong devotion of the staff to our roles. However, we recognize this type of self analysis is apt to fall into a positive analysis of our own. Fortunately, we had the opportunity to have the results of our review evaluated by specialists from outside GSJ. An evaluation committee whose members cover the resources industry, geo-technology industry and the academic community, was organized. Tadashi Sato, professor emeritus of Tsukuba University, took the chair of the committee (Table 2).

Shigeo Aramaki	Professor, Nihon University
Masao Hayashi	Director & General Manager, Idemitsu Oil & Gas Co. , Ltd
Eiji Isawa	Professor, Kyusyu University
Hakuyu	Professor, Kyusyu University
	President, Geological Society of Japan
Yoshimitsu Okada	Director-General, National Research Institute for Earth Sciences and Disaster Prevention, STA
Satoru Oya	President, Oyo Co. , Ltd
Tadashi Sato	Professor Emeritus, Tsukuba University
	Chairman of the Committee
Tomohiko Taira	Professor, Marine Research Institute, Tokyo University
Toshiki Takagi	Senior Managing Director, Sumitomo Metal Mining Co. , Ltd
Hiroshi Wakita	Professor, Tokyo University

Table 2. Evaluation Committee Members for GSJ's review.

BRIEF SUMMARY OF EVALUATION

The summary of the evaluation committee was made in the manner of recording oral and written comments of the members [2].

Comments included strengths and weaknesses, research activity and information service, propriety of important subjects selected by GSJ, quality of products, flexibility of organization and so on. The following is a brief summary:

1) Activity and Product
The work and products of GSJ are useful to specialists but need to be more "user friendly" particularly to non-specialists. GSJ needs to make an effort to give information to industry and policy makers, as well as to the general public.

2) Function
It is hoped that GSJ will be the exclusive solid earth science information center in our country.

3) Research Subject
The selected subjects are reasonable. The importance of other subjects such as water resources, tectonic setting, global resources information and so on are also pointed out.

4) Quality
Scientific excellence is important and GSJ should maintain impartiality in assessing our land. The establishment of a reliable assessment method for natural hazards is awaited.

5)Flexibility
Efforts in the interchange of personnel with industry, universities, and other government organizations including those from overseas in highly recommended.

6) Others
It is effective to have the chance periodically to hear the views form outside GSJ, including those of foreign specialists.

FUTURE SCOPE

We foresee that in our near future we will have to take into account domestic factors.

Generally speaking, the geological community surrounding GSJ is not altogether favorable. It will be represented by the statement of Professor Akiyama on the occasion of his inauguration as the new president of the Geological Society of Japan [3]. The geology departments at many universities are changing their curricula and renaming their courses to Earth and Planetary Sciences and so on, and a geological education based on field observations is shrinking. This tendency will be a significant influence for GSJ in the long run.

But if we anticipate our near future, it can be said to be more favorable because of the following two particular reasons.

1) Increased awareness of the environment and natural hazards by the general public
The before mentioned big earthquake in 1995, which served as a reminder of the public fear of earthquakes and renewed concerns of the ground (geology) they live on. This is reflected in the increase of funding for active fault surveys by both central and local government bodies. The importance of this is woven into the future science and technology policy as described below.

2) Establishment of new science and technology promotion policy
The collapse of the "bubble economy" and the rapid loss of industrial competitiveness is causing a nationwide sense of crisis. As a result, the voices urging strong remedial measures by strengthening the present situation of science and technology became louder. Under such a background, the government established the "Science and Technology Promotion Statute" in November, 1995. This is a primary and constitutional statute and the action plan was set as the "Science and Technology Fundamental Plan" in July, 1996.

This will be the basis of our activity for at least the next several years. According to it, the strong investment of government funding is intended for.

· creative and innovative research fields which promote and enable the expansion of economic frontiers.

· research fields contributing to sustainable development by overcoming such global problems as environment, foods, energy and so on.

· research contributing to safe and stable society by promoting such works as overcoming of diseases, prevention or mitigation of disasters and so on.

The promotion of basic science is denoted as important to support the development of the research areas mentioned above.

The Plan seems to provide a favorable climate for the promotion of earth sciences. GSJ belongs under the umbrella of AIST of MITI, and the chance to receive their benefits will depend largely to what extent our proposals are appropriate in the light of the policies of our government organizations.

CONCLUSION AND ACKNOWLEDGEMENT

GSJ has shifted some of its original functions to other semi-government corporations in the past and is at present strengthening its character as a research organization. We have been reviewing our activities and the results were evaluated by specialists from outside GSJ. Many of the problems and situations for GSJ and the comments pointed out by the evaluation committee members are very similar to those detailed by Cook [1].

Reviewing works included the selection of important subjects for GSJ. We find there are many regional to global scale subjects in our selection of important subjects whose understanding and solutions can not be achieved solely by GSJ. We realize the importance of international cooperative work.

The social conditions of our country at the moment are not necessarily favorable for GSJ. But if we take into account recent factors and moves, we can conclude that our scope for the near future is rather positive. We would like to be active in our contributions to regional and global scale problems.

The International Forum of Geological Surveys (ICOGS) meeting provided us with an excellent opportunity to understand the situations many world geological organizations have faced. We also noticed cooperative work was the common key word for most of us. This means that the common subject of our concern such as the mitigation of geological problems both in other countries and on a global scale will be achieved more successfully by role shares between the many geological survey organizations. If such a subject is the one which gains consensus in our Survey and if it conforms with the content of our "Science and Technology Fundamental Plan", its implementation will be much easier.

We express our cordial thanks to Messrs. Chris Findlay and Henk Schalke who have made a great effort to focus our attention on our common subject. Thanks are extended to Mr. Xionglin Wang who provided an excellent opportunity for further exchange of opinions among us.

The discussions and analyses of GSJ for more than two years have been guided by our Research Planning Office under Dr. Tetsuro Noda, the Director at that time. We express our hearty thanks for his tireless endeavor.

REFERENCE

1. Peter J. Cook. The role of the geological surveys in the 21st century, *Episodes* **17:4**, 106-110 (1994).
2. Anonymous. *Report of the Evaluation Committee for the Geological Survey of Japan.* Geological Survey of Japan (1996) (in Japanese with data file).
3. Masahiko Akiyama. Acceptance of President, *Jour. Geol. Soc. Japan* **102:5**, i-ii (1996) (in Japanese).

Proc. 30th Intern. Geol. Cong., Vol. 26, pp. 197-204
Wang *et al.* (Eds)

Adam Sedgwick and Lakeland Geology (1822-24)

DAVID OLDROYD
School of Science and Technology Studies, The University of New South Wales, Sydney, New South Wales 2052, Australia

Abstract

Using Adam Sedgwick's original notebooks as the main source of information, an account is given of his work in the Lake District, England, during the years 1822-24. Attention is focused on the way in which he worked in an area that had received little attention previously from established geologists. But it is pointed out that he received considerable help from local amateurs, particularly Jonathan Otley. Sedgwick ranged over the country, mapping it according to rock types previously recognised by Otley, who also taught Sedgwick how to distinguish between bedding, joints, and cleavage. Sedgwick's hand-coloured geological map of Cumberland has been identified. Sedgwick did not give much attention to fossils, but he took careful note of dips and strikes, and structural features in general. He mapped a band of "Transition Limestone" (Coniston Limestone) across The Lakes, and showed how it had suffered considerable faulting, the faults being correlatable with some of the major valleys of the region. Sedgwick sought to subsume his observations under the general theory of Elie de Beaumont. Thus Sedgwick's "catastrophism" was grounded in a firm empirical base, and was not, by any means, mere "Biblical" geology. Consideration is given to the differences between the achievements of an amateur geologist such as Otley and a professional such as Sedgwick.

Keywords: Sedgwick, Otley, Elie de Beaumont, Lake District, fieldwork, catastrophism

One of the great founders of British geology was the Cambridge professor, Adam Sedgwick (1785-1873). Aspects of his work have been intensively studied by Rudwick [1] and Secord [2], regarding his role in the establishment of the Devonian and Cambrian Systems respectively. Earlier, Sedgwick was the subject of a two-volume Victorian "Life and Letters" by Clark and Hughes [3], and there is a short non-technical biography by Speakman [4]. But Sedgwick's studies in the Lake District, which provided the conceptual and methodological foundation for much of his subsequent work, have not been closely studied.

In fact, despite Sedgwick's importance in the history of geology, there has been little detailed examination of just how he worked as a young man. I have therefore examined how Sedgwick carried out his investigations with the help of a study of his field notebooks for his work in the Lake District for 1822, 1823, and 1824. From these notes, we can learn something about how early nineteenth-century geologists went about their work; and how one such as Sedgwick began to study the geology of a complex region where little systematic fieldwork had been undertaken previously.

Sedgwick came from Dent in the Yorkshire Dales, to the south-east of the Lake District. His father was the local parson, and the family seems to have been reasonably well-off. Adam was the third of seven children. He went to a nearby school of some reputation, Sedburgh, and on to Trinity College, Cambridge. There he studied mathematics and theology with great diligence, and in 1808 he was fifth in the University for mathematics (5th Wrangler). This led to a fellowship (by further examination) at Trinity in 1810, and then to the geology chair in 1818. The present geology department at Cambridge has its museum named in Sedgwick's honour, and his private notebooks and some manuscript maps are held there.

The story of Sedgwick's appointment has often been told. According to a statement that Sedgwick himself originated [3], there were two applicants, and although Sedgwick knew no geology he got the position since what the other candidate knew was wrong! This story cannot be entirely true. Sedgwick's fragmentary autobiography, preserved among his papers at Cambridge University Library, tells us that as a lad he became intrigued by the rocks and fossils that could be seen near Dent and he realised how the structure of the strata along the sides of the valley could be understood. Sedgwick suffered a breakdown in his health in 1813 and it is surely relevant to the present enquiry that he recuperated by taking a walking holiday in the Lake District. He was actively geologizing on the Continent in 1816, and was "introduced" at the Geological Society that year. He was elected Fellow of the Society in 1818 and FRS in 1821, with Sir John Herschel heading the list of twelve who nominated him, as I am informed with thanks by Mary Sampson, archivist at the Royal Society. No doubt the Cambridge chair smoothed the path to the Fellowship.

Immediately after being elected professor, Sedgwick began fieldwork in a serious way. He went to Derbyshire, Somerset, and Cornwall in 1818-19. He went to Wiltshire, Somerset, and Dorset in 1820, to the Yorkshire coast and Teesdale in County Durham in 1821; and in 1822 he worked his way north through Nottinghamshire and Lancashire, and into Cumberland, thus initiating his Lakeland work proper. In 1823, he went again to Yorkshire and to Cumberland; and in 1824 he spent the whole season in the Lake District. He returned in 1833, 1835, 1845, 1851, and 1857, but his major survey work in the region was done in the years 1822-24 and his later efforts will not be examined here.

Lakeland geology is difficult, compared with that of southern Britain. The terrain is tough, and the rocks are very confused-in most places quite different from the nicely layered fossiliferous strata of the Yorkshire Dales where Sedgwick had grown up. So how did Sedgwick go about his task?

First let me say something about Sedgwick's financial situation, as background to his work. His income in 1822 from his chair and his College fellowship was £407 6s 8d, but, depending as it did on Trinity's financial well-being, it declined in the following two years to £283 1s 0d and £232 15s 8d, according to Jonathan Smith of the Trinity Library, who kindly informed me, but he probably had additional income from private pupils. I have no evidence that Sedgwick obtained financial support for his fieldwork, but being unmarried he would, I think, have had sufficient money to be self-financing. He was in the field from about the beginning of June until mid-October each year.

Sedgwick travelled by carriage, on horseback, or on foot. His obituarist John Phillips stated [5] that Sedgwick travelled chiefly on horseback when in the field-but this cannot have been the case for many of the localities that he visited in The Lakes. Sedgwick had a portable writing desk, a leather specimen bag, and an assortment of hammers, which are on display in the Sedgwick Museum. He also used a portable laboratory, though his notes say little or nothing about the results obtained with it. No doubt he had an acid bottle though.

For accommodation, Sedgwick stayed with friends or at local inns. There is no record of his having made arrangements in advance of his arrival. So I cannot say whether he just "turned up", or made elaborate prior "bookings". But one is struck with the way everything slotted into place so far as accommodation was concerned, despite the fact that he seemed to leave little time to organise his domestic arrangements, being out in the field all day and every day, except for Sundays, when he attended church. Often he was in the field for fourteen hours or more. There is a record [6] that when Sedgwick was visited by his colleague William Whewell in 1824 both were in "breathless haste".

Just as Sedgwick's accommodation arrangements worked smoothly, so too did his transport. There always seemed to be a carriage waiting at an appropriate spot, or a horse. Again, I don't know whether he planned everything carefully beforehand, or altered arrangements from day to day according to the weather and the exigencies of fieldwork. I suspect the latter. However, one thing is certain: he used local people as guides at virtually all stages of his fieldwork, and John Phillips [5] recorded that when he met with Sedgwick by chance in the field in Teesdale in 1822 he had a miner's boy with him as a guide.

Sedgwick seemed to be remarkably little bothered by the weather. There were occasional days when he did not attempt to go out because of rain, but he never mentions problems due to cloud, and he was rarely driven home by rain, despite the Lake District's high rainfall. It may be noted, however, that 1824, which was the year that Sedgwick made his greatest effort in the field in The Lakes, had a an exceptionally dry summer.

The question of maps available to and used by Sedgwick is important. There was a map of the Lakeland region by Thomas Donald (2nd edn, 1810, but originally surveyed in 1771), and we know that Sedgwick possessed a copy, for it still exists in the Sedgwick Museum. But at two miles to the inch Donald's map lacked detail and Sedgwick mentions numerous places and topographic features in his notebooks that do not appear on Donald's map, or any other published map available at that time. It may be noted that Sedgwick complained in his notes on several occasions about the inaccuracy of "the map" and once he mentions Donald by name. One can see, by comparing Donald's work with modern survey maps, how inaccurate his map was.

Significantly improved topographic maps for the Lakeland counties, Cumberland and Westmorland, were published in 1823 and 1824 by C. and J. Greenwood, having been surveyed in 1821-22 and 1822-23 respectively. But even these did not give all the detail that Sedgwick mentions in his notebooks. Therefore, much of Sedgwick's information about local topography must have been supplied by guides.

Sedgwick's geological map of The Lakes was only published as late as 1835, the paper in which it appeared having been read to the Geological Society in 1831. But Sedgwick's manuscript geological map of Cumberland (only), entered on an edition of Donald (1802), is preserved at the Sedgwick Museum. It is interesting to consider how and when this was coloured in.

On his days off in the field, due to rain or the Sabbath, Sedgwick frequently mentions writing up his journal ("journalise", he wrote), but he says little about maps. He probably synthesised his Lakeland work some time in the years 1825 to 1830, and the manuscript Cumberland map probably dates from that period. Interestingly, an annotation in the hand of George Greenough, as identified by John Thackray , to whom I am indebted, to a published map of Westmorland by John Cary (1811), held at the Geological Society, states that "a map of Westmorland on a large scale was published by Hodgson at Lancaster 1828 coloured geologically by Prof. Sedgwick". I have not located Sedgwick's Westmorland map, but the annotation may give a clue to the approximate date of the colouration of his surviving Cumberland map.

Sedgwick's journals for 1822-24, name no less than fifty-one persons in the Lakes with whom he journeyed, stayed, dined, or conversed, or who acted as his guides. These included farmers, miners, the Lakeland poets Robert Southey and William Wordsworth, and local naturalists. His most important scientific contacts were two Keswick residents: the engineer and surveyor, Joseph Fryer; and the local guide, naturalist, meteorologist, cartographer, and clock repairer, Jonathan Otley (1766-1856).

Otley was a man of many parts. Interested in meteorology and an authority on local topography, he knew The Lakes like the back of his hand. In connection with his business as a guide, he published a topographic map of the region in 1818. In 1823, he extended this into a full guide-book, which incorporated the map, and also contained information about the local mineralogy and geology [7]. This contained a stratigraphic subdivision of the Lakeland rocks, which Otley had already published in a local journal [8].

Thus, Sedgwick undoubtedly learned much from Otley, besides how to find his way around in the mountains. Most importantly, there was Otley's division of the Lakeland strata into three main categories. First, there were the dark rocks of the northern Lakes, now called the Skiddaw Slates, forming mountains such as Skiddaw, near Keswick, with generally rounded profiles. South of this unit, there was a more varied group, with slates of blue or green colour, forming the knobbly hills of the central Lakeland. These are now called the Borrowdale Volcanics unit, but Otley called them the "Greenstone division". Third, there was a group forming the gentler region of the southern Lake District. The lowest member of this third unit was a quite narrow band of fossiliferous limestone associated with slaty rock; while other members of the unit (now collectively called the Windermere Group) were true slates. The limestone, which Otley suggested, in Wernerian terminology, might be a "Transition Limestone", ran right across the Lake District, roughly from west to east. Subsequently, it came to be known as the Coniston Limestone.

Otley further noted that the slates of the northern part of the middle division appeared to be generally inclined to the north and those of the southern part inclined south. It was, he said, "as though the mountain ridge, dividing the counties of Cumberland and Westmorland, had acted as a wedge in separating them" [8]. So virtually the first scientific paper on Lakeland geology laid down a broad stratigraphic classification that has lasted until the present. Moreover, an important suggestion about general structure was proposed. Additionally, Otley distinguished between bedding, jointing, and cleavage-features that may easily be confused in The Lakes. Sedgwick himself said that he did not at first understand the difference properly; and it was Otley who taught him.

In his published description of Lakeland geology, based on his fieldwork of 1822-24, Sedgwick [9] described the outcrop of Otley's "Transition Limestone" (a term that Sedgwick himself deployed in his notebooks) as if he had traced it in the field from west to east. He described how the outcrop would continue steadily for a while, and then be displaced to the north or south for perhaps up to three miles. Sometimes a large lake or valley could be found associated with the line of displacement.

On the basis of his published paper, it would appear that Sedgwick followed the outcrop of limestone, using it as a marker band. But in practice this wasn't what he did, as his notebooks reveal. Rather, he would stay in one spot for three or four days and range out from there in several directions. In fact, Sedgwick's "journals" provide enough information to enable one to trace his movements approximately on modern 1:25,000 maps. When this is done, one can see how he ranged over the country so that after three years' work he had covered, albeit hastily, most of the Lakeland terrain. And when his observations were transferred to his map, perhaps back in Cambridge, the outcrop of the "Transition Limestone", and other important features such as the granites near Eskdale and Skiddaw, or Otley's three types of slates, would have emerged as the various areas of the map were coloured in. In fact, not all the Limestone's outcrop was covered even in the same year. It is true that for some days Sedgwick would become particularly interested in the Limestone and he was evidently following its course; but he did not do so consistently. Actually, he very likely knew what to expect, for even in his short paper of 1820 Otley had indicated in words the general outcrop of the "Transition Limestone".

Sedgwick's notebooks contain much information about different rock types, the lines of their outcrop and the boundaries between them, and, most particularly, information about dips and strikes ("ranges" as he called them); also the bearings of principal topographic features. His interest was evidently lithological and structural, and in his early Lakeland work he gave little attention to fossils.

There is so much numerical information about dips and strikes in Sedgwick's journals that one wonders how he remembered it all. Were rough notes made during a day's excursion, and then entered up in the journal in the evenings? Or did he just have a phenomenal memory? I have no means of answering this for certain, but I believe that there must have been rough notes used to compile the surviving notebooks. So we have several stages of knowledge filtration and construction: 1. Observations in the field; 2. Writing of rough notes; 3. Composition of the "Journal" and the destruction of the original notes; 4. Preparation of geological maps from the "Journal"; 5. Composition of a paper for public presentation and evaluation–perhaps in the light of higher geological theory or some ongoing theoretical controversy; 6. Defence of the paper at the meeting of some learned society (usually the Geological Society); 7. Revision of the paper, and its refereeing; 8. Further revision and eventual publication.

Sedgwick's journal/notes contain a few sketches which are not impressive. His procedure seems to have been to cover as much ground as possible in a fairly theoretically neutral or positivistic way, and fill in a map. His categories for this were very simple, but that was doubtless to his advantage in dealing with such a difficult area in pioneering fashion. By contrast with the impoverished petrographic categories, there was a wealth of detail regarding dips and strikes of beds and of cleavage. Little immediate attempt was made to draw sections (in contrast to the notebooks of his later friend, and eventual enemy, Roderick Murchison).

As said, Sedgwick's petrographic vocabulary was impoverished. He often referred in his journal to a rock type by its locality or approximate appearance. Thus we have the "Wallow Crag" type; the "Barrow rock"; "the concretionary" which appears to correspond to the ignimbrites of the Borrowdale Volcanics, not recognized in The Lakes until the 1950s; "the porphyry"; or even "the blue". Interestingly, some of these terms appear in Otley's writings, so Sedgwick probably got them from him. But he did not use them in his published work.

Back at Cambridge, Sedgwick's positivism dropped away to some extent as he developed grander theoretical pictures. For example, when his map was finished it was obvious from its outcrop that the "Transition Limestone" had been affected by massive faults that cut across the outcrop, wrenching it into a number of severed fragments; from which it appeared that the whole region of The Lakes had been upheaved and fractured. Then the great valleys, such as those of Windermere or Coniston, had subsequently been carved out along the lines of weakness produced by the faults. Later, when he published his work, Sedgwick [9] suggested that the movements were caused by the intrusion of the great masses of igneous rock, such as the granites near Eskdale or under Skiddaw.

Since Sedgwick made no effort to squeeze the history of the district into a Biblical time-scale, there was no problem about the erosion of the lines of fracture taking an indefinitely large time. But the fractures, he thought, were produced by what were called at the time (by William Whewell) "catastrophic" earth movements, acting for a relatively short time, before the subsequent unconformable deposition of the New Red Sandstone round the margins of the Lake District. Thus the arguments of Reijer Hooykaas [10], that catastrophism could be scientific and empirically based are well supported by the example of Sedgwick. His catastrophism was not just a form of physico-theology. It was grounded in, and warranted by, an enormous effort of fieldwork.

Over and above Sedgwick's catastrophism, he sought to account for what he had seen in The Lakes with the help of the overarching theory of Elie de Beaumont [11], though with some caveats. In agreement with the Frenchman's doctrines, Sedgwick asserted that the main lines of mountain chains in the Southern Uplands of Scotland, The Lakes, Wales, the Isle of Man, and Cornwall, were approximately parallel, and were thus generated at the same epoch. The elevation of the Pennines, running N-S, could be attributed to a different (later) epoch. Said Sedgwick in his Presidential Address to the Geological Society in 1831: "The investigation of the faults and dislocations interrupting the continuity of our secondary deposits is becoming, daily, a subject of increasing importance; and we are now called upon, not to regard them as solitary phenomena, but to trace them through whole regions, and to examine their relations to each other" [12]. Further, the uplift and fracture of central Lakeland might, Sedgwick suggested, be an exemplification of the theory of "craters of elevation" of Alexander von Humboldt and Leopold von Buch. Thus what started off with observations about the obscure outcrops of calcareous rocks in southern Lakeland-which rocks fizzed with acid and which did not-could, with the help of the maps constructed on the basis of his fieldwork, turn into grand speculations about the history of the globe. Thus it was that the endless notes about dips, strikes ("ranges"), and cleavage were used to underpin grand theoretical models. Ultimately, it seems, this was the point of Sedgwick's preoccupation in the field with all the structural details of the Lakeland rocks.

If we compare Sedgwick's work with that of Otley and other amateurs, we can recognise significant differences. Otley was born and lived in The Lakes, and knew them intimately. Besides discovering the outcrops and showing them to Sedgwick in a number of important cases, he produced two significant theoretical advances, and moreover taught them to Sedgwick-namely the tripartite division of the Lakeland rocks, and the distinction between stratification and cleavage, the understanding of which was essential to the success of Sedgwick's work-otherwise his "structural" observations would have been hopelessly confused and in error. But while he knew many of the exposures, Otley did not ascend to any higher theory. Indeed, by virtue of his profession as a part-time guide, he was, I believe, more interested in topography and meteorology than geology. He sometimes sent Sedgwick topographic information or beautifully drawn topographic maps (there are examples preserved at the Sedgwick Museum), but I have no evidence that Otley sent geological information in the form of maps, though he did produce some.

Other amateurs were also at work. There was Sedgwick's friend Joseph Fryer of Keswick, who is thought to have been the anonymous author of the very first geological paper on The Lakes [13, 14]. There was the cobbler from Kendal, John Ruthven. He was employed by Sedgwick as a fossil collector in some of his later Lakeland work, and published his own geological map of The Lakes in 1855. Ruthven had, I'm sure, covered the ground over a period of years, though he wouldn't have had time or money to spend months at a time in the field like Sedgwick. Perhaps like another remarkable amateur, John Bolton of Furness in western Lakeland, he would have had to sleep out on the fells to do his work [15].

There lay the difference. Sedgwick could approach the task in a professional, full-time manner; and with the advantages of education, connections, and libraries, he could ascend to higher levels of geological synthesis. His structural information was put at the disposal of a grand theoretical model, that of Elie de Beaumont. This theory was soon shown to be unsatisfactory; but such efforts towards higher levels of synthesis were what were required of the geological elite. And the elite players had to be unremitting in their efforts towards reading the literature, and constructing and defending their views against all comers. They had to have money or position; and preferably both. By contrast, the "peripheral" amateurs, whose work was usually confined to a restricted region, had perforce to remain as collectors, and helpers of the "central" players. If, as sometimes happened, they published grand schemes in local journals, no one took much notice.

Of the English geologists of the day, it was only William Smith who achieved immortality from a position of weakness. And he was given a hard time. But given different circumstances, I dare say Jonathan Otley might have been just as distinguished a professor as Adam Sedgwick. Certainly, he taught him a good deal. And Sedgwick surely appreciated the value of his humble friend, visiting him in Keswick not long before he died at the age of ninety, and being reduced to tears at the sight of his dying former companion in the field. I know of no similar case in the history of geology.

REFERENCES

1. M. J. S. Rudwick. *The Great Devonian Controversy.* Chicago University Press, Chicago and London (1985).
2. J. E. Secord. *Controversy in Victorian Geology: The Cambrian-Silurian Dispute.* Princeton University Press, Princeton (1986).
3. J. W. Clark and T. McKenny Hughes (Eds). *The Life and Letters of the Reverend Adam Sedgwick.* Cambridge University Press, Cambridge (1890).
4. C. Speakman. *Adam Sedgwick: Geologist and Dalesman, 1785-1873.* Broadoak Press/The Geological Society of London/Trinity College, Cambridge, Heathfield (1982).
5. J. Phillips. Sedgwick, *Nature* **7**, 257-259 (1873).
6. I. Todhunter. *William Whewell: An Account of his Writings*, 2 vols. Macmillan & Co., London **1**, 32 (1876).
7. J. Otley. *A Concise Description of the English Lakes, the Mountains in their Vicinity, and the Roads by which they may be Visited; With Remarks on the Mineralogy and Geology of the District.* The Author, Keswick, London and Kirkby Lonsdale (1823).
8. J. Otley. Remarks on the Succession of Rocks, in the District of the Lakes, *Lonsdale Mag.* **1**, 433-435 (1820).
9. A. Sedgwick. Introduction to the General Structure of the Cumbrian Mountains; With a Description of the Great Dislocations by which they have been Separated from the Neighbouring Carboniferous Chains, *Trans. Geol. Soc. Lond.* **4** (series 2), 47-68 (1835) (read Jan. 5, 1831).
10. R. Hooykaas. *Catastrophism in Geology, its Scientific Character in Relation to Actualism and Uniformitarianism.* North Holland, Amsterdam (1970).
11. L. Elie de Beaumont. Researches on some of the Revolutions on the Surface of the Globe; Presenting Various Examples of the Coincidence between the Elevation of Beds in Certain Systems of Mountains, and the Sudden Changes which have Produced the Lines of Demarcation Observable in Certain Stages of the Sedimentary Deposits, *Phil. Mag.* **10** (new series), 241-264 (1831).
12. A. Sedgwick. Address to the Geological Society, Delivered on the Evening of the 18th of February 1831, by the Rev. Professor Sedgwick, M.A., F. R. S. &c. on Retiring from the President's Chair, *Proc. Geol. Soc. Lond.* **1**, 281-316 (1831).
13. Anon. [J. H. Fryer]. A Geological Sketch of a Part of Cumberland and Westmorland, *Phil. Mag.* **47**, 41-45.
14. H. S. Torrens. Joseph Harrison Fryer (1777-1855): Geologist and Mining Engineer, in England 1803-1825 and South America 1826-1828: A Study in Failure. In: *Geological Sciences in Latin America: Scientific Relations and Exchanges.* S. Figueiroa and M. Lopes (Eds). pp. 29-46. INHIGEO, Campinas (1994).

15. J. Bolton. *Geological Fragments Collected Principally from Rambles among the Rocks of Furness and Cartmel*. D. Atkinson, Ulverston, and Whittaker & Co., London (1869).

Proc. 30th Intern. Geol. Cong., Vol. 26, pp. 205-215
Wang *et al.* (Eds)

S. S. Buckman (1860-1929), his World-wide Jurassic Biochronology and Work on Chinese Ammonites (1926)

H. S. TORRENS
Department of Earth Sciences, University of Keele, Staffs, ST5 5BG, UK

Abstract

Buckman was an English stratigrapher who worked at the interface between amateur and professional science over a long and active career, spanning 50 years. In 1889 he published a first "milestone" paper, which demonstrated extensive diachronism within lithologically similar Lower Jurassic sands in the south of England. In 1893 he published another, on the Middle Jurassic sediments there, in which he demonstrated, by means of the detailed biochronology of its fossil ammonites, how highly condensed and episodic their original deposition had been. His aim by 1901 was to attempt to construct a biochronological "time table, of worldwide application... a means whereby Jurassic events over a large part of Europe can be exactly dated now; and [for which] there is good reason to think that the same may be said of a far wider field in the future". The value of these papers was immediately questioned by the English geological "establishment" and Buckman came to feel more and more isolated in his opinions. After a breakdown in health in 1904, he abandoned the field work which had been vital to his early work, now published much of his work privately, and started a career as a consultant biostratigrapher. In 1914 he was offered a job for six months as such on the Canadian Geological Survey at $ 2,000 a year. The Survey was busy exploring the more outlying parts of Canada. It was also engaged in a subsidiary role in surveys further afield, including Hong Kong. The First World War caused the invitation to Buckman to be abandoned but his contact with the Canadian Survey was established and continued. In 1923 A.W. Grabau (1870-1946) announced a "discovery of exceptional interest and importance in Chinese geology": ammonites of supposed Lower Cretaceous age. Buckman was now asked, through the Canadian link, to identify further material from Hong Kong and in 1926 he announced that these ammonites were instead of Lower Jurassic age. As an example, his concept of an "ammonite time-table" could be regarded as having been vindicated by this work alone. This paper re-assesses Buckman's work.

Keywords: ammonites, Jurassic, history, Buckman, China, biochronology

INTRODUCTION

Revisionism is an important part of historical research, but has to be pursued with care. A recent historian has commented on the "kind of cynical revisionism that characterizes all the social sciences. The best way to attract attention, get a Ph. D., get a good job, get promoted is to stand things on their head" [1]. But the work of S. S. Buckman has been reappraised by scientists rather than historians and this makes his case all the more interesting [2, 3].

BIOGRAPHICAL SKETCH

Born in 1860, Sydney Savory Buckman was the eldest son of an English naturalist, who lost his Chair in Geology, Botany and Rural Economy at the Royal Agricultural College, Cirencester, in the Cotswolds of England, in 1862. This was for having actively supported Darwinian evolution at the famous Oxford meeting of the *British Association for the Advancement of Science* in 1860 [4]. He was James Buckman (1814-1884), one of the earliest professional scientists in England. The family thence moved to the village of Bradford Abbas in Dorset in the south of England in 1863. Bradford Abbas lies in "one of the most richly fossiliferous districts in the whole world, where ammonites are packed in incredible numbers in the thin beds of the Inferior Oolite formation" of Middle Jurassic age [5]. Here young Buckman went to nearby Sherborne School, from 1871 to 1878, and won the *Dorset Field Club* prize for the best essay by a schoolboy, published in the Club's *Proceedings* in 1878.

Buckman junior now faced the difficult choice of a career. He was first sent to train as a chemist, from 1879 to 1880 at the Chemical Laboratories at Wiesbaden under Karl Remigius Fresenius (1818-1897) [6]. Members of his mother's family were partners in the long established and then well-known firm of pharmaceutical chemists, Savory & Moore in London. But chemistry was not to his taste. Buckman then made unsuccessful attempts to earn his living as a land agent, for which he also trained, and until 1886 as a Cotswold farmer.

From 1887 Buckman turned more and more to geology as a full-time occupation. His Dorset surroundings had clearly had a polarising effect on him. A first attempt at a professional geological career, early in 1887, to gain a post in the Geological Department of the British Museum (Natural History), was unsuccessful. He therefore turned his hand more and more to the "profession of author" [7]. His income-tax return for 1888 shows that 25 % of his income, of about £100, came from this source; in 1889, 23 % from £120. By 1896, 48 % had come from £140 [8].

In 1904 Buckman had a breakdown in health, caused by over-exertion in field-work and on bicycle. He moved first to a house called "Westfield", in Towersey, near Thame and then to "Southfield", also near Thame, in Buckinghamshire in 1910, and both nearer London. Buckman now became more and more of an armchair geologist and became more and more dependent on facts or information that he had not checked in the field. This was in a sense inevitable, the more so as he was now being asked to identify material, ammonites and brachiopods in particular, sent to him from all over the world.

From today's vantage point, Buckman's stratigraphic work up till 1904 had been of magnificent quality and far in advance of its time in its precision, detail and accuracy. But when first read, it had generated very polarised opinions. Many thought that such detail was neither necessary nor desirable, either stratigraphically or taxonomically. The debate culminated in 1907, when his *Monograph of the ammonites of the Inferior Oolite Series* was abandoned by the Palaeontographical Society of London. The criticism of his earliest work, much of it undeserved, undoubtedly caused Buckman to ignore future criticisms, which often were on more solid foundations [9]. His award of the Lyell medal by the Geological Society of London in 1913 acted only as a slight palliative. On the title-page of one of his last volumes, Buckman in 1927 quoted Henry Ford's words: "there are two kinds of people in the world - those who pioneer and those who plod. The plodders always attack the pioneers" [10]. It is clear with which group Buckman here was associating himself!

Buckman's life in geology was always financially precarious, eked out at different times by commercial dealing in fossils and by working as consultant for the British Geological Survey and other overseas surveys. Financial problems seem never to have been far from the surface. Such problems had also to be faced by his father, who also had to earn a living in a world in which science was newly becoming professional. Buckman never held a permanent full-time academic or salaried Government Service position so that the perception and payment for "outside" work was often in flux.

Figure 1. A miniature portrait of S. S. Buckman made by his brother Percy W. J. Buckman (1865-1948), Art Master at Goldsmith's College, London from 1898 to 1930), in 1907 as a silver wedding present (courtesy of Olive Buckman, Australia).

According to letters seen in the 1970's (now lost), Buckman was offered a museum position in the newly expanding Canadian Geological Survey [11] for six months in 1914, at $ 2,000 a year [12]. He had been considering such a job since April 1914, so the first offer must have been made before this. He was to have arrived in Canada in July 1915 but the job had been cancelled, because of the War, by 23 November 1914. Buckman remained however a distant consultant to the Canadian Survey thereafter. This link brought him into contact with Chinese ammonites, to which we now turn.

BUCKMAN'S SCIENTIFIC WORK

Buckman's first prophetic paper was published in 1889. In this he used sequentially ordered ammonite faunas to construct a biochronology and then used this to demonstrate that lithologically similar Jurassic sands in the south and midlands of England occurred on chronologically differing horizons. These became progressively younger, rising from Lower to Middle Jurassic, on going from north to south. It was an early (perhaps even the first) demonstration in this country of diachronism in rocks of uniform sedimentary type and showed how unreliable lithological similarities between rocks could be as indicators of their chronological equivalence. This paper is still quoted today as a classic example [13]. But its profound implications were not then recognised. The Rev. Henry Winwood (1830-1920), honorary curator of the Bath Institution, noted with regret in the discussion "that lithology was cast to the winds if we accepted the conclusions of Mr Buckman's paper" [14]. As an old friend of Buckman's later recorded in an obituary "it is difficult to realise the blank opposition that this paper faced" [15].

Such detailed work demanded detailed new descriptions and labels for many of the ammonites and others were unhappy at the resulting proliferation of new generic and specific names that Buckman felt obliged to introduce. C.D. Sherborn (1862-1942), who had just started on his major work *Index Animalium*, received a characteristic letter from Buckman in October 1891 which noted: "H. B. Woodward rallies me that your labours will be lengthened if I continue in the generic manufacturing line. I tell him I am truly sorry" [16]. This has left us another historical problem in evaluating Buckman's work. W.J. Kennedy and W.A. Cobban commented in 1976 that recently the "recognition of widespread variation in ammonites [had] led to widespread simplification of nomenclature and reduction of numbers of taxa". Their most impressive example was quoted to be the reduction in 1966 by G.E.G. Westermann of over sixty nominal species of Buckman's sonninid ammonites into the range of variation of but a single macroconch/microconch pair [17].

Such a simplification in the comfort of the library of one part of Buckman's work, that of ammonite taxonomy, goes too far in overlooking the other part, his detailed biochronology based on field-work. The sonninid ammonites in question are now in fact known to range through a considerable number of successive faunal horizons, even Zones and so are not coeval in the "single Subzone" that Westermann claimed. Buckman is thus vindicated in having written, "it was very probable that only about one-third of the 71 species of Concavum zone *Sonniniae* were sufficiently contemporaneous to live during what I have for chronological purposes designated "a hemera" [18]. It other words, much of the supposed "variation" is due to what we would today call evolutionary change.

Buckman's second classic paper was read in 1893 and has been recently reviewed [19]. It discussed the fossil-rich beds of the Inferior Oolite round Sherborne in Dorset, with which he had been familiar since childhood. 17 sections were correlated, in an incredible detail for that period, over a distance of seven and a half miles, against a finely-divided series of 12 time-divisions, or "hemerae" as Buckman newly termed them, instead of the previous three standard zones introduced in 1858 by Oppel.

Buckman in this paper showed first, that time-correlations were now possible with very high precision; second, by focusing on the crucial distinction between time and strata, that the ammonite biochronology revealed the existence of many gaps in the preserved sequences of the rocks. These he showed could have resulted either from erosion or from original non-deposition. Third, Buckman showed from this how highly condensed some of these deposits were. This paper has been recently called "one of the most important stratigraphical papers ever written"[21].

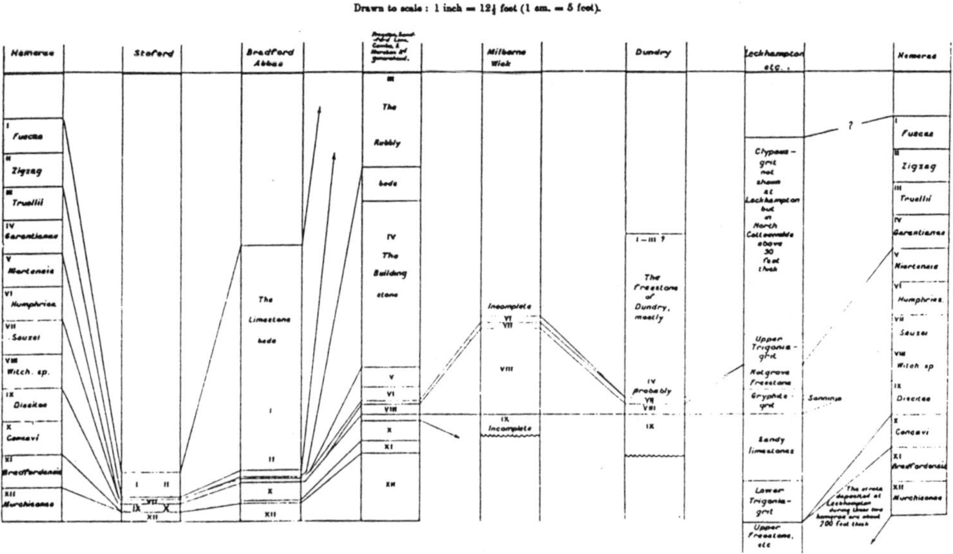

Figure 2. Table III from Buckman's 1893 paper showing the correlations of six sections against his standard scale of 12 hemeral divisions. Hemera, the new term which Buckman had coined, meaning "moment", came from the same root as "ephemeral" [20].

Once again Buckman's 1893 paper was much criticised. In the discussion, J. F. Blake (1839-1906), Professor of Geology at Nottingham University from 1880 to 1888, called "attention to the fact that the hemera [sic, a time-term] of *Ammonites humphresianus* was claimed as only 3 inches of sediment... so that any single specimen would have some difficulty getting into its own zone" [22]. Buckman responded, and one can now imagine with what frustration, "by showing an ammonite which had actually been eroded down to the thickness of its zone, the thinness of which was in this case due to subsequent denudation" [23]. H. B. Woodward (1848-1914), Fellow of the Royal Society from 1896, geologist on the British Geological Survey since 1867 and author of three volumes on *The Jurassic Rocks of Britain (Yorkshire excepted)* of 1893 to 1895 [24], soon noted in his Presidential Address to the Geologists' Association in 1894, that "with the differences that constitute minor zones [of the sort that Buckman had revealed] one need not be at all concerned"[25]. Blake in 1895 supported this, noting "a *reader* [of this paper] will scarcely find in these observations sufficient distinctness to convince him of the necessity of separating these hemerae... it is to be feared that if these hemerae are substituted... we may lose the broader facts by over-specialisation" [26].

On 20 April 1893, nearly three weeks before this paper was read to the Geological Society in London, Buckman had written to its President, W.H. Hudleston (1828-1909), in response to some of the criticism; "I am sorry you think the paper too long and detailed... the detail is the important part of the paper and it is just the detail which I find foreign correspondents want and complain of us for not giving... Much the same has been said about my Cotteswold Sand paper [1889] - the same details have however been found in France... and it was this paper that taught them to look for them [there].... if

H.B. Woodward's advice were followed the details which we — not myself only by any means — want and take such pains to get, would be entirely falsified. I am strong for details. I expect the time I spent in an analytic laboratory is answerable for that" [27].

In these two papers Buckman was one of the first to demand the proper separation of sedimentary and chronological records in stratigraphy. In 1893 he had pointed out how ammonites "species may occur together in the rocks, yet such occurrence is no proof that they were contemporaneous... If they have been proved to be successive species, then their occurrence together only shows that the deposit in which they are embedded accumulated very slowly" [28]. In 1910 he memorably noted how "a schoolboy once defined a net as a series of holes strung together, and the Dorset Inferior Oolite might be defined as a series of gaps united by thin bands of deposit". He then re-emphasised how "the amount of deposit can be no indication of the amount of time [involved]" and observed that "one can hardly view the few feet of Inferior Oolite at Burton Bradstock [on the Dorset coast]... and imagine that it represents an interval of time equal to a quarter or a fifth of the whole Jurassic Period - a time during which thousands of feet of strata were laid down. But this is because we do not allow sufficiently for the gaps" [29]. Buckman's two papers of 1889 and 1893 were based on a prodigious amount of field-work. They demonstrated how highly diachronous, condensed and episodic the rock record could be. They were far ahead of their time in stratigraphic thinking and have been entirely vindicated.

A BIOCHRONOLOGICAL TIME-SCALE

In 1901 Buckman first wrote of his vision in extending such detailed polyhemeral work over a much larger geographical area. He now wanted to attempt a biochronological "time-table... of world wide application. There is no local limit to time, and there can be no local limit to a time-table. Whether the records of the rocks in distant localities may be sufficiently perfect to enable their dates to be stated with as great exactitude as in my time-table, is another matter. But the table of strata which I have given in connection with this time-table... is a means of exactly dating Jurassic events; therefore it has much more than local value... the time-table is a means whereby Jurassic events over a large part of Europe can be exactly dated now; and there is good reason to think that the same may be said of a far wider field in the future" [30].

John Callomon has rightly noted how "towards the end of his life Buckman's knowledge of Jurassic ammonites was unrivalled. At recognising stratigraphically important species he was adept, so that the choice and number of hemerae in his final table [1930] is not very far out. But their sequences are largely incorrect" [31]. This was in consequence of the newer working methods that Buckman had to adopt after his breakdown in health in 1904. He was after this unable to do much fieldwork and his later work suffered as a result. But his vision of what might be achieved by means of ammonite biochronology never diminished. It is well demonstrated by Buckman's study of Chinese ammonites.

BUCKMAN'S WORK ON CHINESE AMMONITES

Early in 1923 a physician, pathologist and bacteriologist then based in Hong Kong who was also an amateur geologist, Dr Charles Montague Heanley (1877-1970) [32], discovered ammonites on the north side of the Tolo Channel there. These were the first invertebrate fossils to be found there [33]. They were soon identified by A. W. Grabau (1870-1946), Professor at the National University in Beijing and Chief Palaeontologist to the Chinese Geological Survey [34]. Grabau had moved to China in 1920, having like so many suffered from the anti-German feeling then prevalent both in the United Kingdom [35] and in the United States during and after the First World War. He identified Heanley's ammonites as Lower Cretaceous forms, *Hoplites (Blanfordia) wallichi* (Gray) var. *hongkongensis*

Grabau nov., compared them with forms from the Spiti shales of the Himalayas and called them a "discovery of exceptional interest and importance in Chinese geology" [36].

Figure 3. Grabau's 1928 illustration of the Lower Jurassic ammonites from Hong Kong which Buckman had first correctly identified in 1926.

Buckman had been offered a position in the Canadian Geological Survey in 1914, but the First World War had put a stop to these plans. By 1923 the Canadian Geological Survey, then busy exploring the more outlying parts of Canada, was however continuing to use Buckman's services as a consultant. He published a number of papers on Canadian fossils [37]. The Canadians had also become involved before the War in a project to survey the territory of Hong Kong under its then director, R. W. Brock (1874-1935). After the War this project was renewed, and from 1923 on a team from the University of British Columbia came to investigate the Geology of Hong Kong [38]. In 1924 one of the team, Dr Merton Yarwood Williams (1883-1974) [39], re-excavated Heanley's original ammonite-bearing locality under Heanley's direction. Heim has discussed Heanley's work, the "pioneer observer", in Hong Kong and published photographs of Heanley at this original locality and details of his work there [40].

The further ammonites found were submitted to Buckman. The material was not of high quality, but in 1926 Buckman re-identified it as being infact Schlotheimiid ammonites of Lower Jurassic age, to which he gave the new generic name *Hongkongites* [41]. He detailed the many significant differences he observed between *Hoplites wallichi* and the new Hong Kong material in a report dated 19 November 1926. His identification demonstrated that this fauna, then the only Lower Jurassic known in south-east Asia, was much older than had previously been thought and that Grabau's assignment of the ammonites to the Cretaceous was in error.

Grabau was sent Buckman's *Report* by Williams and who accepted his re-determinations. In 1928 Grabau published and illustrated the ammonites as of Lower Jurassic, Coroniceratan age [today lowest Sinemurian] and accepted Buckman's new generic name *Hongkongites*[42]. This was the earliest formal publication of Buckman's new genus and, because of his incomplete application of the Rules of Nomenclature, Grabau has wrongly received credit for having invented it.

Buckman's *Report* was not published until 1943 and thus only appeared 15 years after his death on 26 February 1929. It was delayed as a consequence of the Great Depression, which perhaps understandably diminished interest in a Canadian publication on Chinese ammonites identified by an English palaeontologist. Then came the Second World War, which caused further delay.

CONCLUSIONS

Buckman's determinations have been amply confirmed in the 70 years since. This Lower Jurassic ammonite fauna today assigned to the ammonite *Sulciferites hongkongensis* (Grabau)[43], is nowadays widely known on both sides of the Tolo Channel, in Hong Kong [44] and in the Guangdong province of south-east China. It remains the oldest Jurassic ammonite fauna known there.

Buckman's early work had demonstrated, over one hundred years ago, how very incomplete the stratigraphic record could be. A paper re-emphasising this was presented to the 30th *International Geological Congress* in Beijing in 1996 [45]. It pointed out, as Buckman had, that the incompleteness of the stratigraphic record had very important implications for sequence stratigraphy and geochronology. There are other lessons from Buckman's work on such geographically distant ammonites. In their recent paper Wang and Smith merely note that "the geological age of the original "*Hongkongites*" was not clear (cited initially as Cretaceous and later interpreted as Jurassic by Grabau 1928), but [that] the Guangdong material demonstrates its position within the Lower Simemurian" [46]. This account one can only regard as "economical with the truth", particularly in how it ignores Buckman's remarkable contribution. History has an important role in revealing such economies of truth [47].

Acknowledgements

I thank the following for much help and encouragement: Olive Buckman (Canberra), John Callomon (London) who kindly read and commented on a first draft, Anne Dickason (Geological Survey of Canada, Ottawa), Charles Heanley (Henfield), Lee Cho Min (Hong Kong Polytechnic), Jeffrey S. Murray (Canadian Government Archives, Ottawa), David Oldroyd (Sydney), Raynor Shaw (Hong Kong Geological Survey), John Thackray (London), and Wang Hongzhen (Beijing) for organising such an interesting INHIGEO session for us there. Documents recorded as in "the author's collection" will be deposited in the archives of the British Geological Survey, Keyworth, UK.

REFERENCES

1. D. Landes. The Fable of the Dead Horse; Or, the Industrial Revolution Revisited. In: *The British Industrial Revolution: An Economic Perspective*. J. Mokyr (Ed). pp. 132-170. Westview Press, Boulder (1993).
2. J. H. Callomon. Time from Fossils: S. S. Buckman and Jurassic high-resolution geochronology. In: *Milestones in Geology*. M. J. LeBas (Ed). pp. 127-150. Geological Society Memoir **16**, London, UK (1995).
3. P. Doyle. *Understanding Fossils*. Wiley, Chichester (1996).
4. S. S. Buckman. MSS Autobiography 1860-1880, author's collection.
5. W. J. Arkell. *The Jurassic System in Great Britain*. Oxford University Press and R. B. Chandler and D. T. C. Sole. The Inferior Oolite at East Hill Quarry, Bradford Abbas, Dorset, *Proceedings of the Dorset Natural History and Archaeological Society* **117**, 101-108 (1996).
6. H. Fresenius. *Geschichte des Chemischen Laboratoriums zu Wiesbaden....* Kreidel's Verlag, Wiebasben (1898).
7. Buckman's two novels, *John Darke's Sojourn in the Cotteswolds and elsewhere* (1890) and *Arcadian Life* (1891), were published by Chapman & Hall, London.
8. S. S. Buckman's copy letter book for 1884-1896, pp. 549, 579, 630 and 643, author's collection.
9. W. J. Arkell. op. cit. 5, pp. 17-37.
10. S. S. Buckman. *Type Ammonites*. Volume 6. Wheldon & Wesley, London (1927). The words are from Ford's *Today and Tomorrow* (1926).
11. M. Zaslow. *Reading the Rocks*. Macmillan Company of Canada, Toronto (1975).
12. This statement is based on letters in the possession of Peter Buckman of Dry Sandford, England seen by me in the 1970's but which have since been 'disposed of.
13. P. D. Taylor (Ed). *Field Geology of the British Jurassic*. London, Geological Society (1996).
14. S. S. Buckman. On Cotteswold, Midford and Yeovil Sands, *Quarterly Journal of the Geological Society of London* **45**, 440-474 (1889).
15. A. Morley Davies. The Geological Life-Work of Sydney Savory Buckman, *Proceedings of the Geologists Association* **41**, 221-240 (1930).
16. British Library. Add MSS 42579, f. 176.
17. W. J. Kennedy and W. A. Cobban. *Aspects of Ammonite Biology*, London, Special Papers in Palaeontology **17**, (1976).
18. S. S. Buckman. *A Monograph of the Ammonites of the "Inferior Oolite Series"*, London, Palaeontographical Society Monograph (1887-1907).
19. J. H. Callomon and R.B. Chandler. A Review of the Ammonite Horizons of the Aalenian-Bajocian stages in the Middle Jurassic of Southern England, *Memorie descrittive della Carta Geologica d'Italia* **40**, 85-112 (1990).
20. The archives of the Geological Society show only that the paper's referee was the past president, Robert Etheridge (1819-1903), but his Report has not survived (Geological Society archives, Com. P 3/1).

21. Taylor. op. cit. 13, p. 62 see also B. M. & J. E. Conkin. *Stratigraphy; Foundations and Concepts*. Stroudsburg, PA, Hutchinson Ross Publ. Co. (1984).
22. S. S. Buckman. The Bajocian of the Sherborne District, *Quarterly Journal of the Geological Society of London* **49**, 479-522 (1893).
23. A. Morley Davies. op cit. 15, p. 227.
24. published London, Her Majesty's Stationary Office.
25. H. B. Woodward. Geology in the Field and in the Study, *Proceedings of the Geologists' Association* **13**, 247-273 (1894).
26. J. F. Blake. *Annals of British Geology*. Dulau and Co., London (1895).
27. MSS in Cambridge University Library. Add 7652 V 2, letter dated 20 April 1893.
28. Buckman. op. cit. 22, p. 518.
29. S. S. Buckman. Certain Jurassic (Lias-Oolite) Strata of South Dorset; and their correlation and Certain Jurassic species of Ammonites and Brachiopoda, *Quarterly Journal of the Geological Society of London* **66**, 52-110 (1910).
30. S. S. Buckman. Jurassic Brachiopoda, *Geological Magazine,* Series 4, **8**, 478 (1901).
31. J. H. Callomon. Palaeontological Methods of Stratigraphy and Biochronology: Some Introductory Remarks, *Geobios*. **17**, 16-30, p. 25 (1994).
32. I thank his son, Dr Charles Heanley of Henfield, Sussex, for much valuable assistance in helping me to unravel his father's complex life. Dr C.M. Heanley later moved to Salisbury, Rhodesia where he died on 3 January 1970 and was buried.
33. C. M. Heanley. Notes on some fossiliferous rocks near Hong Kong, *Bulletin of the Geological Society of China,* **3**, 85-89 (1924).
34. H. W. Shimer. Memorial to Amadeus William Grabau, *Proceedings of the Geological Society of America Annual Report for 1946*, 155-166 (1946).
35. H. S. Torrens. Ernest Feuerheerd & his Rotary Lobe Pump of 1919, *Icon.* **1**, 37-69 (1996).
36. A. W. Grabau. A Lower Cretaceous Ammonite from Hongkong, South China, *Bulletin of the Geological Survey of China* **5:2**, 199-207 (1923).
37. S. S. Buckman. Report on Ammonites from the Firth River. In: *Report of the Canadian Arctic Expedition 1913-18* Part A, pp. 14A-15A. Ottawa (1924) and S. S. Buckman. Jurassic Ammonoidea, *Bulletin of the National Museum of Canada* **58**, 1-27 (1929).
38. Zaslow. op. cit. 11, p. 283.
39. V. J. Okulitch and H.V. Warren. Memorial to Merton Y. Williams, *Geological Society of America Memorials* **6**,1-5 (1977).
40. A. Heim. Fragmentary Observations in the Region of Hong Kong, Compared with Canton, *Annual Report Geological survey of Kwangtung & Kwangsi* **2**, 1-32 (1929).
41. S. S. Buckman. [1926 Report on Ammonites.] In: M. Y. Williams. The Stratigraphy and Palaeontology of Hong Kong and the New Territories, *Transactions of the Royal Society of Canada*. Third series, Section 4, **37**, 93-117 (1943).
42. A. W. Grabau. *Stratigraphy of China: Part 2, Mesozoic*, Peking, Geological Survey of China (1928).
43. D. T. Donovan and G. F. Forsey. Systematics of Lower Liassic Ammonitina. *University of Kansas Paleontological Contributions* **64**, 1-18 (1973).
44. C. M. Lee *et al*. Discovery of *Sulciferites hongkongites* (Grabau) on the South Shore of Tolo Channel... *Geological Society of Hong Kong Newsletter* **8:3**, 23-33 (1990) and In: M. J. Atherton *et al. Report on the Stratigraphy of Hong Kong*. Department of Civil & Structural Engineering, Hong Kong Polytechnic (1990).
45. M-P. Aubry. On the incompleteness of the stratigraphic record: Implications for sequence stratigraphy and geochronology, *Abstracts of the 30th International Geological Congress, Beijing, China 1996* **2**, 10 (1996).
46. Yi-Gang Wang and P. L. Smith. Jurassic ammonites from China, *Journal of Paleontology* **60**, 1075-1085 (1986).

47. Cho Min Lee. History of Geological Work in Hong Kong. In: *Useful and Curious Geological Enquiries beyond the World.* D. F. Branagan and G. H. McNally (Eds). pp. 76-84. Springwood, NSW (1994).

Proc. 30th Intern. Geol. Congr., Vol. 26, pp. 217-224
Wang *et al.* (Eds)

In Memory of Professor Peter Misch

JOSEPH A. VANCE
Department of Geological Sciences, University of Washington, Seattle, WA 98195-1310, U. S. A.

LI WENDA
Nanjing Institute of Geology and Mineral Resources, Nanjing 210016, R. P. China

Abstract

Peter Misch (1908-1987) was born in Berlin, Germany and received a Ph. D. in geology at Goettingen in 1933. Following teaching at Goettingen, and geologic studies in the Alps, Pyrenees and Himalayas, he migrated to China where he taught and carried out geologic field studies through the World War II years. After the war he went to U. S. A., accepting a professorship at the University of Washington in 1948, where he served as a teacher until 1980 and remained active as Emeritus Professor until his death. His legacy to science includes pioneering geologic studies in west Yunnan, China, and in the North Cascades of Washington State and in Great Basin region, U. S. A., and the training of more than 100 M. S. and Ph.D. students whom he supervised. His example lives on in the work of many outstanding geologists whom he trained both in China and in U. S. A.

Keywords: Peter Misch, regional metamorphism, structural geology

Peter Misch died on July 23, 1987 after a long period of declining health. With his passing away the geologic community has lost a colourful and controversial personality, a productive scientist, and an inspiring teacher. His life was disrupted by the two great wars and had more than the usual allotment of hardship and turmoil. Certainly, his constant and passionate dedication to geology was a major force in helping him through the dislocations of the second World War and the recurrent health problems of his last two decades.

Peter was born in Berlin, Germany on August 30, 1908 into a university family. His maternal grandfather was Wilhelm Dilthey, a philosopher whose work is still highly regarded. His father, Georg Misch, also a philosophy professor, taught at Berlin and Marburg before settling in Goettingen in 1916. Several of Peter's lifelong interests developed very early. His love of the outdoors grew during boyhood rambles through the countryside around Goettingen and led to a fascination with natural science and with rocks and fossils in particular. During these walks he encountered students of the noted tectonicist, Hans Stille, under whose influence he soon fell. As a boy Peter began painting water color landscapes, an interest he was to pursue the rest of his life. His artistic talents served him well in later years when he would cover the blackboard with splendid cross-sections and sketches in his structural and regional geology courses (The illustrated pictures inserted in the text are greeting cards made by Peter himself in the winter of 1981-82. It displayed his artistic talent and pursuit). During visits to the Alps, first on family vacations as a teenager and later as a university student, Peter took up mountaineering. He became an avid alpinist and skier, acquiring skills which would later enable him to work in difficult terrain in several parts of the world.

After graduation from the Gymnasium where he followed the classical studies program Peter began geologic studies at the University of Goettingen under the guidance of Hans Stille. He served as field assistant to advanced students working in the Palaeozoic and Alpine orogenic belts of Europe. He soon revealed himself to be a master of reconnaissance geologic mapping. Field work in complex structural settings was his greatest passion and he was to spend almost every summer in the field until slowed by a serious skiing injury to his leg in his mid-twenties. Peter's geologic interests were wide-ranging. V. M. Goldschmidt and C. W. Correns were also at Goettingen and Peter spent a semester in Graz studying metamorphic petrology with Franz Angel. This exposure to geochemistry and petrology laid the groundwork for Peter's interest in the role of metamorphism in orogenic belts and helped him bring a felicitous multidisciplinary approach to the solution of geologic problems. Peter's doctoral research was on the structures and metamorphism of an area in the central Spanish Pyrenees. Here he slept on the ground without a sleeping bag, rose at first light and worked until dark.

He completed his Ph.D. dissertation in 1932, taught for a year at Goettingen, and then joined the 1934 German Himalayan climbing expedition to Nanga Parbat. As a member of the scientific team he carried out a geologic reconnaissance of a large area on and around the mountain and began to develop the ideas on granitization which he was to publish many years later. The expedition ended tragically with the death of a climbing party trapped at high camp in storms and heavy snows. Peter and another member of the scientific team made a valiant but unsuccessful attempt to reach this high camp.

Upon his return to Germany Peter began work on his Nanga Parbat materials. However, the political climate under the Nazis was becoming intolerable. Raised as a Lutheran, Peter learned that his mother was Jewish. He was now married and the safety of his family was threatened. He never recovered from the shock of learning that he was unwelcome in his homeland. In 1936 he left Germany with his young wife to accept a teaching position in China. The Nazi regime had attempted to confiscate his Nanga Parbat materials, but he was able to smuggle his notes, thin sections, and small rock slices out of the country with him. From this time on, he largely cut himself off from German culture, indeed it was with great reluctance that he would speak German, even with old friends.

Peter's ten years in China were a turbulent period dominated by Japanese invasion starting in 1937. Peter and his family were forced to flee his first teaching post at the National Sun Yat-Sen University (Zhongshan University) in Canton (Guangzhou). Before this, Peter was first introduced to Dr. Zhu Jiahua, a high rank officer in China, by Hans Stille, whom was an old acquaintance of Zhu. Zhu invited Peter as a professor of Sun Yat-Sen University. Here he got his Chinese name (米士) which was said to be chosen by Zhu. In 1938 he reluctantly allowed his wife and small daughter to return to Germany and Peter himself entered into the then Shamian Concession to seek asylum. When Canton was occupied by the Japanese army, Peter moved to Chengjiang, Yunnan Province, where the University had migrated to. At the end of 1940, Sun Yat-Sen National University was removed to Pingshi, Guangdong Province, Peter accepted a professorship of the Geological Department of Peking University, with salary no more than his Chinese colleagues, which was then associated with the Qinghua and Nankai Universities as the National Southwest Associated University in Kunming, Peter also became a staff member of the latter.

Following the defeat of Japan, Peter received an offer to continue his teaching duties in Peking University, but the unstable political situation and steadily deteriorative economic condition of

China persuaded him to leave in 1946. He visited India where he met Dr. D. N. Wadia, the director of India Geological Survey, and then went to America. At first he lectured on both coasts, finally settling at Stanford University where he met his sister and began to study the Upper Palaeozoic and Triassic fossils suits that he had collected in China. At this time he also began work on his three provocative papers on granitization, in which he argued that some sedimentary and igneous rocks had been transformed into granitoid rocks by metasomatic metamorphic processes. While the granitization hypothesis is now generally rejected, examination of these papers reveals the carefully reasoned argument and clarity of expression which characterized Peter's work. While living in the San Francisco Bay area Peter met and married the delightful and talented Nicoletta (Niki) Rosenthal.

Peter Misch (1908-1987)

(This photo was taken around 1952)

In 1947 Peter met George Edward Goodspeed, then chairman of the Geology Department at the University of Washington. Goodspeed, an enthusiastic granitizer, asked Peter to join the department. Attracted by the proximity of the wild and geologically unexplored North Cascades, Peter accepted. After the resolution of some difficult visa problems Peter moved to Seattle. He had at last found a home.

1981 1982

Greeting Cards made by Peter Misch

Peter's six years in Kunming were occupied with teaching and field works with geological surveys in west Yunnan. Some of his field work was done under trying conditions, occasionally accompanied by armed guards to protect the party against bandits. He also made detailed field studies in central and east Yunnan. His studies on the Sinian stratigraphy, the Carboniferous bauxite deposits, the red beds and marine Triassic deposits, as well as the discovery of Upper Permian (Lopingian) in west Yunnan were fruitful. His study of the Sinian stratigraphy of east Yunnan provided a further basis for correlation and also a supplement to the section of the Sinian type locality in Yangtze gorge. When he worked in Cangshan-Jinshajiang-Shigu area of northwest Yunnan, through his arduous field observations, he proposed a concept of red-bed metamorphism and secondary retrogressive metamorphism; this more or less inspired the later research work although his hypothesis is not completely correct. Based on his field observations of geological structure from west to east, he summarized the transitional relation of Alpino-Germano and Sino-type structures of the young orogenic deformation. He pioneered the research of young dynamo-metamorphsm and Alpine orogeny in west Yunnan. Among three types of young orogenic deformation, Sinotype is a replenished concept to the classification of Hans Stille in light of the evidence of west Yunnan. When he studied the Sinian and Carboniferous stratigraphy, he also gave some attention to applied aspects of Sinian phosphate and Lower Carboniferors bauxite deposits respectively, which greatly inspired the later prospecting work. His systematic study on the "Metasomatic granitization of batholithic dimemsions" covered Northwest Himalayas and northwest Yunnan. He as a transformist played an important role on the development of the history of granitization school although the granitization theory nowadays has been rejected.

When he worked in National Southwest Associated University, besides teaching general geology, petrology, structural geology and geological mapping, he initiated courses on regional metamorphism, and Eurasian orogenesis for senior students and post-graduates. Under his training, most of his Chinese students could be independently responsible for geological mapping and surveying right after graduation. His teaching course and study works on regional metamorphism in west Yunnan greatly motivated his Chinese students and young colleagues, for example, the doctoral theses of Zhang Bingxi and Chi Jishang in U. S. A., Dong Shenbao in France and Ma Xinyuang in Scotland, are mostly concerning regional metamorphism. After they returned to China, they all took on professorial duties and academic activities which greatly promoted the study of regional metamorphism in China.

In Washington University Peter taught undergraduate courses in field geology, metamorphic petrology and structural geology (what we would now call tectonics) and graduate courses in the regional geology and tectonics of Europe and Southeast Asia, as well as courses on metamorphic processes. He was a stimulating lecturer who motivated many students to continue on to graduate work. Although Peter would often deplore the rigid European university system and the authoritarian professor with whom one could not argue, he could himself be dogmatic. He was a forceful personality with strong opinions both about people and geology. Peter was known to change his mind, but usually only after long discussion and subsequent reflection.

Peter demanded much of his students, but gave generously of himself and his time. As early as in his years in China, when he taught his students how to map in the field, he spent almost all his Sundays with his students in the remoter outskirts of Kunming city. He asked his students to observe and describe every outcrop and to argue about the complex phenomena. It seems that he was a positivist, the principle of practice, experiment and feedback was his working hypothesis. He together with his young Chinese assistants led the undergraduate students to field practice. During those days of the Japanese invasion, Chinese people lived a hard and bitter life. Peter and his students suffered all kinds of hardships in the field, they slept in the farmer's thatched cottages or old dilapidated houses. They ate noodles simply boiled in water, rarely with fried eggs. Peter never complained of his hard life. He was happy and active and often made jokes on his students, and thus inspired a happy and vigorous atmosphere among his students. When he taught in University of Washington, he asked his graduate students to prepare weekly thesis. The weekly thesis conference was a unique experience and could be traumatic, if you were not well prepared. Peter would carefully look over your thin sections asking questions to make certain that you had not missed any important minerals or textures. Thesis chapters were edited and revised word by word. He instilled a rigor in thought and writing, indeed he taught a generation of graduate students to write clear, concise, well organized English. Many of Peter's students were skiers or climbers, and classes were not scheduled on Thursdays of the winter quarter in order to permit "downhill research" at nearby ski areas. Over the years more than one hundred M.S. and Ph.D. theses were completed under his supervision. In his later years Peter sometimes wondered whether he had sacrificed too much research time to work with his students. He came to realize, however, that their success in reaching influential positions in academy and industry fully justified his efforts. Indeed, his students may be his greatest legacy to geology. Peter is well remembered by his former students, many of whom are now prominent geologists. After the "cultural revolution "in China Peter corresponded with some of these Chinese geologists who have visited him at his home. In view of his contribution to Chinese geological science and education, in 1982 the Geological Society of China sent a letter inviting him to visit China and attend the 60th Anniversary Meeting of G.S.C. as an honorable guest. But because of his poor health he failed to fulfill the sincere expectation of his former Chinese friends and students.

In 1948 Peter embarked on his pioneering studies of the pre-Tertiary crystalline rocks of the North Cascades, a task which would occupy him the rest of his life. At that time this rugged, brushy range, difficult of access, was almost unknown geologically. By the mid-1980's Peter had completed a detailed reconnaissance of some 5,500 km^2 of the range and in 1966 he published "Tectonic Evolution of the Northern Cascades of Washington State: A West-Cordilleran Case History", a summary and synthesis of his findings and those of his graduate students who had worked in the area. This paper names and defines the principal stratigraphic, metamorphic and plutonic units of the area and describes their structural relations. He outlines the major Cascade thrusts and high-angle faults and the metamorphic and plutonic episodes. Peter's Cascade work bears his characteristic stamp, the combination of careful field observation with the study of small scale structures, down to the thin section level, and their integration to develop the scheme of tectonic evolution. Unfortunately, in his later years Peter became somewhat agitated about the arrival of other workers in the North Cascades which by then he considered his own private domain. It is unclear whether he felt that his conclusions might be proven wrong, whether less able geologists would incorrectly reinterpret the geology, or whether he was simply frustrated that his declining health would no longer permit him to continue active field work. Although modern geochronology, more detailed mapping, and the reanalyses of the area in the light of plate tectonics require reinterpretation of his work, recent studies largely confirm his basic conclusions. His Cascade study remains the foundation for all future work in the area.

Peter's other major contribution to regional geology was in the Great Basin. He spent several field seasons in the 1950's as a consultant for Union Oil Company. Working with John Hazzard he carried out regional stratigraphic and structural studies in a complex terrane which combined regionally developed metamorphic rocks and low-angle faults. These structures are what we call metamorphic core complexes. Although Peter interpreted these structures as a classic compressive fold thrust system, his careful mapping and description foreshadowed the development of current thinking on the subject. In 1980 Peter was honored when the Geological Society of America Memoir on "Cordilleran Metamorphic Core Complexes" was dedicated to him. Peter's death marked the passing of a classical geologist *par excellence*. This exceptional man has been missed, but his example lives on in the work of the many outstanding geologists whom he trained both in China and in U. S. A.

Acknowledgements

Special thanks are due to Dr. R. V. Fisher, professor of the Department of Geological Sciences, University of California, Santa Barbara, and Dr. S. S. Szetu, consulting geologist of Canada Exploration Consultants Limited, Ontario, for providing informations about Peter; and also due to Drs. Zhang Bingxi, Dong Shenbao, and Wang Hongzhen for their critical reading of the manuscript.

Major Works of Peter Misch

1935 The Scientific Work of the German Himalayan Expedition to Nanga Barbat, 1934, Geologieal

Studies, *Alpine Journal for 1935* 49-52.

1937 Himalayan Research, *The Earth*. National Sun Yat-Sen University, Canton (in Chinese).

1939 The Crystalline Rocks of Tali, West Yunnan, *The Earth*. National Sun Yat-Sen University, Canton (in Chinese).

1942 Sinian Stratigraphy of Central East Yunnan, *National University of Peking, Contribution of the Department of Geology* **4**, 30 pp.

1945 Sinotype Structures in East Yunnan, *Academia Sinica, Science Records* **1**, Nos. 3-4, 534-540.

1945 Young Dynamometamorphism and Other Alpinotype Structures in West Yunan, *Academia Sinica, Science Records* **1**, Nos. 3-4, 541-548.

1946 On the facies of the Carboniferous of the Kunming Region, East Yunnan, With Special Reference to the Bauxite Deposits, *Bulletin of the Geological Society of China* **26**, 1-64.

1946 On the Discovery of Upper Permian (Lopingian) in West Yunnan, *Bulletin of the Geological Society of China* **26**, 65-82.

1947 Observation on Red Beds and Marine Triassic of Yunnan, *Academia Sinica, Science Records* **2:1**, 106-111.

1947 (Wtih Carl O. Dunbar) Fusulina bearing Permian Rocks of Northwestern Yunnan, *Bulletin of Geological Society of China* **27**, 101-110.

1949 Metasomatic Granitization of Batholithic Dimensions. Pt. I. Synkinematic Granitization in Nanga Parbat Area, Northwestern Himalayas, *American Journal of Science* **247**, 209-245.

1949 Metasomatic Granitization of Batholithic Dimensions. Pt. II. Static Granitization in Sheku Area, Northwestern Yunnan (China), *American Journal of Science* **247**, 372-406.

1949 Metasomatic Granitization of Batholithic Dimension. Pt. III. Relationships of Synkinematic and Static Granitization, *American Journal of Science* **247**, 673-705.

1952 Geology of the Northern Cascades of Washington, *The Mountaineer* (Annual of the Seattle Mountaineers, Inc.) **45**, 4-22.

1953 (With John C. Hazzard *et al.*) Large Scale Thrusting in North Snake Range, White Pine County, East Nevada, *Geological Society of America Bulletin* **64**, 1507-1508.

1956 Tectonic Patterns of the Western United States, *American Institute of Mineral Engineers*. Seattle Meeting, Out of invited paper, 3pp.

1957 (With K. F. Oles) Interpretation of Ouachita Mountains of Oklahoma as Autochthonous Folded Belt: Preliminary Report, *American Association of Petroleum Geologists Bulletin* **41**, 1899-1905.

1959 Sodic Amphiboles and Metamorphic Facies in Mount Shuksan Belt, Northern Cascades, Washington, *Geological Society of America Bulletin* **70**, 1736.

1960 Regional Structural Reconnaissance in Central Northeast Nevada and Some Adjacent Area: Observations and Interpretations. In: *Intermediate Association Petroleum Geological Guidebook to the Geology of East Central Nevada*. pp. 17-42.

1962 (With John C. Hazzard) Stratigraphy and Metamorphism of Late Precambrian Rocks in Central Northeastern Nevada and Adjacent Utah, *American Association of Petroleum Geologists Bulletin* **46**, 289-343.

1963 (With Gerald M. Miller) Early Eocene Angular Unconformity at Western Front of Northern Cascades, Whatcom County, Washington, *American Association of Petroleum Geologists Bulletin* **47**, 163-174.

1964 Stable Association Wallastoneite Anorthite, and Other Calc Silicate Assemblages in Amphibolite Facies Crystalline Shcists of Nanga Parbat, Northwest Himalayas, *Beitr. Mineral u. Petrogr.* **10** (Corren Volume), 315-356.

1965 Radial Epidote Glomeroblasts Formed Under Conditions of Synkinematic Metamorphism — A New Mechanism of Collective Crystalloblastesis, *Geologische Rundschau* **54** (Bederke Volume), 994-956.

1966 Tectonic Evolution of the Northern Cascades of Washington State — A West Cordilleran

Case History. In: *A Symposium on the Tectonic History and Mineral Deposits of the Western Cordillera in British Columbia and Neighboring Parts of United States*. N. C. Gunning (Ed). pp. 101-148. Canadian Institute of Mining and Metallurgy. Spec. Vol. **8**.

1968 Plagioclase Compositions and Non-anatectic Origin of Migmatitic Gneisses in Northern Cascade Mountains of Washington State, *Contr. Mineral. and Petrology* **17**, 1-70.

1969 Paracrystalline Microboudinage of Zoned Grains and Other Criteria for Synkinematic Growth of Metamorphic Minerals, *American Journal of Science* **267**, 43-63.

1970 Paracrystalline Microboudinage in a Metamorphic Reaction Sequence, *Geological Society of America Bulletin* **81**, 2483-2486.

1975 (With Jack M. Rice) Miscibility of Tremolite and Hornblende in Progressive Skagit Metamorphic Suite, North Cascades, Washington, *Journal of Petrology*, **16:1**, 1-21.

1976 (With A. C. Onyeagocha) Synplectite Breakdown of Ca-rich Almandines in Upper Amphibolite Facies Skagit Gneiss, North Cascades, Washington. A Study of Chemical Exchanges and Imperfectly Attained Successive Equilibria, *Contrib. Mineral. Petrol.* **54**, 189-224.

1976 (With Bernard W. Evans) A Quartz Aragonite Talc Schist from the Lower Skagit Valley, Washington, *American Mineralogist* **61**, 1005-1008.

1977 Bedrock Geology of the North Cascades. In: *Geological Excursions in the Pacific Northwest: Geol. Soc. Amer.* 1977 *Ann. Meet. Seattle Guidebook*, pp.1-62.

HISTORY OF GEOLOGY
GEOSCIENCE DISCIPLINES

Proc. 30th Intern. Geol. Congr., Vol. 26, pp. 225-235
Wang *et al.* (Eds)

The Methodology of the Researches of Geological Science — Petrology as an Example

DONG SHENBAO
Department of Geology, Peking University, Beijing 100871, P. R. China

Abstract

Geological phenomenon represents a kind of eternal motion of matter in nature that embodies different geological processes with immense space and eternal flux of time and shows interconnection with each other. Being controlled by this uniqueness, the study of geological science belongs to a higher form of motion of matter with its own geological constraints, and cannot be replaced by the lower form of motions, yet, their physico-chemical constraints are necessities to reveal the innate nature of geological process. The study of geological process relies fundamentally on the observation in nature, whereupon the method of approximation is generally admitted as the working principle, which constitutes the alternations from inference to observation in approaching the truth. Herein, the abduction adopted by C. S. Peirce as the inference to the best explanation both by inductive and deductive methods, is best suited to this approach. In the history of the investigation of geological science, the controversy of the conceptual idea generally occurs among the important topics with vigour. In reality, this debate largely images the opposites of the contradiction of the geological process, and proceeds to elaborate them to a new unity for further debate to reach finally its verity. It is considered as the impetus in promoting the development of geological science. Some examples of petrology are also discussed.

Key words: motion of matter, petrology, law of unity and opposite, method of approximation, uniformitarianism

PREAMBLE

The researches of geological science aim at the elaboration of the geological processes in nature, the investigation of their interconnections and mutual constraints in order to guide the necessity of the existence and the living conditions of human being throughout the whole world.

Geological science is a kind of natural science wherein theories and practice are closely connected. Its researches can only be done by bearing in mind that they belong together and supplement each other. Since the very beginning of the historical records of mankind, narrations on the floods, earthquakes, volcanic eruptions as well as ore findings had abounded in various countries and became one of the pillars of the antiquities with pristine simplicity in dialectics notably in Greek's philosophy. These are the vivid streams converging into the founder of geological science that Lyell [1] first brought sense into the forbidden zone enmeshed in theology by substituting the realistic principle for a slow transformation of the earth.

As a whole, feedbacks of practice — theory — practice prevail alternatively through the progress of the history of geological science since Lyell's time through the establishment of diverse branches of geological science during the lower half of nineteenth century up to the modern researches on the dynamic perspectives of the interconnection between different spheres of the earth. The repeated processes of observation and inference, analysis and synthesis, induction and deduction, all reverting to practice and theory, invoke the recognition of change and transformation of quantity and quality of each geological process at a comparative high stage of development, and will attain the full development from casuality to causality of theory of knowledge in the future. If we make a survey on the development of history of geological science, we may find that there has been a great progress in a broad field encountered from the surface features to the nature of mantle and their crust-mantle action; from continental crust to oceanic crust and their interaction as to form the theory of global tectonics; from temperature and pressure evaluation of the origin of rocks to the thermal perturbation and relaxation of heat flow under depth with the initiation of grand tectonics; from static observation mainly qualitative to quantifying geodynamic processes, thus leading to the gradual approach to the very essence of the geological processes in nature. The mighty advance of the mode of geological thinking during century has hitherto been the guide of ever-increasing welfare of nations and averting the unduly loss of people's lives. If a nation that wants to keep pace with modern industry cannot possibly manage without geological theories flourished from its own land.

THE FORM OF MOTION OF MATTER OF GEOLOGICAL PROCESS

Geological science embodies the totality of earth as its object. The main objective reality of the investigation lies upon the geological processes operated within the accessible part of the earth from lithosphere to atmosphere. By geological process, it is understood as the natural process operated by the motion conceived as the inherent attribute of inorganic matter. It comprehends various modes of action, broadly named as endogenic and exogenic actions that exist through immense space and endless flux of time showing multiple geological constraints from each other and passing away as eternally moving and ever-changing matter in accordance with some natural laws. Based on this uniqueness, studies can only be done through observation in nature in attempting to understand the scattered geological records performed by the grand laboratory of Nature that exist only as the ancient relics partly disintegrated by the successive geological events and obliterated through weathering. The inadequacy of observation is then supplemented by experiment and relevant principle to form inference and is returned as feedback to nature to verify the results for further investigation. Thus the method of approximation that represents the gradual passage for the recognition of the nature of geological processes is generally adopted as the working method that essentially differs from those of basic sciences.

Geological process is characterized as the higher form of motion of matter in a series of forms of action such as magmatic, metamorphic, diagenetic orogenic, weathering *etc*. Each of them possesses its own inherent attribute, that forms bodily the geological science as a whole, and that passes into one another during the ceaseless motion. The geological process expressed as the higher form of motion of matter is obviously different from the simple or lower form of motion of physics and chemistry, and can neither be treated explicitly as the functional relations of physical variables nor replaced by the latter. Geological constraints have become the principal factors governing this form of motion, whereas the subsidiary physico-chemical actions behave as the basic constraints without which it cannot be considered as an exact science. For instance, the high temperature-high pressure experiments have led to the thermodynamics as the foundation in igneous petrology. Nevertheless, if the geological constraints of a magmatic complex such as the attainment of equilibrium, the influence of the source rocks, the mechanism of the motion of magma, the change of fluid condition during crystallization *etc.* are lacking, they can only tell us what cannot be happened rather than what could be happened.

Different forms of motion of matter with inherent attribute ascertain precisely the classification of sciences. Geological process as a science has its own unique form of motion both with geological and physico-chemical constraints. Neither these constraints can be confused with each other, nor be ignored. During the studies of geological sciences, any violations, either indulging into the empirical experience to spurn all experiments and principles of basic science, or simply using some physico-chemical formulations to replace the complexity of geological process, can never be averted from the impunity from mysticism and the notion of immutability of nature, characterized by Newton and Linnaeus in the later half of eighteenth century.

Viewing back through the history of geological science, it is not astonishing that during the history of geological science, the defects and lingering caused by such violations cannot be avoided in the early period of its development, even the famous geologists could not avert the pitfalls, and their very errors may indeed stimulate investigation on the part of other side. However, it would be a damage, that some of them may remain unchanged and even exist as the sole guide in some modern researches.

For instance, opinion that claims the dominate action of surface manifestation of certain structural types such as decollement structure or ductile shear zone in governing the magma genesis or inducing the regional metamorphism, has gained some popularity in certain modern researches. However, these are only some local phenomena of what belong to the studies of heat flow, wherein the main topics focus upon the thermal perturbation and relaxation of crust-mantle action and their relations with tectonics. 'That plate motions are the surface manifestation of a terrestrial heat engine operating by the subtle interaction of physical and chemical processes and transformation of thermal energy to mechanical energy,' [2], and that the P-T-t paths of metamorphic rocks that are linked by the physics of heat transfer and the thermodynamics and kinetics of possible transformation of rock masses tells something of that orogen's history [3] may represent the modern thought in this respect.

Another defect currently met with is the undue interpretation of the time of major magmatic or metamorphic thermal crystallization events by means of geochronological dating without careful examination of their context of geological environment even up to the ad hoc assumption in selecting such geological facts. Naturally, confusion is to be expected with endless arguments, and in fact violates its development. In the majority of cases, the simple assumption that the precise analytical data relate to major geological events such as metamorphic terranes and magmatic complexes, is much more questionable. A good deal of evidences has suggested that many of analytical data must be treated with caution, that they do not relate to the time of major crystallization as supposed, rather do they relate other subordinate thermal events in its history before their closure. "Unless geologists and geochronologists squarely face up to the importance of the correct interpretation of analytical dates, the whole subject is in danger of deteriorating into an elaborating number-game." [4].

Diverse opinions have been resulted in the study of geological science as the consequence of the implications of the complexities of geological process and the imperfection of field observations due to dismembering and weathering of rock units by the successive events. Grown up amid these diversities, the conceptual notion, a kind of sensuous knowledge, named as "multiple solutions of geological science" has been formed. It claims that unlike the strictness of the sciences of physics and chemistry, within each field of geological science there exists several solutions in solving its nature. This statement implicitly signifies that certain degrees of uncertainty have been inherited in the geological science whose nature cannot be exactly defined by certain laws.

Actually, this notion cannot be treated as a concept with much more significant idea, since it does not account for the innate nature of geology, rather does it account for its superficial features. Strictly speaking, much diversity of opinions has been originated from the observation of the contradiction of opposites inherited from the geological process. Its appearance can be ascribed as the consequence of necessity, a necessary step toward the approaching of the nature of geological process. As the geological process has been implicated by mutual constraints of actions and their interchanges during an eternally changing motion, we can only approach them by the observation from different side of its actions or its particulars in contrast to the whole view. In general sense, such controversy on the diversity of opinions is a means to disclose the opposite of the contradiction of things and their transformation in the aim of attaining a new insight or unity to deepen our knowledge on the geological process and for an ensuring contradiction, that again invokes another controversy. It is a universal law named as "Law of Unity of Opposite". For example, in physics, the nature of light was debated as ray propagation *versus* wave motion through a longer duration since Newton's time. It is by this controversy that some fundamental laws of physics and chemistry have been discovered as motion of particles in quantum state within the microscopic world. Herein, this law appears even more important as a driving force to guide the investigation through all fields of geological science (see later).

Another appeal to the "multiple solutions of geological science" comes largely from the subjective consciousness of workers whose prejudices have controlled the selection of facts. They could be taken as suppositions, yet, some of facts may be worthy to stimulate thinking.

HISTORICAL DEVELOPMENT OF THE RESEARCHES OF GEOLOGICAL SCIENCE

At the very beginning, geological science had been flourished by people's narratives on the observations in nature such as floods, earthquakes, volcanic eruptions *etc.* as to form a materialistic basis in nature. Subsequently the ancient natural philosophy notably the Greek's, became substantially as the guide of thinking in explaining the geological phenomena despite the obstruction exerted by the mysticism or religious habit. It was to Lyell, who analyzed these narratives in determined opposition to mysticism, and brought reasoning into geology by the realistic principle and historical comparative method in the middle of nineteenth century, that promoted the rapid development of geological science by the method of analysis and synthesis, classification and correlation in every field of it. At the end of nineteenth century, the interaction of basic sciences through the application of laboratory experiments and technical knowledge into the geology has actively supported its study to such an extent that the necessity of classifying it in each separate branches in accordance with its inner connection has became imperative. Thus, the general outlook (first comprehensive outlook) expressed by Lyell's view with pristine simplicity had to be yielded to another mode of outlook of their particulars based in each branch of geological science.

The advancement of the branching sciences is the necessity of the development of geological science in compare with those general outlook of geology coming directly from contemplation. However, during their course of studies, some departure from nature has been encountered by the entanglement of the puzzles of its own field, that it blocked its own progress from an understanding of the part to an understanding of the whole, to an insight into the general inter-connection of things. Herein lies potentially the danger, on account of which it cannot avert the metaphysical concept of immutability of nature. In this period, the division of geological science as a complete system composed of different geological processes reached a perfection, yet, its whole outlook seemed rather weakened. This is the reason why we are often compelled to refer Lyell's principle, though imperfect may it be, it contains in embryo the later modes of outlook of geology.

Modern researches of geological science has been propagated through several lines of approach. Great achievements have been carried on in the branching of sciences especially in breaking through the imposed constraints and to fill the gaps between two branches of sciences to create a new field as boundary science. Among them the interaction between geology and physics or chemistry such as geophysics and geochemistry occupies and important place leading to the deep understanding of the interior of the earth, a forbidden area in tradition geology like oceanic crust and mantle, and the theoretical consideration of geological phenomena. They may be exemplified in the acknowledgment of physical nature of lithosphere and upper mantle and their interactions in the geophysics, and the geochemical process operated by the determined electronic configuration and isotopic changes of elements in the geochemistry. The creation of the boundary sciences in the geological sciences marks a course of jump toward the realization of the nature of the geological process.

The advancement of geophysics and geochemistry have had much effects in causing a basic change of geological science from the pure descriptive narration of facts toward the quantifying of them through relevant data and related principles, thus, leading to the evaluation of some important problems of geology in geodynamic perspectives. Physico-chemical process was largely documented by analytical and experimental data with considerable success. However, they can only be considered as a first approximation of the recognition of the natural process, since some of its constraints are yet to be determined. Studies of geodynamics of the geological process have been enlightened by the global programs since the seventies of the century, among which the heat flow studies in the earth and their bearing on problems of evolution of lithospheres, plate tectonics, magma genesis, P-T-t paths of metamorphic rocks, *etc.* are of assistance in revealing their dynamic behavior, though there remains large areas of uncertainty.

As a whole, modern researches have achieved a great success in (1) quantifying the geodynamic perspectives of the geological process that initiate the recognition of the forms of motion of intrinsic nature in opposing to those of immutable outlook of nature; (2) performing a second comprehensive outlook of geological science under a restricted analysis of dissected scientific knowledge in supplementing the Lyell's realistic simplicity of nature.

As regards to the unique nature of geological science, the method of investigation is different from those basic sciences as physics or chemistry that a simple hypothesis, a formula even a deduction nomological explanation can be taken as the principal guide within certain realm, that most of the interconnections can be deduced from them. Relied on the observation of geological facts as the chief source, geological study is essentially an epistemological study, wherein the methods of approximation generally helds as the optimistic one to approach the consequence of geological process through the repeated feedbacks from inference to observation. As inference or hypothesis in this respect must be developed, improved, and for its improvement there is as yet no other means than to verify it again in the observation in nature.

Method of approximation usually begins with the induction of empirical knowledge. Through analysis by means of principle, it forms a claim, an assertion with simplicity as first approximation. Further verification of claim in nature constitutes the inference, a conclusion drawn from a set of reliable information with a theoretical background. Feedbacks of claim and inference may eventually reach a kind of hypothesis, an advanced status of scientific inference as starting-point for further investigation of known facts. As a whole, their method of thinking may be ascribed to abduction a term proposed by Peirce, C. S., that signifies a pattern of reasoning of facts inferred from a hypothesis or inference which offer the most plausible or satisfactory explanation of the evidence through both inductive and deductive methods [5].

In his well known essay, "The method of multiple working hypotheses", Chamberlin (1897) [6] discussed various modes of attack upon scientific problems. He expressed his worries on the danger of a theory hastily advanced in explanation of any groups of observations. To prevent the danger in working hypothesis from gliding with the utmost ease into a 'ruling theory', the method of multiple working hypotheses is urged, "where one can find all reasonable processes that might lead to an observed relation, carefully considered the full consequences of each process envisioned, impartially compared these deduced inferences with the re-examined facts, and thus, reached a consequence as to the possible process." [6]. This opinion has gained much support from many famous geologists, such as Bowen, Oxburgh *etc*.

To establish an advanced hypothesis or a corroboration, it might be tested through falsification, a view advocated by Popper (1972) [7]. He claims that for an inference to be scientific, it must specify which observation would falsify it. A hypothesis can never be verified to any degree, and that science progresses by systematically attempting to falsify previously advanced hypothesis. In contrary, if one attempts to verify a hypothesis *a priori*, to all conceivable events and not to refute it by falsification, the hypothesis itself can hardly be developed. It may be exemplified what Popper (1972) [7] has called a "conventional stratagem" of which he remarks as:

> Some genuinely testable theories, when found to be false, are still upheld by their admirers by introducing ad hoc some auxiliary assumption or by re-interpreting the theory *ad hoc* in such a way that it escapes refutation. Such a process is always possible, but it rescues the theory from refutation only at the price of destroying, or at least lowering, its scientific status. (p.37). Again he also emphasizes that "A theory which is not refutable by any conceivable event is non-scientific. Irrefutability is not a virtue of a theory (as some people think), but a vice." (p. 36).

It may be exemplified by referring the current hypothesis of mantle -magma with the plate tectonics. The currently obvious intention has been to find some behavior of basalt that is unique to a particular tectonic setting, such as plate tectonics. However, this correlation can be done without any effort in re-examination of the facts. In his concluding remark of the petrologic assessment of the mantle-magma system, Carmicheal says (1974) [8]:

> "One of the most intriguing aspects of both mantle-magma and global tectonics models is at the same time a current source of weakness.... More rigorous mutual constraints by which current hypotheses may be tested by attempted falsification are needed before we shall see any enduring general synthesis of a viable mantle-magma system with the phenomena of global tectonics." (p. 664).

THE IMPETUS AND OBSTRUCTION OF DEVELOPMENT OF GEOLOGICAL SCIENCE

Viewing from the development of geological history, most important consequences of the recognition of the objective of geological process have been achieved through the controversy of the conceptual idea from opposites. The sequential development of such elaborated disputes has been unceasingly deepening the knowledge of the interconnections of geological process. It is generally known that the Law of the Unity of Opposites can be applied in whole of the universe, but herein lies the utmost importance as an impetus to guide each field of geological science.

Generally speaking, each one's opinion in study of geological science may be considered as some meditated images that partly reflect the objectivity of geological process exclusive of certain *ad hoc* assumptions. Restricted by one's particular and obstructed by the control of pre-possessed concept, prejudice cannot be averted especially in the beginning stage of development. Very errors though the prejudice may possess, it stimulate the investigation on the other side, and can form a scientific conception to get rid of its unreasonable part through the joint effort in attacking a problem. Any scientific conception can be treated as the thought that reflects the contradiction of the essence of geological process, and thus, invokes controversy with others. Among them opposites are common, and even cannot be reconciled yet, unless their transformation attains a new unity. During the granite controversy in the fortieth of the century, Bowen, himself as a great magmatist, has even urged to adopt Chamberlin's method of multiple working hypotheses (see earlier) to end the conflict of granite problem. Though sceptical in his mind, he still hoped:

> "Each of us is, of course, quite sure that he himself had done what Chamberlin recommends, but he is sure that the other fellow has done no such thing. ... The situation is deplorable, but may not be hopeless. It may be that the mutual conflicts of prejudices will dull the rabid edge of each, and that the communal petrologic mind can eventually reach the truth by the sorry-method as surely, if not directly, as Chanberlin's ideal individual mind." [9].

It seems that Bowen's appeal of multiple prejudices cannot be otherwise than a new thought of unity that will break through the controversy and emerge.

The controversy of conceptual idea of the studies of geological process cannot merely be considered as some reflections of geological phenomena in different aspects. It aims at the finding of the opposites within a contradiction of matter and their condition of transformation in order to reach unified conception of deep understanding of the nature of geological process. The unity thus attained serves as evidence for another conflict of opposites in a new form of contradiction, and invokes another controversy in the conceptual idea. Although it follows the line of thought of their predecessors in main direction, yet, forms another alternative differ from the old ones.

In the investigation of geological process, the controversy prevails over all the branches of geological science, such as neptunic *vs.* igneous; diagenesis *vs.* burial metamorphism; crystallization *vs.* deformation; rock association *vs.* tectonic setting; continental crust *vs.* oceanic crust; geosynclinal theory *vs.* plate tectonics; hydrothermal deposits *vs.* strata-bounded deposits *etc.* As a whole, they have made a tremendous effect in promoting the recognition of geological science.

The progress of petrology has been greatly facilitated by the controversies of the conceptual thought along some main themes. The controversy between the neptunist and magmatist known as "great debate" each with its own extreme, since the later half of the eighteenth century marked the beginning of the history of petrology. This debate was ended by Lyell's fourfold classification of rocks as aqueous, volcanic, plutonic, and metamorphic rocks [10] by the introduction of their transformation as metamorphic and plutonic action. Afterwards, the shift of "great debate" turned to the origin of granite, a universal debate of magmatism *vs.* migmatization from the lower half of ninetieth to the fiftieth of the centuries. The former accentuated the magmatic behavior of granite as the differentiates of basaltic rocks or as the contamination of them by Rosenbusch and Bowen; while the latter emphasized the migmatic behavior of granite as anatexis or diffusion from the continental crust from the French tradition to Sederhom and Read. There have been also minor debates under this category such as wet *vs.* dry granitization, migma *vs.* magma, metasomatism *vs.* fractional crystallization, arterite *vs.* veinite *etc.*, the composite of which formed the bitter controversy that has lasted nearly hundred years with their adherents distinctly grouped as magmatists and migmatists between which there has been passage from the view of extreme to intermediates. Through the enduring efforts of the experimental studies of formation of granitoid magmas from diverse crustal sources since sixties, the conflict has subsided into the intermittent debate. It approached a consensus view of anatexis of granitoid magma from continental crust in a consequence that the migma occurs at the initial and the magma essentially at the developing stages of the formation of granitoids. Under this agreement, modern controversy have largely shifted to the contention of the hypothesis of restites *vs.* syntexis, *i.e.* the partial melting of continental crust against the mingling or mixing of basaltic rocks with the previously formed anatexitic granitic magma. This debate comes essentially as the descendant from the conflict of migmatist and magmatist, but opines toward a more broad field wherein types of granite, characterized a variety of processes including fractional melting, fractional crystallization and metasomatism and involved different source rocks can be the very image of their tectonic environment.

In the history of the study of metamorphism, there has been a long-enduring controversy nearly a hundred years ago since Daubree's time (1860), whether thermal or stress action is the main cause governing the regional metamorphism. Through an obstinate and harsh debate, however, it is generally admitted, that neither of them can by any means be considered as the principal source to conduct such varied types of metamorphism from burial or oceanic metamorphism essentially of epeirogenic character to high temperature granulite terrain, rather they represent only the surface manifestations, the interconnection of which is controlled by some innate quality under depth.

The evaluation of thermodynamic P-T condition of the mineral assemblage of metamorphic rock has ever been implicated with the contention of the course of metamorphism *per* static *ad* dynamic perspectives. The contention from the Barrow zone, the testimony of the depth zone, to the metamorphic facies concept marks the embryo of the transfer of heat-flow related to certain types of metamorphism. Next change commences from the quantifying the P-T path through modern geothermobarometry to the P-T-t study backing by the forward modelling in connection with the orogenesis. Thus, the line of evidences of the geodynamic process of certain orogenesis in connection with the thermal perturbation and relaxation of heat-flow in the evolution of metamorphic history may be treated as an approach toward the elucidation of their interconnection.

The philosophy of the law of opposites and unity that exists as a guiding thought governing the hollistic approach of geological science, such as uniformitarianism *vs.* non-uniformitarianism; fixism *vs.* mobilism, *etc.* prevail unconsciously in every investigation of geological science. Generally, these guiding thought belongs to the realm of superstructure, the advancement of which relies on the achievements of the recognition of the behavior and their inter-connection of all geological processes, Before the fifties of the century, the uniformitarianism and fixism, expressed as the sameness of the geological phenomena and the cyclic process from past to present, predominated as ruling theories guiding the working hypotheses of each field of geological science. However, since the seventies, the appearance of plate tectonics and the subsequent debates on its application in the Precambrian tectonics as well as some implications of petrological problems such as ophiolite and andesite have greatly altered such situations

.

The obstruction which forms the opposite with the impetus in the contradiction of geological thinking, is usually present during the development of geological science. This thought has rooted from the concept of the immutability of nature, that some great geologists could not exempt from this pitfall. Even Lyell, one of the founders of geological science, merely emphasized the importance of the historical correlation of geological phenomena and discarded entirely the possibility for the transformation of interconnection of geological processes through their evolution. As a rule, the obstruction usually attempts to interpret the complicated geological process with an ever-changing mode by an over-simplified generalized scheme, that can be shown as: (1) replacing the rational analysis and inference by perceptual and empirical knowledge alone; (2) replacing multi-factor analysis by one-sided analysis; (3) replacing the geological constraints of higher form of motion of nature by the simple physico-chemical constraint of lower form of motion of matter without recoursing of former, (4) replacing the re-examination of facts assuredly by the interpretation of certain current theory *a priori* even with some *ad hoc* assumption. In fact, any conceivable theory is only a benchmark in the eternal flux of knowledge to verity. Neither can it be stopped, nor can its tract of progression be ignored.

It is noteworthy to review some important obstructions in the development history of petrology. Before the controversy of neptunist against magmatist, the theory that all rocks on earth have been originated from sedimentation process by Werner, became a ruling theory over a long interval of time. Since then, in the beginning of the century, in the sphere of magmatic rocks, the Heidelberg school headed by Rosenbusch, has made a great contribution in the study of magmatic rocks by their petrographical descriptions, but it also prevented its further progress, by its insistence that the cause of crystallization of all magmatic rocks depends mainly on their features of petrography.

Since the appearance of plate tectonics, one principal impetus is to relate the rock association with the tectonic setting of plate tectonics in order to verify its course through individual process, as is exemplified by the categorization of the granite types having an available group of mineralogical and geochemical parameters with distinct generative process and source in connection with geological environment of Phanerozoic epoch [11]. However, there has been another current tendency that opined at the interpretation of ancient geological setting simply through the use of various petro-chemical diagrams of some magmatic rocks evidenced by the empirical data in a known region. This treatment has notably been applied in the petro-chemical diagrams of basaltic rocks with recognized geological context of plate tectonics in the attempt in modelling the ancient plate tectonics from them [12], and has been supported in such a wide extent even extended to Precambrian epoch. As we know, many distinctions of these diagrams are for rock suites rather than for identifiable tectonic settings. In reality, they neither rely on the laboratory experiments conducted under controlled conditions to provide unambiguous data relevant to the origin of basaltic rocks [13], nor possess the theoretical consideration with their geological context. Empirical observation alone can never adequately prove necessity. This consideration also assumes lack of variation of the composition of basaltic rocks through Precambrian, an idea of uniformitarianism, which is invalidate in plate tectonics. With these respects, it seems that this approach may impede further progress along this direction, if it were not to be modified.

From these brief accounts, it is to be noted that the obstruction is inevitable from the impetus in the investigation of geological science. We might pay more attention to it especially in the coming flourishing development of geological science.

SOME CONCLUDING REMARKS

Geological phenomenon represents a kind of eternal motion of matter in nature that embodies different geological processes with immense space and endless flux of time, and shows interconnection with each other. Based on these premises, study can only be done through observation in nature and forms inference by induction and analysis. This deduced consequence is compared with the re-examination of facts in nature and thus reached a conclusion as to the probable process for further identification. Thus, the method of approximation is best suited as the working hypothesis in approaching the objective of geological science.

Geological process is characterized as the higher form of motion of matter in a series of actions such as magmatic, metamorphic, diagenetic, orogenic, weathering *etc*. Each of them possesses its own inherent attribute, and passes into one another during the ceaseless motion. Thus, it is obviously different from the simple or lower form of motion of physics or chemistry. The geological constraints become the principal factors, whereas the physico-chemical constraints are subordinate to, yet, as an important and basic one in ascertaining the nature of geological process. Two sides of potential danger either indulging into the empirical experience to spurn all the principle of basic sciences, or simply using some physico-chemical formulations to replace the geological constraints, must be averted.

During the study of geological science, the feedback from inference to observation is a necessity toward the approaching of its truth. Herein, the method of thinking may be ascribed as **abduction**, a term proposed by Peirce, C. S., that signifies a pattern of reasoning of facts inferred from a hypothesis or inference which offer the most plausible or satisfactory explanation of the evidence through both inductive and deductive methods.

To an advanced hypothesis, it might be tested through falsification, a view adopted by Popper, who claims that for a hypothesis to be scientific, it must specify which observation would falsify it. In contrary, if one attempts to verify a hypothesis *a priori* to all conceivable events, and not to refute it by falsification, the hypothesis itself can hardly be developed.

In the history of the investigation of geological science, the controversy of the conceptual idea generally occurs among the important topics of the geological process with vigour. In reality, the debate largely images the opposites of the contradiction of the geological process, and proceeds to elaborate them to a unity for further debate until reaches its truth. Thus, it is considered as the impetus that promotes the development of each branch of geology, such as neptunic *vs*. igneous; magmatist *vs*. migmatist; diagenesis *vs*. burial metamorphism; deformation *vs*. crystallization; continental crust *vs*. oceanic crust; geosynclinal theory *vs*. plate tectonics; hydrothermal *vs*. strata-bounded deposits *etc*.. Controversy of philosophical abstraction governs the hollistic approach of geological science. Its contention like uniformitarianism *vs*. non-uniformitarianism; fixism *vs*. mobilism *etc*. exists in each of workers that no one can escape from it. Recently, the appearance of plate tectonics and its subsequent contention in the Precambrian tectonics have actively supported the view of mobilism and non-uniformitarianism.

The obstruction which forms the opposites with the impetus in the contradiction of geological thinking is usually present in the development of geological science. This thought has deeply rooted from the concept of immutability of nature. This potential danger can neither be ignored nor imitated.

REFERENCES

1. C. Lyell. *Principles of Geology*, 1st edition (1830).
2. E. R. Oxburgh. Heat flow and magma genesis. In: *Physics of Magmatic Processes*. R. B. Hargraves (Ed). pp. 161 - 199. Princeton University Press, Princeton (1980).

3. R. A. Haugerod and E-an Zen. An essay on metamorphic path studies or Cassandra in P-T-t space. In: *Progress in Metamorphic and Magmatic Petrology, A Memorial Volume in Honour of D. S. Korzhinsky*. L. L. Perchuk (Ed). pp. 323 - 348. Cambridge University press (1991).
4. S. Moorbath. Recent advances in the application and interpretation of radiometric age data, *Ear.-Sci. Rev.* **3**, 111 - 133 (1967).
5. R. Boyd, P. Gasper and J. D. Trout (Eds). *The Philosophy of Science*. MIT Press, M. I. T., Cambridge, Massachusetts (1991).
6. T. C. Chamberlin. The method of multiple working hypotheses, *Jour. Geol.* **5**, 155 - 165 (1987).
7. K. Popper. *Conjecture and Refutations*. 4th. edition. Routledge and Kegen Paul (1972).
8. I. S. E. Carmichael. *Igneous Petrology*. MacGraw Hill Book Company (1974).
9. N. L. Bowen. The granite problem and the method of multiple prejudices, *Geol. Soc. Am. Mem.* **28**, 79 - 90 (1948).
10. C. Lyell. *Elements of Geology*, 12 edition (1838).
11. W. Pitcher. A kind of conclusion: a search for order among multifactorial processes and multifarious interactions. In: *The Nature and Origin of Granite*. W. Pitcher (Ed). pp. 292 - 297. Blackie Academic and Professinal (1993).
12. J. A. Pearce. Basalt geochemistry useed to investigate past tectonic environments on Cyprus, *Tectonophysics*, **25**, 41 - 67 (1975).
13. P. J. Wyllie. Petrogenesis and the physics of the Earth. In: *The Evolution of the Igneous Rocks - Fiftieth Anniversary Perspectives*, 2nd. edition, H. S. Yoder (Ed). pp. 481 - 520. Princeton University Press (1980).

Proc. 30th Intern. Geol. Congr. , *Vol. 26*, pp. 237-249
Wang *et al.* (Eds)

The Role and Future of Geology in Modern Integrated Environmental Research and Decision Support

GYÕZÕ JORDÁN AND ANDREA SZÛCS
Uppsala University, Institute for Earth Sciences, Norbyvägen 18B, Uppsala S-75236, Sweden
present address: Hungarian Geological Survey, Stefánia út 14, Budapest 1143, Hungary

Abstract

Geology is the only interdisciplinary geoscience which studies systematically the hierarchically organized biotic and abiotic systems and their interactions from the micro-scale of biogeochemical processes to the macro-scale of continental and global processes. The concepts and approaches of plate tectonics, environmental facies and hierarchical systems view, three-dimensional and temporal modelling, and modern integrated stratigraphy and basin analysis provide it with both the theoretical and practical foundations to analyse and model the spatio-temporal evolution of these systems. Environmental sciences provide data and information for decisions on the control of the integration of human social systems and natural systems. Geology has been yielding basic data on geological systems and indispensable expert knowledge on natural hazard prediction and environmental risk assessment. However, only the scientific tools and methods of geology have been applied which is insufficient to meet challenges of integrated environmental research related to the problems of temporal and spatial modelling, analysis of complex systems and extrapolation between scales. It is argued that geology has much more to offer for modern environmental analysis by the use of its concepts and approaches. The modelling techniques and methodologies of geology are discussed and a comparative case study is presented to demonstrate the application of geological concepts in environmental research and management.

Keywords: decision support, environmental geology, environmental research, geochemistry, geology, modelling

INTRODUCTION

As a result of technological development the interaction between human society and its natural environment has become extremely complex and environmental problems have become a major concern. The need for practical solutions has drawn the attention of scientific research to problems related to the modelling of complex systems [1-4], the unresolved methodological and theoretical problems in extrapolating fine-scale knowledge to larger regional or global scales [5-7] and unconnected databases [8].

It has been realized that the complexity of these problems can be approached and solved only by the integration of related scientific disciplines. One of the greatest challenges before geologists is to find the place and role of geology in this new, integrated environmental science. The challenge has been recognized and environmental geology has developed to meet the requirements of environmental research [9]. This new branch of geology is primarily concerned with geological hazard assessment (earthquake and volcanic activity predictions, landslides, *etc.*) and environmental geochemistry. Although they both originate from geology, neither of them takes a full advantage of the unique concepts and approaches of geology which could make them more powerful in solving environmental problems.

The fast development and complexity of environmental research has resulted in a great diversity, and often discrepancy, of related terminology. For this reason, prior to the discussion of geological concepts and their use in modern integrated environmental research and decision support, it might be useful to establish a common ground and ensure consistency within the text by giving first a brief review what is understood here by geology and geosciences, environment, environmental research and environmental decision support.

Geology and Geosciences

Geology is unique among geosciences because it is the only discipline which has a holistic view on all aspects of geoscientific problems. Geology is defined as the science that aims to describe and understand the genesis of materials that make up the solid Earth and to study its past and present transformations [9]. Here the definition of geology is confined to the "classical" hard rock geology and its traditional sub-disciplines. The distinction between geology and geosciences is stressed because, in contrast to other geosciences, such as climatology, hydrology, biology, geography, *etc.*, geology has the capability to approach environmental problems in their entire complexity and to provide methodologies for integrated problem solving. Despite this fact, experience shows that all too often geology is not applied when it should be and geological expertise and knowledge is often not involved in environmental projects and impact studies.

Environment and the Scope of Environmental Research

By definition, environment is everything outside a system. Human social system together with its surrounding biotic and abiotic systems form nature [1]. However, the interaction of man and his environment has become complex and as a result environments have developed which are hard to separate from human existence. The word "environment" also implies this involved inter-relationship. Bearing these in mind and having no better alternatives, the idea of "natural systems" is used here to denote systems in which human influence is absent or minimal (*e.g.* pre-cultural states of landscapes).

The scope of environmental research is best approached from the historical perspective of the evolving relationship of the two fundamental systems, the hierarchically organized human social systems and natural systems (Fig. 1). These systems are interacting through the flow of matter, energy and information. Prior to the industrial and technological revolution humans and human societies were mostly subdued to natural systems, and the relationship of the two systems was characterised by the one-sided dependence of humans on natural processes.

As a result of industrial and technological development man has gained substantial influence and increasing control on natural systems [10]. For example, human activity can decrease biodiversity and trigger even global climatic changes by pollution and erroneous use of land resources which, in turn, feed back adversely to human systems. The ultimate environmental problem is whether it is possible to establish harmony with natural systems while maintaining development, that is to reach sustainable integration of the two systems and form the Total Human Ecosystem, which is man and his total environment forming one single whole in nature [11]. Or, by the destruction of natural systems, the two systems will collapse and disintegrate. Modern environmental research aims at addressing this problem and studies the integration of human social systems and natural systems.

What is Environmental Decision Support?

Population growth and desire for higher living standards lead us to either change our environment or change our attitude toward the environment by taking actions in the form of controlling the flow of matter, energy and information. There is a controversy between economic development through the exploitation of natural resources and the restriction of exploitation to avoid environmental deterioration. Finding a compromise is the main task of environmental management and decision support.

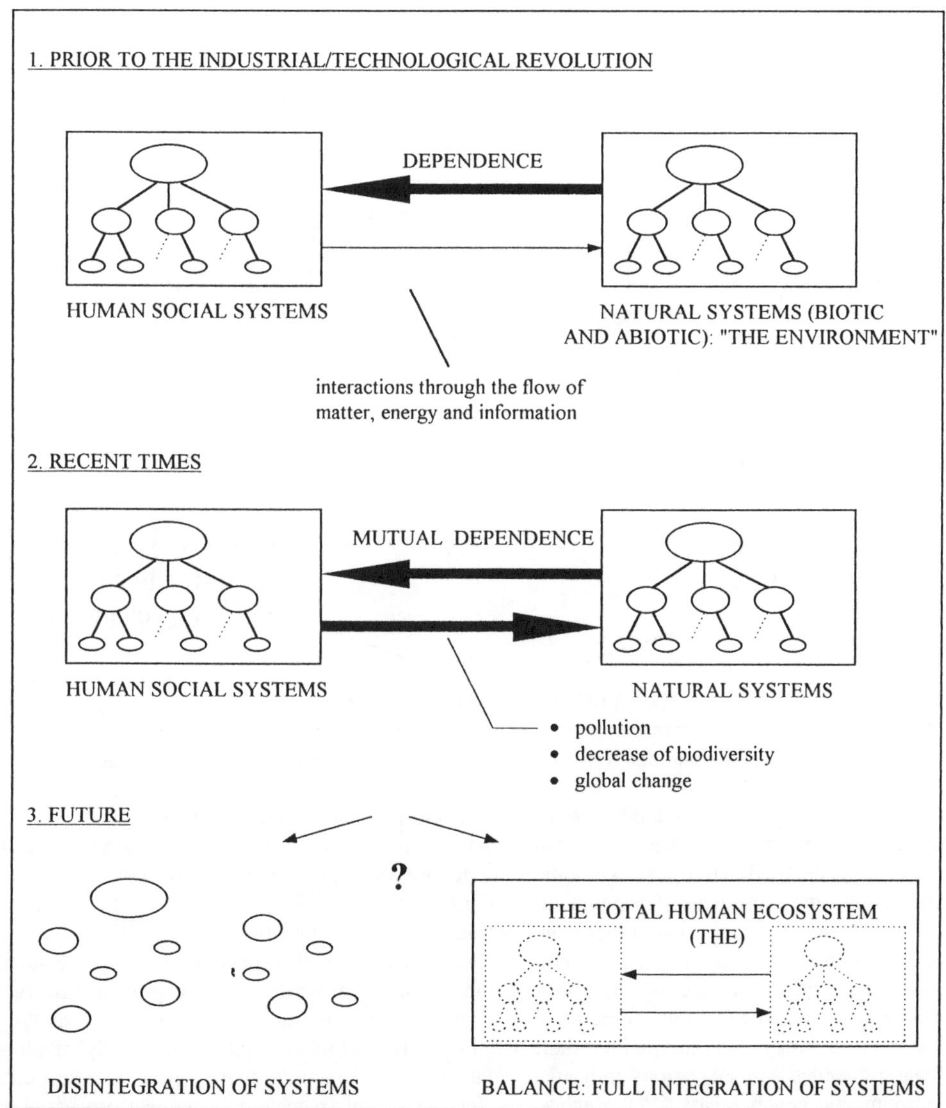

Figure 1. The scope of environmental research from a historical perspective: the analysis of the integration of Human Social Systems and Natural Systems.

The effective management and control of the environment by social systems is possible only through understanding the underlying processes and mechanisms. The essence of good planning

and decision-making is founded on the basis of relevant information. The data acquisition and interpretation is the decision support role of sciences (Fig. 2). The decision-making process has to be analysed as a whole which calls for interdisciplinarity among natural and socio-economic sciences. The following deficiencies of natural sciences limit the process of decision-making:
1. Lack of interdisciplinarity (as opposed to multidisciplinarity) which results in isolated case studies and unconnected databases;

2. The hierarchical aspect of interacting systems is often neglected, which leads to either detailed micro-scale analyses or broad generalisations on macro-scales;

3. The analyses are often limited in their spatial and temporal view.

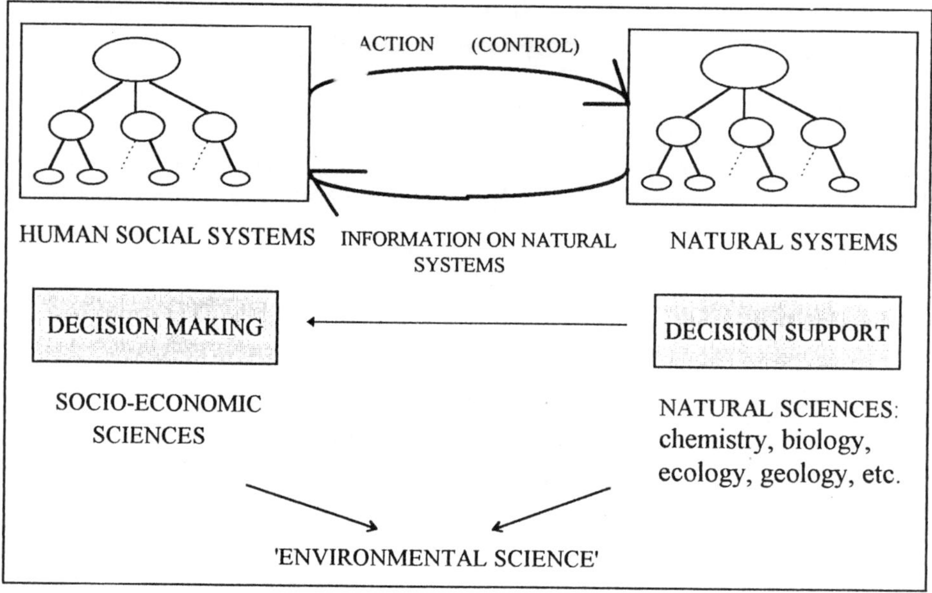

Figure 2. The environmental decision-making process and the decision support role of sciences.

How could Geology Contribute Better to Environmental Research and Decision Support?
At present, geology contributes to environmental research in two ways. The first is related to the traditional geological activity of exploration, modelling and exploitation of earth (energy, mineral and groundwater) resources. The input into environmental analysis is limited to basic geological information in this case. The other contribution of geology has two major fields which are collectively termed as environmental geology. The first is geological hazard and risk assessment, which deals with the prediction of seismic, flood, coastal and marine hazards, ground movement hazards and volcanic activity forecast [9]. Most recently, an other subdiscipline of geology, Quaternary geology, has got an important role in environmental research by the study of global climatic changes and glaciations indicated in the geological records of near past. The second is environmental geochemistry [12] which either focuses on micro-scale geochemical processes, or, primarily based on exploration geochemical experience, concentrates on broad scale geochemical mapping to support land-use management and health decisions.

Although these fields use the scientific tools and methods of geology extensively they take no advantage of geological concepts and approaches. In the following sections the concepts and approaches of geology will be briefly summarized and their potential use in integrated environmental research and systems analysis will be shown.

GEOLOGICAL CONCEPTS AND APPROACHES

Plate Tectonics: the Fundamental Theory

Plate tectonics or global geodynamics is the fundamental theory of geology like the Darwinian evolution theory of biology. The theory explains a great number of phenomena with great simplicity, such as the various aspects of the geological history of the Earth's surface, including the formation and distribution of oceanic and continental crust, mountain ranges and sedimentary basins and gives explanation for climatic variations, formation of volcanic rocks and the successive geographies encountered in the course of geological time. Most importantly, it helps geologists to recognize the interrelationships between geological processes, such as tectonics, petrological and geochemical, sedimentological and palaeontological processes. Beyond teaching to understand processes in their relations, and to appreciate that the history of Earth is a permanent "Global Change", the theory helps also to recognize the evolution of interacting earth systems, such as oceans, continents and climate, in the past, fundamental for the understanding of present environmental processes.

Concept of Environmental Facies

This concept enables geologist to think in terms of complex abiotic and biotic processes and systems in interaction. The idea of facies is much similar to the ecosystem concept of ecology [13] with the important additional emphasis on abiotic processes and system evolution. Most significantly, the interaction of sub-facies and their integration into larger facies or environmental systems is well established in contrast to the ecosystem concept in ecology [14, 15].

Hierarchical Systems View

In common geological practice the simultaneous analysis and integration of processes on various scales is also inherent: modelling often has to range from the micro-scale thin-section and geochemical analysis, through the local geological reconstructions, to the emplacement of the studied area in regional and global geological relationships. The scale of observation spans from a crystal, to a continent, to the Earth as a whole, and the geologist must therefore manage to synthesize complex and multiform data. This practice is fundamental for the understanding of the processes in the hierarchically organized environmental systems.

Three-dimensional and Temporal Modelling

Natural processes take place in the four-dimensional spatio-temporal space and should be studied and modelled accordingly. *Stratigraphy* embodies this idea and provides four-dimensional spatio-temporal modelling techniques for geologists. A recent development of these techniques is *basin analysis* which places special emphasis on integrated systems analysis.

By the very nature of geology, geologists think in terms of three-dimensional spatial models and produce maps and cross-sections accordingly. Moreover, time and temporal processes are considered explicitly in geology like in no other natural science. Geologists have to consider different time scales simultaneously from the "geological time periods" that may be as long as millions of years, to the instant catastrophic event of a meteoric impact. Only the study of geological records can reveal the mechanisms and hazard of very rare events and give direction for the analysis of the long-term stability of Earth systems. This spatio-temporal view is essential for the understanding of present and future Earth system processes and it enables geologists to consider systems in their dynamics and evolution.

Interdisciplinary Approach

As it was mentioned earlier, environmental research calls for interdisciplinary approaches. The need for appropriately trained scientists has long been recognized and higher education courses are developed to meet the needs for experts with broad environmental knowledge. Geologists study and use the integration of physical sciences, such as geophysics, sedimentology, petrology, mineralogy and geochemistry, and biological sciences, such as palaeontology and palaeoecology and receive a wide environmental training; on this basis, geology could have a central role in environmental education and problem solving.

Comparing Results of Geology and Environmental Geology

The potential of modern geological concepts in environmental systems analysis and the fundamental difference between modern geology and present day environmental geology is demonstrated by comparing related maps. A typical early geological map is characterised by giving information only about the distribution of rock types. A modern environmental geochemical map prepared in 1994 (Fig. 3) also shows that, despite the much higher-level technical implementation, conceptually it is at the same level as a geological map made fifty years ago: it is descriptives, showing only the distribution of parameters, in this case chemical concentrations.

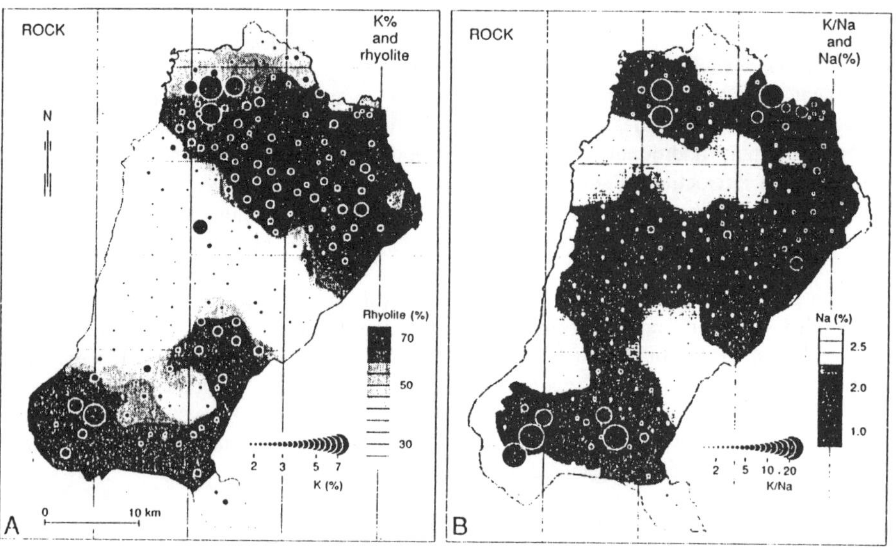

Figure 3. Modern environmental geological map. It is descriptive showing the distribution of (chemical) parameters (after Hartikainen *et al.*, 1992).

As a result of the enormous progress made in geology in the last decades due to the application of the new scientific concepts and approaches discussed previously, three-dimensional and dynamic maps have become common which show geological systems and their relationships. In Fig. 4 a landscape geochemical map is shown which attempts to reach the modelling sophistication of a modern geological map. The map is based on the integrated systems approach of geology: 1. it shows systems homogeneous with respect to geochemical character; 2. it reflects their relationships (by surface matter transport in this case); 3. it is three-dimensional, because, similarly to geological practice, the map is encompassed by a cross-section and 4. it is dynamic, that is, it indicates not only the state of the environment but processes. These make the map predictive, similar to modern geological maps and models. It is noted, that by the application of GIS techniques, the distribution of chemical concentrations can also be superimposed on the map (Fig. 4).

CASE STUDY

The objective of the case study is to demonstrate geological concepts and approaches used in geological model-building and their application in environmental analysis through the stepwise comparison of two case studies. The first is a complex geological investigation aiming at the facies analysis and modelling of the Early Cretaceous sedimentary succession in the Gerecse Mountains in Hungary [17]. The other case study is from the Slaettberg area in Central Sweden and its objective is the modelling of control mechanisms of heavy metal retention in peat sediments, in order to estimate the long-term metal retention in a wetland environment receiving acid mine drainage [18]. It is not intended here to give in-depth explanations in either case; the aim is to indicate the possible use and benefits of geological concepts and approaches in environmental analysis.

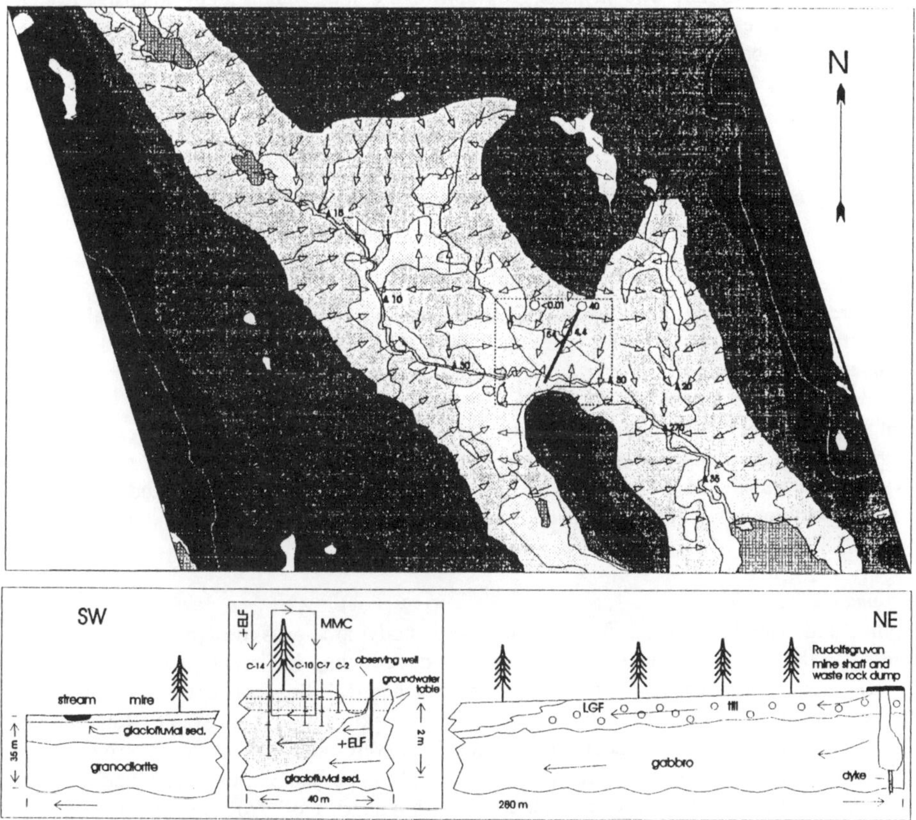

Figure 4. Landscape geochemical map and cross-section [18]. It is based on the integrated systems approach of geology: three-dimensional and dynamic. White parts on the left and right sides are watershed boundaries. Grey levels in the order of decreasing shading: eluvial, trans-eluvial, eluvial-accumulative and super-aqual landscapes.

Step 1. Hierarchical Systems View

An hierarchical systems view helps in building the general prior knowledge and model of the studied system. In hierarchical systems the functioning and key properties of local-scale levels are determined by systems at higher hierarchies. This is studied by the emplacement of the study area in the relevant global and regional scale situations. This procedure also enables the comparison and analysis of the relationships of systems located in different physico-geographical areas.

Geological case study
The Gerecse Mountains are situated in the North Pannonian and Western Carpathian Unit and are found in the North Eastern edge of the Transdanubian Mountains ("Bakony Unit"). The latter has a South Alpine affinity and it can be paralleled with the Belluno trough east (or south-east) to the submerged Trento plateau [19].

Environmental case study
In order to analyse regional matter flow patterns which determine the general geochemical processes in the Slaettberg area, regional scale geological, physical geographical maps and climatic data were overlain as shown in the inset in Fig. 5. From this composite map it can be seen that the area has a background geochemical composition characteristic of island-arc igneous rocks and associated ore deposits [20]. The physical geography and climatic data suggest intensive leaching in the area. Wind directions hint that the major source of heavy metal contamination in top layers of soils throughout Scandinavia is the atmospheric output of continental and Southern Scandinavian industrial regions. This was confirmed by the observation of N-S metal concentration gradients [21].

Step 2. Observation: sampling and sample analysis
The next step in geological analysis is changing scale from local, through site, to micro-scales, and taking observations and analysing samples accordingly.

Geological case study
In the Gerecse case study the physical and sedimentological properties of geological formations were first described. Marls with varying carbonate contents, laminated-graded sandstones, conglomerates, breccias and folded sandy marls were identified and their spatial relationships were determined. Transport directions were indicated by lineations on the foot wall of turbidites, fold axis of slid beds and the orientations of belemnitids and plant fossils. Rock samples were analysed for carbonate content, carbon isotope ratios and oxygen isotope ratios. For the palaeoecological analysis of the ancient environment the fossil content of the sediments was studied and various belemnites, ammonites, continental plant fragments, trace fossils, crinoids and radiolarids were identified.

Environmental case study
Sampling and sample analysis for the Slaettberg study area was designed and implemented according to the needs of geochemical analysis. First, following landscape geochemical principles [22, 23], areas homogeneous with respect to geochemical flow patterns were identified as elementary landscapes, such as reducing wetlands and oxidising stream water. Relationships by matter flow between adjacent geochemical landscapes were analysed by the GIS overlay of the classified elementary landscape map, watershed model, slope and slope-break model and run-off model derived from digital elevation model. The boundaries of geochemical landscapes were interpreted as landscape geochemical barriers and were classified according to Perel'man's scheme [24]. Based on the above, sample locations were identified at the reducing gley geochemical barrier where acid mine drainage discharges into the mire. Fifteen peat sediment cores were taken and analysed at different depths for heavy metal (Pb, Ni, Cu, Zn, Fe and Mn) content in five fractions in order to study the retention of metals by the recipient peat sediments.

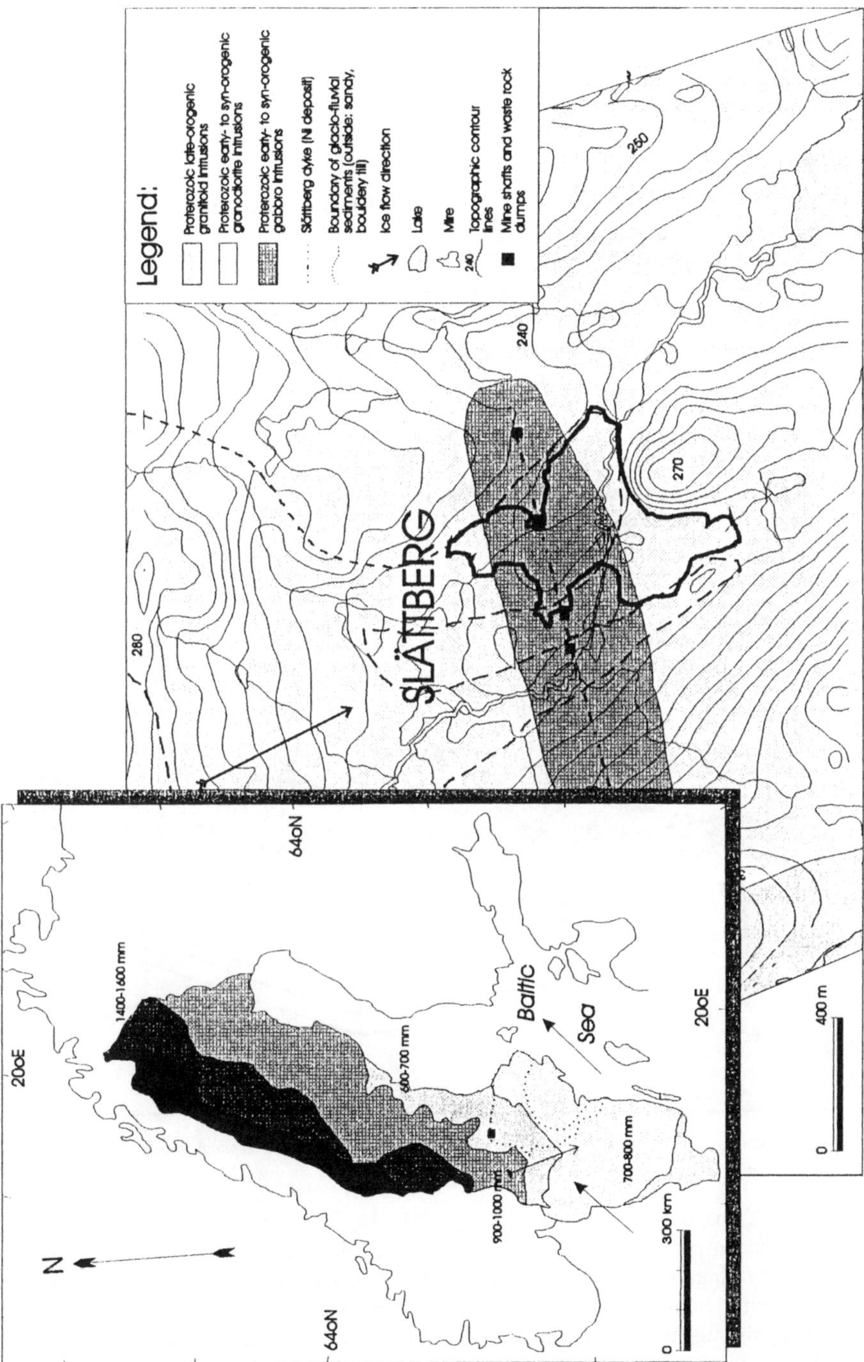

Figure 5. Hierarchical systems view. Location of the Slaettberg study area. *Inset:* the study area (solid box) is situated in the "southern boreal zone" of Scandinavia and falls into the region of the Svecofennian island arc lithologies of Early Proterozoic age (dotted line). Arrows indicate wind directions in winter; figures are average annual precipitations in millimetres. The single arrow of NW-SE direction shows regional ice movement during the last glaciation.

Step 3. Systems Modelling
At this step interdisciplinary knowledge of geology is applied for the spatio-temporal modelling of the studied systems. This is done by using integrated stratigraphy and basin analysis techniques of geology.

Geological case study
Differential dissolution before burial places the Early Cretaceous depositional environment in the Gerecse between the Aragonite and the Calcite Compensation Depth. The specific depth is also indicated by ammonites, significant planktonic/bentonic ratio and the presence of trace fossil *Zoophycos*. The analysis of the samples suggest that sedimentation rate was about 1 cm per thousand year: one or two magnitude lower than that of deep-sea fans. These features most likely indicate a mud or silt-dominated submarine slope with linear feeding systems (Fig. 6). Progradation of the slope is demonstrated by a coarsening and thickening upward sequence. The tectonics and sedimentological analysis indicate that slumps and debris flows of ophiolitic composition were deposited on a slope determined by a thrust front. Turbidites derived from the shelf edge are exposed by a dextral strike-slip fault cutting the thrust belt. Two significantly different transport routes were active: one is shown by debris flows and slumps with orientation to the SW and the other is represented by turbidites directed to the W.

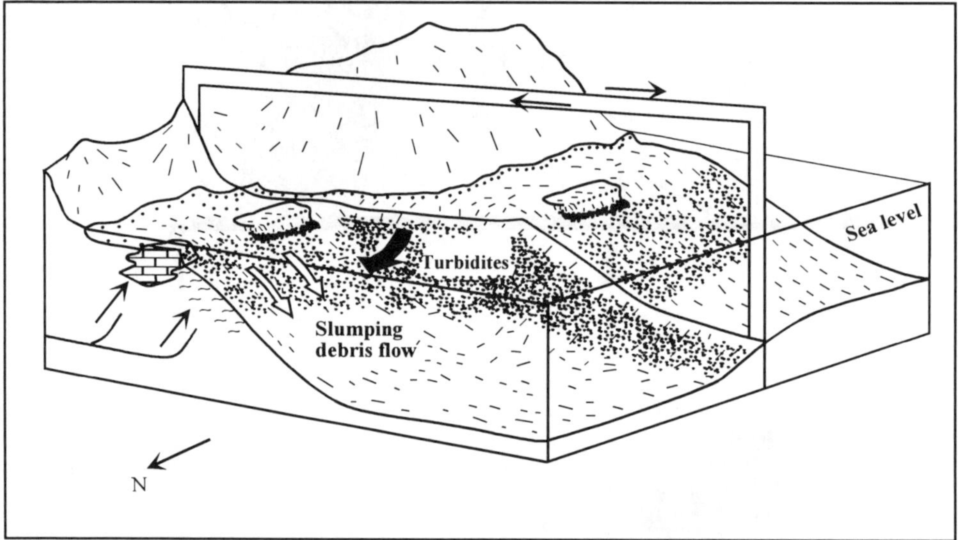

Figure 6. Geological systems modelling. Sedimentological and palaeogeographical model of Early Cretaceous sediments in Gerecse.

Environmental case study
Geochemical data interpretation and modelling was carried out using the robust techniques of Exploratory Data Analysis to treat multimodal data populations and accommodate information inherent in outlying values. The interpretation of results was enhanced by Multivariate Statistical Analysis. Analysis suggests that the main controlling phases of trace metal retention are organic matter and amorphous sulphides. Where AMD discharges into the mire a hydrogen sulphide-adsorption barrier forms in the peat sediments. The analysis and modelling of geochemical data revealed that under the acidic conditions in the bog only the most stable copper sulphides may precipitate within a relatively short distance of plume propagation (about 10 m). The observed phenomena can be most readily interpreted by the assumption of microsites in peat sediments where,

despite the bulk geochemistry of the sediments, microbial sulphate reduction and copper sulphide formation can take place. Where migrating groundwater becomes depleted with respect to Cu, H_2S remains in gaseous form and metal retardation is due to organic complexation. The studied metals exhibit markedly different mobility and adsorption stability in the acidic reducing peat sediments, Cu can be precipitated as stable sulphides, Pb is mostly organic bound while Mn and Ni are found mostly in the more labile exchangeable and water soluble phase, which shows that these environments are highly selective.

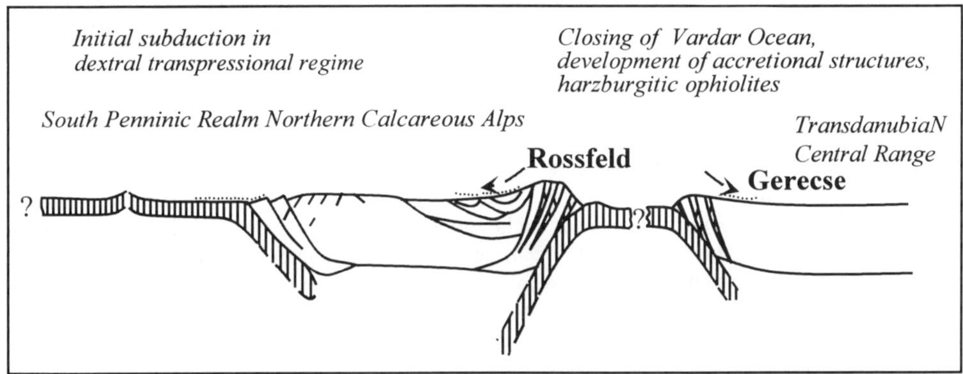

Figure 7. Integrated modelling of systems evolution. Position of Gerecse in the closing Vardar Ocean system (after Faupl *et al.*, 1992).

Step 4. Integrated Modelling: Spatio-temporal Systems Evolution

As the final phase of model formulation, testing and verification of the geological model is implemented by emplacing the modelled system in broader-scale situations and by analysing its functional relationship with neighbouring systems. If the studied system's spatio-temporal evolution can be interpreted in relation to other systems the results can be used for dynamic and predictive modelling.

Geological case study

The studied environmental facies or system in the Gerecse was interpreted as a prograding mud or silt-dominated submarine slope most likely with linear feeding systems. Other evidence on the ophiolitic composition of the turbidites, the location of the Gerecse in the Bakony Unit and the tectonic setting of the area lead to the conclusion that the modelled system belonged to the closing Vardar Ocean in the Early Cretaceous [25] and its evolution was controlled by the related macro-processes (Fig. 7).

Environmental case study

Based on the results gained from the Slättberg case study, it seems that despite the naturally high adsorption capacity of organic-rich sediments, they are selective towards different metals usually found together in mine waste waters. Wetlands are extremely vulnerable to dewatering introduced either by human activity or naturally, such as changes in climate or local drainage systems. Drying is not balanced by natural feedback mechanisms and leads to the irreversible destruction of the mire-peat sediment system. Since the metal retention of wetlands is based almost exclusively on that they are reducing waterlogged environments, their long-term metal retention potential and stability depends on only a single variable. Upon drying the sediments are eroded, organic matter and sulphides are quickly oxidised and released to the hydrological system [26].

CONCLUSIONS

The objective of modern integrated environmental research is the analysis and modelling of the integration of human social systems and natural systems. This requires an interdisciplinary approach and complex analysis of systems behaviour and processes. Up to the present, geology has provided its scientific tools and methods which, despite their importance in environmental research and management, cannot address the challenge to the solution of modern integrated environmental problems. It has been argued in this paper that geology could contribute to environmental research and decision support much more efficiently if its unique concepts and approaches were implemented in environmental analysis.

It has been shown that geological methodologies are applicable and needed in complex environmental modelling. In fact, geology is the only interdisciplinary science which offers well-established techniques for complex systems analysis, integration of processes on various scales and spatio-temporal environmental modelling. What geologists can contribute especially is the evaluation of geologic elements of the models and the overall co-ordination in combining the specialised expert fields. In order to implement geology in environmental practice there is a need for geologists in environmental research, who are fully aware of the concepts and approaches of geology and are capable of applying them in the solution of environmental problems.

Acknowledgements

We thank associate Professor Ulf Qvarfort at the Uppsala University for providing facilities for the geochemical analysis and Emõke Jocháné Edelényi, István Horváth and György Tóth at the Hunagrian Geological Survey for making possible the preparation of this paper. Suggestions for the paper from László Fodor are greatly appreciated. Scholarships were provided by the Swedish Institute, The Soros Foundation and the Szádeczky-Kardos Elemér Foundation. Their assistance is gratefully acknowledged.

REFERENCES

1. Z. Naveh and A. S. Liebermann. *Landscape ecology, theory and application.* Springer-Verlag, Inc., New York (1984)
2. R. T. T. Forman and M. Godron. *Landscape ecology.* John Wiley & Sons, Inc., New York (1986).
3. J. A. C. Fortescue. *Environmental geochemistry: A holistic approach.* Springer-Verlag, Inc., New York (1980).
4. R. J. Bennett and R. J. Chorley. *Environmental systems: Philosophy, analysis and control.* Methuen, London (1978).
5. M. G. Thruer, V. H. Dale and R. H. Gardner. Predicting across scales: Theory, development and testing, *Landscape Ecology* **3**, 245-252 (1989).
6. S. E. Jorgensen, B C. Patten and M. Straskraba. Ecosystems emerging: toward an ecological of complex systems in a complex nature, *Ecological Modelling* **62**, 1-27 (1992).
7. C. A. Wessman. Spatial scales and global change: bridging the gap from plots to GCM grid cells, *Annual Review of Ecology and Systematics* **23**, 175-200 (1992).
8. J. A. C. Fortescue. Landscape geochemistry: retrospect and prospect - 1990, *Applied Geochemistry* **7**, 1-53 (1992).
9. G. L. Lumsden (Ed). *Geology and the environment in western Europe.* Clerendon Press, Oxford (1992).

10. W. H. Schlesinger. *Biogeochemistry: An analysis of global change*. Academic Press Inc., San Diego, California (1991).
11. F. E. Eagler. *The way of Science: a philosophy of ecology for the layman*. Hafner, New York (1970).
12. J. A. Plant and R. Raiswell. Principles of environmental geochemistry. In: *Applied environmental geochemistry*. I. Thornton (Ed). pp. 1-40. Academic Press (1983).
13. A. M. Schultz. The ecosystem as a conceptual tool in the management of natural resources. In: *Natural resources: Quality and quantity*. S. V. Cirancy-Wantrup and J. J. Parsons (Eds). pp. 1339-148. University of California Press, Berkeley(1967).
14. I. S. Zonneveld. The land-unit - A fundamental concept in landscape ecology, and its applications. *Landscape Ecology* **3**, 67-86 (1989).
15. R. V. O'Neill, D. L. DeAngelis, J. B. Waide and T. F. H. Allen. *A hierarchical concept of ecosystems*. Princeton University Press, Princeton (1986).
16. A. Hartikainen, I. Horváth, L. Ódor, L. Ó. Kovács and J. Csongrády. Regional multimedia geochemical exploration for Au in the Tokaj Mountains, Northeast Hungary, *Journal of Applied Geochemistry* **7**, 533-545 (1992).
17. A. Fogarasi. Sedimentation on tectonically controlled submarine slopes of Cretaceous age, Gerecse Mts., Hungary—working hypothesis, *Általános Földtani Szemle* **27**, 15-41 (1995) (in Hungarian with English abstract).
18. G. Jordán, A. Szûcs, U. Qvarfort and B. Székely. Evaluation of metal retention in a wetland receiving acid mine drainage, *Proc. 30th Intern. Geol. Congr.* **19** (1997) (in Press).
19. M. Kázmér. Lower Cretaceous facies zones in the Bakony Unit, *Annales Universitatis Scientiarium Budapestiensis de Rolando Eötvös Nominatae, sectio Geologica* **28**, 161-168 (1988).
20. G. Gaál. Tectonic styles of Early Proterozoic ore deposition in the Fennoscandian Shield, *Precambrian Research* **46**, 83-114 (1990).
21. A. Ruhling and G. Tyler. Regional differences in the deposition of heavy metals over Scandinavia, *Journal of Applied Ecology* **8**, 497-507 (1971).
22. M. A. Glazovskaya. On geochemical principles of the classification of natural landscapes, *International Geological Review* **5**, 1403-1430 (1963).
23. A. I. Perel'man. *Landscape Geochemistry*. Vysshaya skhola, Moscow (1972) (in Russian).
24. A. I. Perel'man. Geochemical barriers: theory and practical application, *Applied Geochemistry* **1**, 669-680 (1986).
25. P. Faupl and H. Wagreich. Cretaceous flysch and pelagic sequences of the Eastern Alps: correlations, heavy minerals, and palaeogeographic implications, *Createceous Research* **13**, 387-403 (1992).
26. W. Shotyk. Review of the inorganic geochemistry of peats and peatland waters, *Earth-Sciences Reviews* **25**, 95-176 (1988).

Proc. 30th Intern. Geol. Congr., Vol. 26, pp. 251-260
Wang *et al.* (Eds)

Science Across Cultures: The Case of Vertebrate Palaeontology in China

PATRICIA KOMAROWER
Department of Ecology & Evolutionary Biology, Monash University, Clayton, Vic. 3168, Australia.

Abstract

At the beginning of the twentieth century, the hypothesis that central Asia was the cradle of mammalian evolution was proposed by Osborn (1910) and Matthew (1915). A multinational effort set out to prove or disprove this hypothesis and achieved much more. Collaboration with the Geological Survey of China led to the discovery and description of a substantial fraction of the fossil vertebrate record of northern China and the discovery of the Peking Man site in Zhoukoudian.

The five field seasons of the Central Asiatic Expeditions launched by the American Museum of Natural History resulted in the discovery of a rich and varied dinosaurian and mammalian fossil fauna and confirmed Osborn's and Matthew's hypothesis. A cooperative venture between the Geological Survey of China and the Rockefeller Foundation was set up in order to carry out a systematic study of the Zhoukoudian site and led to the foundation in 1929 of the Cenozoic Research Laboratory which became a focus of interaction between the Geological Survey of China and the international palaeontological community and provided an institutional training ground for Chinese vertebrate palaeontologists. The development of vertebrate palaeontology in China highlights the nature of the factors controlling the dissemination of scientific disciplines across cultures. The dominant factor in this case was cognitive in nature: the urge to explore the fossil record of central Asia and China in order to answer some central questions about the origin, evolution and dispersal of mammalian faunas, including man. Other factors, of a historical, economic and socio-cultural nature, have influenced the modes of interaction between the different scientific groups at work in the region.

Keywords: history of palaeontology, China, Peking Man, Cenozooic Research Laboratory, Central Asiatic Expeditions, Teilhard de Chardin

INTRODUCTION

The first flourishing of vertebrate palaeontology in China occurred during the 1920s and 1930s as a result of the significance of the fossil record of the region for the study of the evolution of major mammalian groups. Following the multinational exploration of this record facilitated by the Geological Survey of China, considerable information about the biogeographical evolution of the region was collected, the Middle Pleistocene Peking Man site in Zhoukoudian was discovered and a set of institutions specifically dedicated to the study of vertebrate palaeontology was developed in China.

This episode in the evolution of Chinese science has theoretical significance for historians of science as it is characterised by a complex network of factors, both external (sociological) and internal (epistemological) in nature. This paper will address the circumstances of the development of vertebrate palaeontology in China from a palaeontological and historical point of view in order to address the tension between these internal and external controlling factors.

CHINA AT THE TURN OF THE 20TH CENTURY: AN HISTORICAL OVERVIEW

During the late 19th century and the early 20th century which saw the end of the Qing dynasty and the transition to a republican state, Chinese political life and, more specifically, its interaction with outside powers, was framed by the tension between internal rebellions and outside aggression. In response to this double challenge, the Self-Strengthening Movement was initiated by Feng Guifen and implemented by Zeng Guofen and Li Hongzhang who had defeated the Taiping and Nian rebellions respectively [1].

According to Feng and Fairbank [2], the road to self-strengthening started from the inclusion of foreign languages, mathematics and science in the education curriculum and led to the use of modern military technology and the need to develop modern communications, arm manufacture and primary resources such as mining to provide supplies. However, this program had to face fervent nationalism from many quarters which led to incidents such as the Boxer Rebellion in 1899. The end of the Qing Dynasty and the transition to republicanism did not solve these tensions but led to a period of civil war and intellectual ferment fuelled by these tensions.

The geological institutions that were to be such an important partner in the development of vertebrate palaeontology in China were born from the need to assess resources and train geologists. According to Lee [3], the Geological Survey of China was inaugurated in 1916: its main function was to produce national geological maps and survey mineral resources. By 1931, it comprised several units including a Laboratory of Palaeontology dedicated to the study of invertebrate fossils and the Cenozoic Research Laboratory which focused on the study of vertebrate palaeontology and palaeoanthropology. The first university department of geology was created in 1918 at Peking University and was soon followed by several others. However the unsettled conditions that prevailed during the and the 1920s and 1930s made geological and palaeontological field work quite difficult and the Japanese invasion in 1937 put a temporary hold on it.

VERTEBRATE PALAEONTOLOGY AT THE TURN OF THE 20TH CENTURY: THE SIGNIFICANCE OF THE FOSSIL RECORD OF CENTRAL ASIA AND CHINA

As a result of the publication of Darwin's *Origin of Species* in 1859, the questions of the early record of life, the nature of intermediate forms (or missing links) and the special question of the evolution of man became of central importance to the evaluation of Darwin's thesis. The simultaneous developments in collecting, transportation and preparation techniques allowed the exploration and interpretation of distant fossil faunas and the gradual development of biogeographical studies.

Interest in the fossil record of man had been fuelled by the discovery, in 1856, of the fossil remains of Neanderthal man (*Homo neanderthalensis*) which was recognised, after much discussion, as a human ancestor and that, in 1891, of Java man (*Pithecanthropus erectus*, later renamed *Homo erectus*) which was originally interpreted by its discoverer Dubois as a genuine missing link between man and ape. This discovery suggested that Asia, rather than Europe, might have been the cradle of human evolution. The subsequent discovery in 1921 of the first stone artefacts in northern China confirmed this hypothesis and was instrumental in generating an intense interest in the Asiatic fossil record.

Post-Darwinian biogeography, according to Greene [4] was a synthesis between geology and palaeontology. It assumed the contraction hypothesis according to which the main geological features of the Earth, such as mountain ranges and the distribution of continents and oceans could be explained by the gradual contraction of the earth's surface as it cooled. The work of Suess, De Launay and Chamberlin in the early part of this century highlighted the tectonic nature of the factors controlling marine transgression cycles, climate change and the past and present distribution of faunas and floras. The subsequent emergence of alternative hypotheses such as continental drift and the permanentist model created a tension which informed the work on the origin and distribution of vertebrate groups by W. D. Matthew and H. F. Osborn who became interested in the Central Asiatic fossil vertebrate record as the potential cradle of mammalian evolution.

Henry Fairfield Osborn [5] published his seminal review of the mammalian faunas of the world, *The Age of Mammals*, in 1910. In his closing chapter, he emphasised the need for more information on the Asiatic fossil vertebrate record in order to corroborate his conclusions regarding the close affinities between mammalian faunas of the Holarctic region (which includes North America, Asia and Europe).

One of Osborn's colleagues at the American Museum of Natural History, William Diller Matthew [6], tackled the problem of the past and present distribution not only of fossil mammals, but of other vertebrate groups, invertebrates and plants (Matthew, 1915). Matthew demonstrated that the central Asiatic region was an important centre of evolution and dispersal for man as well as other mammals such as horses, antelopes and elephants. The two overwhelming conclusions of Matthew's work were that Central Asia had been a major centre for mammalian evolution and that further exploration, especially of the Tertiary fossil record, was needed in order to refine the conclusions drawn from an often incomplete and poorly documented record.

THE DISCOVERY OF THE CENTRAL ASIATIC AND CHINESE FOSSIL VERTEBRATE RECORD: AN INTERNATIONAL EFFORT

The Peking Union Medical College

As a result of the investigation into Chinese medicine carried out by the Rockefeller Foundation in 1908, a medical college, the Peking Union Medical College, was established in China with the aim of extending the training of a number of Chinese doctors whose training in Western medicine until then was restricted to missionary schools of unequal standards and conditions.

In 1919, Davidson Black, a Canadian medical graduate from the University of Toronto, was appointed to the Chair of Neurology and Embryology at the Peking Union Medical College. In the course of his career, he had become interested in comparative anatomy and anthropology and had been strongly influenced by Matthew's (1915) theory about the importance of central Asia for the dispersal of mammals. As will be shown further in this paper, Black's appointment was critical in establishing the importance of the Zhoukoudian Peking Man site which resulted in the setting up of the Cenozoic Research Laboratory, a cooperative project between the Peking Union Medical College and the Geological Survey of China.

The Central Asiatic Expeditions

The initial proposal, planning, organisation of the Central Asiatic Expeditions were very much the brainchild of Roy Chapman Andrews, although it had the full support and backing of Henry Fairfield Osborn, the Head of the Department of Vertebrate Palaeontology and the President of the American Museum of Natural History. Osborn received the plan enthusiastically for several reasons: (1) it was a chance to confirm the biogeographical ideas he had developed in *The Age of Mammals*, (2) the palaeoanthropological dimension of the project opened new opportunities for Osborn, and it was likely to have enormous popular appeal which, in turn, would make the promotion and funding of the Expedition much easier and (3) many of Osborn's business friends and patrons were interested in the opportunities offered by Asia in the context of the establishment of the Open Door policy in China in 1899 [7].The five field seasons of the Central Asiatic Expedition of the American Museum of Natural History took place in 1922, 1923, 1925, 1928 and 1930. They were centred on the Gobi Desert, as an agreement had been made between the Geological Survey of China and the Central Asiatic Expeditions that "the Expeditions would not enter upon geological, palaeontological or archeological explorations in northern China".

Johan Gunnar Andersson: An economic geologist with a difference

The Swedish presence in China at the beginning of the twentieth century did not originate from a primary interest in the fossil record of the region, but from a cooperative agreement between the Chinese and Swedish governments aiming at the exploration and exploitation of Chinese ore deposits. J. G. Andersson, the Director of the Geological Survey of Sweden, arrived in Peking in 1914, where he had been invited as a mining adviser by the Chinese government. Andersson's activities, however, soon expanded from the field of economic geology to include palaeontology, palaeoanthropology and archaeology. Andersson had read Schlosser's review of the mammalian fossil record of China [8] and he wrote to Christian missionaries and foreign residents, asking for help to locate "dragon bone" localities. He became interested in dating the loess deposits first identified by Von Richthofen and suggested using the vertebrate fossils they contained to date them. He contacted Carl Wiman, a vertebrate palaeontologist at the University of Uppsala, Sweden. Wiman undertook to work on the material collected by Andersson and sent him an assistant, Otto Zdansky, who arrived in 1921. The collection and transport of the fossils was financed by the China Research Committee created by Axel Lagrelius, a rich Swedish industrialist. The collaboration between Andersson and Zdansky was to be very productive, although tense at times, and led to the discovery of the Peking Man site at Zhoukoudian, southwest of Peking.

The French Jesuits: Emile Licent and Pierre Teilhard de Chardin

The last, but not least, element in this multinational community of scientists at work on the fossil vertebrate record of central Asia was the pair of French Jesuits, Emile Licent and Pierre Teilhard de Chardin, originally attached to the Hoangho–Paiho Museum in Tianjin which was part of the Jesuit College there. Fr. Licent was an entomologist and the curator and founder of the Hoangho–Paiho Museum. After his arrival in China in 1914, Licent went on many field trips, mostly along the Huang He (Yellow River) valley to collect geological, zoological and botanical specimens for his museum. In 1921 he discovered two localities in the Qingyang region (Gansu Province) which contained Late Pleistocene stone artefacts. These were the first palaeoliths found in China, and their discovery corrected the view that no stone age human had lived in China. This inspired Licent to proceed with his search for human fossils.

In 1922, Licent travelled to the southeastern corner of the Ordos Plateau in Inner Mongolia. In a bend of the Yellow River, he found a great variety of vertebrate fossils. Licent requested the expertise of a vertebrate palaeontologist, and Professor Marcellin Boule from the Museum of Natural History in Paris sent him Teilhard de Chardin, who arrived in China in 1923. Teilhard de Chardin continued to work in China for most of the time until 1946, and he had a most important impact on the development of vertebrate palaeontology and palaeoanthropology in China. Teilhard de Chardin had previously worked on mammalian fossils from the Early Cainozoic of France and

Belgium. In China, he moved up-section into the Late Tertiary and Quaternary. Teilhard was also a geologist and he became interested in the sedimentological, tectonic and igneous history of China as well as its fossils. His Chinese career can be divided into three periods.

The first period (1923-1928) was dedicated to studies on the stratigraphy of the Late Cainozoic, the fossil mammals of the period and the remains of Early Man. Teilhard's work during this first period was closely associated with Licent and the Hoangho Paiho Museum in Tianjin. In 1924, Teilhard de Chardin started work on the Lower Pleistocene mammalian fauna of the Zhoukoudian (Peking Man) site which had been discovered by Andersson and Zdansky.

During the second period (1929-1940) his contributions to Chinese geology and palaeontology led to his nomination in 1929 as a scientific adviser to the Geological Survey of China and his subsequent involvement in geological studies sponsored by the Geological Survey as well as the developing program at Zhoukoudian. He took part in the 1930 Central Asiatic Expedition and in the 1931 Haardt-Citroen Croisiere Jaune Expedition as adviser which provided him with the opportunity to carry out stratigraphic correlation that provided a stratigraphic framework for the search for fossil hominid remains in the region.

The final period (1940-1946) was one of analysis and interpretation of the material he had collected as the war made field work impossible except in the area around Peking. He also established with another Jesuit, Pere Leroy, the Peking Institute of Geobiology and an associated journal *Geobiologia*, whose main aim was to focus on studies of continental evolution, tracing the joint evolution of rocks and organisms within the framework of the Asiatic continent.

PEKING MAN, THE CENOZOIC RESEARCH LABORATORY AND THE DEVELOPMENT OF VERTEBRATE PALAEONTOLOGY IN CHINA

During the second decade of the twentieth century, the fossil vertebrate record of Central Asia had became a focus of interest to the international palaeontological community. The hypothesis that Central Asia was the cradle of mammalian evolution had been proposed by Osborn (1910) and Matthew (1915). A multinational effort set out to confirm or refute this hypothesis and, in the process, achieved much more:

1. The Central Asiatic Expeditions considerably expanded current knowledge of the central Asiatic fossil vertebrate record: the earliest known fossils in 1915 were Late Tertiary in age; by the end of the 1930s, the fossil record of the region had been extended back to the Permian and considerable information about the geological and biogeographical evolution of central Asia had been collected.
2. Osborn's and Matthew's hypothesis was confirmed
3. The Middle Pleistocene Peking Man site in Zhoukoudian was discovered
4. As result of these new discoveries, and particularly in response to the need for the systematic study of the Zhoukoudian site, funds and personnel were poured into the region and cooperative ventures set up with the Geological Survey of China, leading to the foundation of the Cenozoic Research Laboratory and the development of a Chinese palaeontological community.

Let us look at some important dates in the story of the Cenozoic Research Laboratory. It started in 1921, with the arrival in China of Otto Zdansky, a vertebrate palaeontologist sent by C. Wiman from the University of Uppsala in response to Andersson's request for expert help to identify the fossil material he had been collecting. During the summer of that year, they jointly discovered the Zhoukoudian Peking Man site as a result of a report by a local resident: fossil mammal remains, stone tools and the first human tooth were quickly recovered from that site. By the 1920s,

Andersson's ethnographical, archaeological and palaeontological activities were taking up an enormous amount of his time, and a special agreement had to be established with the Geological Survey of China. He negotiated a contract for 1924-1927 whereby he continued in his capacity as an adviser but donated his salary to start a fund dedicated to the description and publication of the material he had collected in a monograph series, Paleontologia Sinica to be published by the Geological Survey of China.

The contract between Andersson and the Geological Survey of China included provisions concerning the conservation of specimens. Fossil vertebrates were to be sent to Wiman at the University of Uppsala and fossil plants were to go to Halle at the Swedish Museum of Natural History in Stockholm. Duplicates of vertebrate and plant fossils were to be made and sent back to China. All fossil invertebrates stayed in Peking, although duplicates were made and sent to Stockholm in exchange for duplicates of Swedish invertebrates. Andersson became the Curator of the Museum of the Geological Survey of China in 1921.

In 1926, a meeting took place at the Peking Union Medical College with the Crown Prince of Sweden. The Prince was the Chairman of the Swedish-China Research Committee which had funded Andersson's work. At the meeting, various papers were read, summarising the progress of work on the Chinese fossil vertebrate record: Teilhard de Chardin on the fossil sites in the Ordos Plateau explored with Licent, Andersson on the work done at Uppsala by Wiman on the Chinese fossil material and Black on the Zhoukoudian molars found and described by Zdansky which were subsequently published by *Nature* and *Science*. Black also applied to the Rockefeller Foundation for funds to finance a 2-year research program at Zhoukoudian. As a result, a cooperative venture for the systematic study of the Zhoukoudian site was set up. The project was jointly funded by the Rockefeller Foundation and the Geological Survey of China and the Cenozoic Research Laboratory was established. A successful series of field seasons at Zhoukoudian brought up a wide variety of human and mammalian fossil remains. Chinese and foreign palaeontologists were associated with this work: Pei Wenzhong, the discoverer of the first human skull and Yang Zhongjian (C. C. Young), Teilhard de Chardin, Davidson Black until his death in 1934 when he was succeeded by Weidenreich. The 2-year Rockefeller grant was extended and work continued until 1936. It resumed after Liberation under the auspices of the Institute of Vertebrate Palaeontology and Palaeoanthropology with special emphasis on dating and palaeoenvironment reconstruction.

What are the main factors that have allowed vertebrate palaeontology to flourish in China after 1920? There is no doubt that the initial success in answering the questions asked from the fossil vertebrate record of the region was the single most important factor in attracting the funds and personnel needed to further explore that record and develop the institutional framework. The regional nature of palaeontology and its links to another regional science, geology, are critical, as there was a substantial geological network already in existence in China by the 1920s. This geological network was very important in interacting with individuals such as Andersson and Teilhard de Chardin (who were employed by the Survey at different times) and negotiating with foreign institutions favourable conditions regarding funding, conservation of specimens and joint projects. These conditions allowed a local network of Chinese palaeontologists to develop more rapidly and to have more control over their research and training.

This growth, however, was cut short by historical events such as the invasion of China by Japan in 1937. As a result, some of the Peking-based institutions were closed because of their foreign connections (such as the Peking Union Medical College) or relocated in Chongqing in Sichuan Province (southwest China) along with most other governmental institutions in what was the Chinese capital during the war.

INTERPRETATION

This first flourishing of vertebrate palaeontology in China has theoretical significance for historians of science who have long been interested in the dynamics of the development of particular branches of scientific knowledge in particular places and at particular times. There are 2 approaches possible to this problem: one, the internalist approach, looks at explanations based on the cognitive and logical structure of scientific knowledge, while the second, the externalist approach, focuses on the social dynamics of scientific activity as well as the social context within which it is embedded. It soon became apparent that, in order to do justice to individual case histories, one needed to combine both in what could be loosely described an ecological approach. Using this third approach, one could produce a multidimensional account of scientific activity and of the epistemological, cultural, sociopolitical and ideological spaces in which it is situated. This is particularly useful in a cross-cultural context.

Basalla [9] was the first to attempt to model the pattern of events taking place when Western-style science became established in various parts of the world. His three-phase diffusionist model comprises an exploration phase, a colonial phase and an independent phase. The exploration phase (which lasted until the 1920s in our example) is characterised by the opening up of a new area which can act as a source of data for Western science. In China, this phase started after the middle of the 19th century with the expeditions of the American geologist, R. Pumpelly, the German geographer, Von Richthofen and the Russian geologist, V. A. Obrutchev. During the second colonial phase (1920s to 1950s in our case), a local scientific community starts to develop but the activities of this community are still very much framed by the interests of the imported Western science. The transition to the third, independent, phase is characterised by the replacement of the external scientific culture by a national one, characterised by the development of national scientific institutions, local education and employment of scientists and the opening up of new fields of study within this scientific community which can start acting as a centre for the diffusion of modern science. The Basalla model has often been criticised for its general character (failing to take into account specific national backgrounds and scientific disciplines) and its superficial treatment of the political context. However it offers a useful framework to approach these issues and is currently used, albeit with some reservations, by some Chinese historians of geology such as Yang [10]. There is also an internalist dimension to the shortcomings of Basalla's model: its lack of cognitive focus, its failure to incorporate the epistemological dynamics of the cross-cultural episode under study.

One of his critics, Macleod [11], in his *Reflections on the Architecture of Imperial Science*, gave an account of the relationship between science and imperialism and the associated cultural and economic interdependence. He attempted to correlate the main stages of imperial development with different aspects of scientific practice, institutional structures, social and political characteristics, economic and technological functions. He recognised five phases, based on his studies of Australia and other parts of the British Empire. Macleod's model is particularly useful when it is charting the political dimension of imperial science, the changing nature of vested interests both in Britain and in the colonies as well as the complex nature of the cultural and economic interdependence associated with the natural history tradition. However, his model is still very much an externalist model, more concerned with the context of science than with its internal dynamics and its cognitive content.

David Wade-Chambers [12] attempted to deal with the tension between the internalist and externalist dimensions of the periodization schemes used by historians of science and the difficulty of choosing a useful historical framework for "colonial science". The rewards of constructing such a framework are great: a better understanding of the process of diffusion of ideas from one cultural setting to another, an easier cross-cultural comparison between various episodes of scientific development, an apprehension of the institutional framework necessary for this development. He concluded that the road to a general framework for a cross-cultural history of science must be grounded in multiple case studies of the multi-dimensional network nature of science.

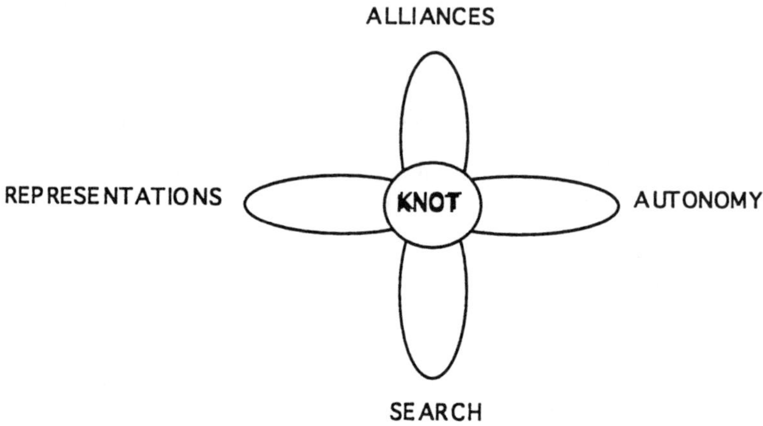

Figure 1. The rose-window model (after Latour and Polanco, 1900).

Latour & Polanco's Rose-Window model [13] attempted to bridge the gap between the internalist and externalist approaches and integrate the contextual and epistemological aspects of scientific knowledge. Their model is based on Polanco's concept of "science-world' which itself is an adaptation of Braudel's concept of 'economy-world'. In contrast to the more widespread concept of 'world-science" which, in spite of its apparent neutrality, often refers to mainstream Western science, the phrase 'science-world' more specifically describes any autonomous section of global science capable of providing for most of its needs (journals, institutions, scientific community) and possessing an epistemological unity. Latour and Polanco's rose-window model (Fig. 1) is a graphic representation of the network aspect of scientific activity; it comprises 5 interlocking circles and resolves the opposition between conceptual and social aspects, scientific content and scientific context. The first circle (the 'search') represents the dynamic interaction between the natural world and science. Here, we have those questions asked from the Central Asiatic and Chinese fossil record and the expeditions, collection and interpretation of the fossil material which produced answers to these questions. The second circle ('autonomy') reflects the gradual emergence of institutions (training, research and publications and the development of relationships with related disciplines. Here, we have the foundation of the Cenozoic Research Laboratory, the development of palaeontological museums (in Tianjin and at the Geological Survey of China). The third circle is the expression of 'alliances' with other social groups such as missionaries, interested laymen (informants) and also rich and influential groups such as the Geological Survey of China, the Swedish China Research Committee, the Rockefeller Foundation, which could facilitate the funding and growth of institutions. The fourth circle, is 'representation' or presentation of science to the public in such a way that ground will be laid for the acceptance of a particular theory or discipline by society at large. This aspect is probably not so important in our case study. The 'knot' in the centre (our fifth circle) is formed out of the scientific results obtained and their success and fruitfulness in terms of the other circles: cognitive, institutional and public. It is the dynamic core of this model whose growth is a reflection of the strength of the discipline under consideration.

CONCLUSIONS

The Latour and Polanco model successfully integrates the network of factors at work in the development of vertebrate palaeontology in China:
— the cognitive drive to explore the central Asiatic and Chinese fossil vertebrate records,
— the regional nature of palaeontology and its association with an existing geological network,
— the multinational interaction, collaboration and competition between individuals and their personal and institutional networks, and
— the historical framing of this First Golden Age of Chinese Palaeontology between the agony of the Chinese Empire and the Japanese invasion.

One of the strengths of this model resides in its account of the multiple dimension of scientific activity and their dynamic interaction. Future comparative studies of the developemnt of vertebrate palaeontology in several non-western countries and of the development of such fields as chemistry, plate tectonics and medicine in China should allow historians of science to understand better the nature of the relationship between these dimensions and to generalise the model beyond this case study of the development of vertebrate palaeontology in China.

REFERENCES

1. J. Spence. *The search for modern China.* W. W. Norton & Company, New York and London (1990).
2. S. Y. Teng and J. K. Fairbank. *China's response to the West.* Harvard University Press, Cambridge, Massachusetts and London, England (1979).
3. Lee, T. S. Y. *Geological Sciences in Republican China, 1912-1937.* A Dissertation submitted to the Faculty of the Graduate School of State University of New York at Buffalo in partial fulfillment of the requirements for the degre of Doctor of Philosophy (1985)
4. M. Greene. Plate tectonics and biogeography in historical perspective, *Earth Sciences History* **4**, 93-97 (1985)
5. H. F. Osborn. *The age of mammals in Europe, Asia and North America.* Macmillan (1910).
6. W. D. Matthew. Climate and Evolution, *Annals of the New York Academy of Sciences* **XXIV**, 171-378 (1915).
7. R. Rainger. *An agenda for antiquity – Henry Fairfield Osborn and vertebrate palaeontology at the American Museum of Natural History, 1890-1935.* The University of Alabama Press, Tuscaloosa and London (1991).
8. M. Schlosser. Die fossilen Saugethiere Chinas nebst einer Odontographie der recenten Antilopen, *Abhandlungen bayer Akademie der Wissenschaften* **22**, 1-221 (1903)
9. G. Basalla. The spread of Western science. *Science* **156**, 611-622 (1967)
10. J. Y. Yang. The transference of modern geology to China (1840-1937) with a discussion of Basalla's model. In: *Interchange of geoscience ideas between the East and the West – Proceedings of the XVth International Symposium of INHIGEO, 1990, Beijing.* H. Z. Wang, G. R. Yang and Y. Y. Yang (Eds). pp 31-38. China University of Geosciences Press (1991)
11. R. MacLeod. On visiting the "moving metropolis": Reflections on the architecture of imperial science. In: *Scientific colonialism – A cross-cultural comparison.* N. Reingold and M. Rothenberg (Eds). pp 217-249. Smithsonian Institution Press, Washington, D. C and London (1987).
12. D. Wade-Chambers. Period and process in colonial and national science. In: *Scientific colonialism – A cross-cultural comparison.* N. Reingold and M. Rothenberg (Eds). pp. 297-321. Smithsonian Institution Press, Washington, D. C and London (1987).
13. B. Latour and X. Polanco. A propos de l'histoire social des sciences: Quelques remarques, le modele de la rosace. In: *Naissance et developpement de la science-monde. Production et*

reproduction des communautes scientifiques en Europe et en Amerique Latine. X. Polanco (Ed). pp. 10-52. Editions La Decouverte, Paris (1990).

Proc. 30th Intern. Geol. Congr., Vol. 26, pp. 261-272
Wang *et al.* (Eds)

Modern Concepts in Palaeontology and Early Life on Earth

WANG HONGZHEN AND WANG XUNLIAN
Department of Geology, China University of Geosciences, Beijing, 100083, P. R. China

Abstract

Traditional evolutionary ideas of gradualism and uniformitarianism prevailing before 1950 were replaced by modern concepts in palaeontology based on biological events and neocatastrophic ideas. The new stage of palaeontology was marked by Schindewolf's work in the middle of this century. After Simpson's monumental treatise on evolution in the forties, fossil breaks and biotic extinctions successively discovered were summarized by Newell in 1967. Gould and Eldredge advanced the doctrine of punctuated equilibria in evolution and of cladistics in classification in the seventies. Alvarez et al. raised the impact viewpoint for the Cretaceous-Tertiary extinction in the beginning of the eighties, and serial papers by Raup and Sepkoski on periodicity of mass extinctions in the geologic past and their extraterrestrial causes appeared in the early eighties. In the eighties Raup reviewed mass extinctions, and Walliser summarized global bioevents and evolution. In recent years a scrutiny and retrospect of these matters has been made by Gould, Raup, and Sepkoski. These new concepts were introduced to China in the early eighties.

Study of the early life history on earth began in the middle of the century. In the late fifties Glaessner described the first known metazoan Ediacara fauna. Cloud in later years discussed the pre-metazoan evolution and biogeologic history. Schopf summarized the earliest life records on earth in 1983. Meso- and Neoproterozoic microbiota were reported in the seventies by Xing in North China. In recent years eucaryote fossils and macroalgae remains were discovered in Mesoproterozoic bearing ages of 1700 and 1450 Ma. Ediacara type of fossils were discovered from Upper Sinian in Hubei, and annelid remains from Upper Sinian in Huainan, Anhui, which may be older than the Ediacara fauna. Of special importance was the discovery of the Chengjiang faunal treasury representing an Early Cambrian biologic explosion, which has been partly worked out by Chinese palaeontologists. The Ediacara fauna, the Chengjiang fauna and other biologic explosions and radiations in earth's history are strong support for the doctrine of neocatastrophism and the idea of punctuated progression in organic and inorganic evolution as well.

Keywords: modern concepts, biologic explosion, uniformitarianism, punctuated equilibria

INTRODUCTION

Here the term "modern concepts" is used in a historic sense and for distinction from the traditional ones, although some of the terms mentioned have been used for years and are not altogether new. Early life on earth covers a very wide scope of study and is of important significance both in palaeontology and geology. In this paper the fossil life discussed is confined to those prior to the Middle Cambrian.

From the seventies of this century onwards, new ideas and concepts in palaeontolgy have been thriving, and the development of theoretical palaeontology in particular has stepped into a new stage. The linkage between palaeontology and biology was emphasized, the term "palaeobiology" was used, and a journal bearing that name was initiated in 1975.

One of the basic characteristics of modern palaeontology consists in the recognition and re-evaluation of the mutual relation and relative importance between gradualism and uniformitarianism on the one hand, and saltationism and catastrophism on the other, in the fundamental processes of biologic evolution. Actually discussion and controversy over this problem have never ceased. In fact, this problem has remained the core of discussion in palaeontology and even in geology as a whole, because it encloses the basic concept and leading thought that direct the tendencies in the whole field of research. One of the major tasks of modern palaeontology has been to establish a new theme in biologic evolution in interpretation of the revolutionary processes of mass extinction, radiation and explosion in the history of life and to explore their controlling factors, including the interactions between the biosphere and other shallow earth spheres.

MODERN CONCEPTS IN PALAEONTOLOGY

The Plate Tectonics Theory that was established in the late sixties is usually regarded as a revolution in geoscience. Modern concepts and ideas in palaeontology flourished in the seventies. This may have partly come from innovation of geological ideas affected by plate tectonics, and partly from renewed interests and discussions on biological evolution among geologists. In the following only some of the outstanding palaeontological ideas are discussed.

Uniformitarianism or Neocatastrophism

Uniformitarianism is a basic idea in geosciences and is not confined to palaeontology. Indeed, the contradiction between uniformitarianism and catastrophism is concerned with the fundamental time-space relation concept in geosciences as a whole. Uniformitarianism has dominated geological thinking since Lyell's time, but has been challenged for over a century and a half. Traditional evolutionary ideas characterized by gradualist and uniformitarian views had been predominant until the fifties of this century. Neo-Darwinism or the modern synthesis of J. Huxley [1], and the theories of evolution of G. G. Simpson [2] in the thirties and forties seemed on the whole inclined to gradualism. It was the embryologist R. Goldschmidt (1940) [3] who first distinguished microevolution at species level from macroevolution at higher taxonomic levels. The distinction between micro- and macroevolution marked the recognition of the importance of catastrophic change in the history of life.

The year 1950 marked a turning point in the history of paleontologic thinking, as it was in that year that Schindewolf related the modern idea of catastrophism in his “Principles”, and used the term Neocatastrophism in 1962 [4]. Meanwhile, Eldredge and Gould, impressed by the discontinuity of fossil records in earth’s history and inspired by the new evolutionary ideas, proposed in the early seventies the concept of “Punctuated Equilibria” in evolution [5, 6] in place of the traditional phyletic gradualism. This was accepted by the majority of paleontologists within a few years. Consequently in classification and taxonomy, evolutionary systematics was also soon replaced by cladistic systematics, as reviewed and summarized by S. W. Stanley [7] in 1982.

The phenomenon of biotal mass-extinctions in the history of the earth has long been known and was excellently summarized by N. Newell [8] in the sixties. The Alvarezs’ [9, 10] renowned papers (1982, 1984) calling for an extratelluric impact cause for biotal mass extinction brought about a high tide in biotal extinction studies. Eight extinction events from Permian to recent at an interval of ca 26 Ma were recognized (Fig. 1) by Raup and Sepkoski [11, 12], and five cycles for the whole Phanerozoic by Sepkoski recently [13] (Fig. 2). As revealed by our recent studies (Fig. 3), more extinction cycles, probably eight, may be discerned in the Phanerozoic, and which are roughly correlatable with our Megasequence in sequence stratigraphy and sea level changes, averaging 60-120 Ma of time intervals [14]. We would like to point out that this is in good accord with House’s early diagram of 1971 [14] for the appearance and extinction of invertebrates, and vertebrates as well as plants. From the eighties onwards, many palaeontologists, notably Walliser [15, 16] and Sepkoski [17], laid special emphasis mainly on extratelluric causes of bioevents in earth's history, and a large number of publications in this respect have appeared. Probably it may be said that Uniformitarianism as a record of fact has to be abandoned, but as an idea and way of thinking it is still and probably will still be influential among earth scientists for a considerable period of time. For our part, we consider that, at present, the uniformitarian concept is a harm rather than a benefit as far as palaeontological and geological thinking is concerned.

Phyletic Gradualism or Punctuated Equilibria

The dominant doctrine in biological classification and evolution had been phyletic gradualism and the modern synthetic theory or neo-Darwinism before the fifties of this century. Phyletic gradualism holds that species arises from very slowly and evenly proceeded changes that result in gradual divergence of forms. On the other hand, punctuated equilibria advocated by Eldredge and Gould [5], regard speciation as a rapid process. Species originate in a very short time span and remain relatively stable after its origination for a rather long period of time. The history of biologic evolution is a story of homeostatic equilibria disturbed and broken by rapid and episodic events of speciation and generation of superspecific ranks. This new concept met with general welcome by palaeontologists, especially as it may interpret the abundant missing links of taxonomy in earth's history, which had long vexed the taxonomist and were usually explained by incomplete records. Although Malmgren [18] and others advanced the idea of punctuated gradualism in interpretation of the comparatively rapid yet gradual speciation of the Neogene pelagic foraminifera, and Mayr [19] still held that the idea of punctuated equilibria is enclosed in the modern synthetic theory, the alternation between a slow and stable phase and a rapid episodic phase in evolutionary processes seems to have been established.

Evolutionary concepts will naturally directly affect ideas in systematic classification. Cladistics or cladistic systematics is closely tied to punctuated equilibria, and evolutionary systematics based on phyletic gradualism was soon replaced by cladistics. The forerunner of cladistic systematics was the idea of phylogenetic systematics initiated by Hennig [20] in 1966, which strengthened an overall analysis of the phylogenetic relationship between the biocharacters, and put special emphasis on the importance of macroevolution. Stanley [7] made a very good review of macroevolution in 1982.

The outburst of modern concepts and ideas in palaeontology occurred in the seventies, and was soon brought to the attention of Chinese palaeontologists. Among the earliest to introduce these ideas to China may be mentioned Chow and Zhang [21] on cladistics and Yin [22] on punctuated equilibria. Later Xu [23] gave a special analysis of their geological significance. Good work using this new concept was done in regard to Palaeozoic brachiopods by Rong and others [24].

DEVELOPMENT OF EARLY LIFE ON EARTH

The early life on earth referred to here comprises the life prior to the Middle Cambrian, whose lower boundary is taken as approximately 540 Ma (see later), and our discussion will be confined to the main bioevents during this duration of time. First we would like to point out that it may be difficult to designate the first appearance of the different levels of life, which is subject to constant changes by new discoveries. But it may be easier to assign the extinction and to a certain extent the radiation epochs of them. In the following we will give a brief discussion on important epochs of biotic profusion and explosion in the order of taxonomic levels and of geological age (Fig. 4).

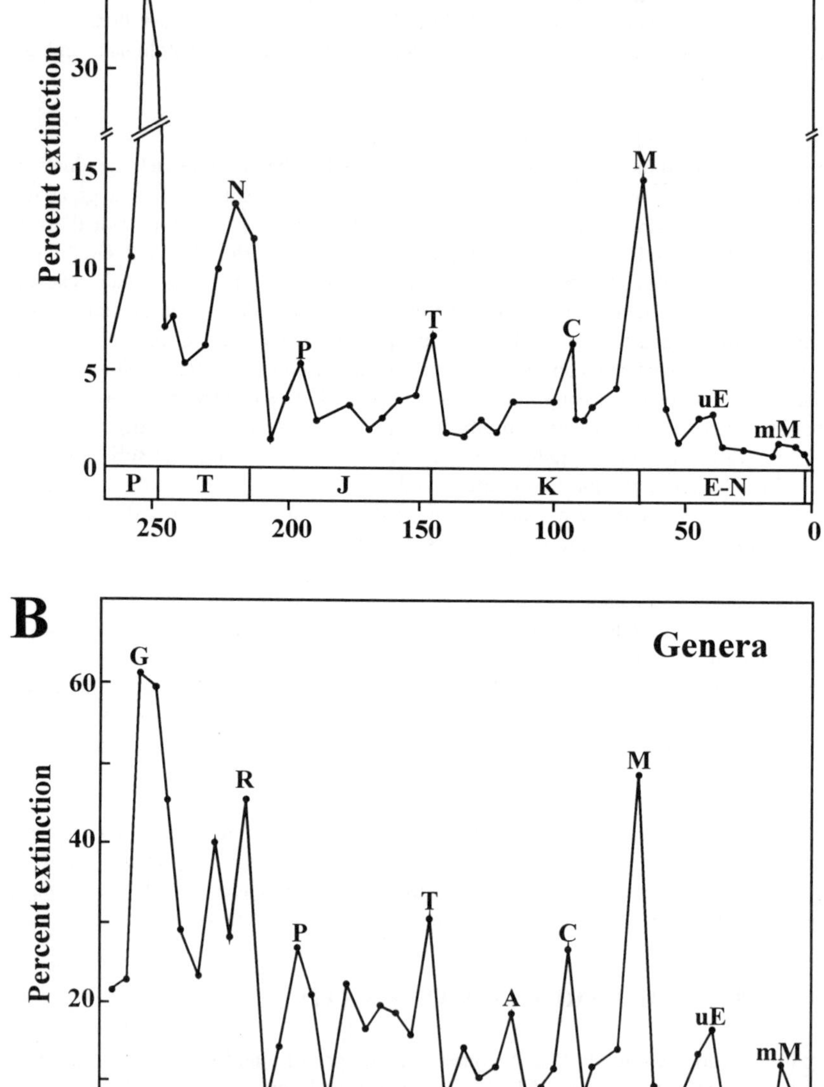

Figure 1. Percent extinction for marine families (A) and genera (B) for the Permian to recent (after Raup and Sepkoski, 1986).

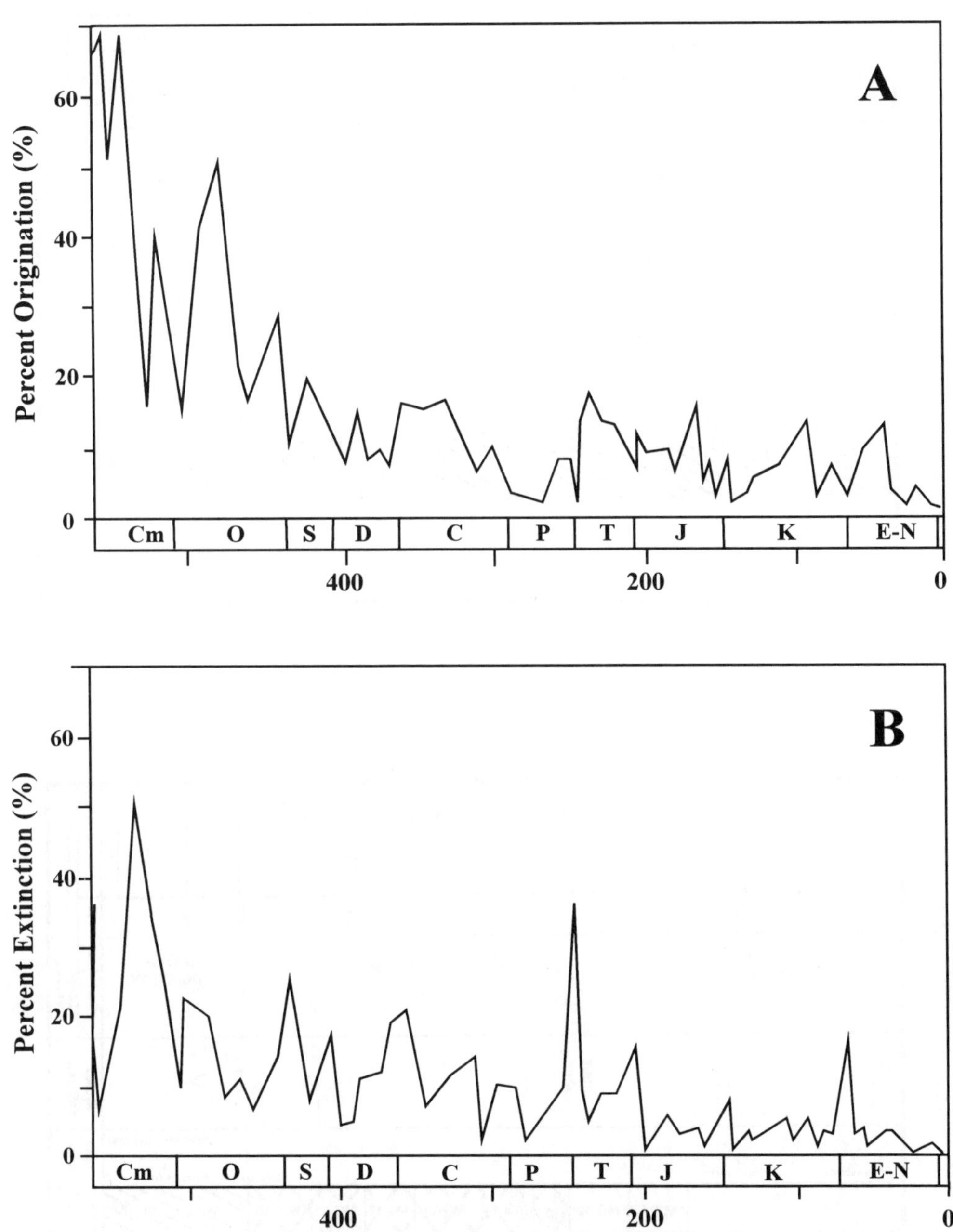

Figure 2. Origination and extinction of organisms in the Phanerozoic (after Sepkoski, 1993).

Geologic Time: Era | Period | Ma

Era: Nz, Mz, Pz

Period: Quaternary, 1.64, Ng, 23.3, Pg, 65, K, 135, J, 208, T, 250, P, 290, C, 362, D, 407, S, 439, O, 510, Cm, 570, Z

Ma: 0, 50, 100, 150, 200, 250, 300, 350, 400, 450, 500, 550, 600

Tectonic Stage: Megastage

INTRA-CONTINENTAL DEVELOPMENT

PANGEA FORMATION

Pangea cycle

Pangea 250

Bio-event

Angiospermae, Aves, Gemnospermae, Insects, Ferns, Brachiopods, Shelly Triobites

Mammals, Primates, Dinosaurs, Diapsids, Reptilids, Amphibids, Agnathids, Anthozoans, fossils, Chordata

Sequence: Mega-Sequence

MG-1, MG-2, MG-3, MG-4, MG-5, MG-6, MG-7

Sea-level change: Rise ⟷ Fall

Plume-related CFB; Biotic explosion; Biotic mass-extinction; Glacigene deposits

Geologic time: Era | Period | Ma

Era: NPT, MPT, PPT, Archean

Period: 570, Z_2, 700, Z_1, 800, Qb, 1000, Jx, 1400, Ch, 1800, Ht, 2100, Ga, 2500

Ma: 600, 700, 800, 900, 1000, 1100, 1200, 1300, 1400, 1600, 1800, 2000, 2200, 2400

Tectonic stage: Megastage

PANGEA FORMATION

PLATFORM FORMATION

PROTO-PLATFORM FORMATION

Pangea cycle

Pangea 800, Pangea 1450, Pangea 1900, Pangea 2600

Bio-event

Metazoans, Trilobites, Ediacara fauna, Macro-algae, Eucaryota, Cyanophyta

Sequence: Mega-Sequence

MG-8, MG-9, MG-10, MG-11, MG-12, MG-13, MG-14, MG-15, MG-16, MG-17

Sea-level change: Rise ⟷ Fall

GS-1 Gigasequence 1; Gigasequence 2; Gigasequence 3; GS-4 Gigasequence 4

Figure 3. Pangea cycles, tectonic stages, bioevents and sequence stratigraphy (after Wang and Shi, 1996).

The First Profusion of the Procaryota (2000-1800 Ma) and the Eucaryota (1450-1300 Ma)

The oldest known fossils are the bacteria and filamentous remains found in the Mesoarchean (3.2-3.6 Ga) rocks in the Barberton Greenstone belt in South Africa [25] and in the Pilbara Greenstone belt in western Australia [26]. They were also discovered in the Anshan Group (>3.1 Ga) in northeastern China. Blue algae associated with stromatolites were known in the Mesoarchaean and are common in the Neoarchean (3.3-2.8 Ga) and Epiarchean (2.8-2.5 Ga) sediments. The first profusion or explosion of the Cyanophyta (Procaryota) may probably be (Fig. 3) taken at 2000 Ma represented by the Gunflint fossils [27]. This occurrence of abundant procaryotes seems to be coincident with the first appearance of continents with shelves on the earth.

Formerly people thought that the eucaryotes appeared as late as in the Mesoproterozoic, but in the eighties they were reported from Palaeoproterozoic sediments 1900 to 2000 Ma of age in several regions of different continents, which may be regarded as primitive forms of Eucaryota. Eucaryotes are common near the base of the Changchengian "System" (1750 Ma) [28], which is included in the Mesoproterozoic in the Precambrian time scale used in China. Leiosphaerida and even the macroalgae *Chuaria* were also identified from this horizon [29]. In the upper portion of the Mesoproterozoic, they are also found in the Gaoyuzhuang Formation, 1500-1400 Ma in age, in North China [30]. In the sixties Barghoorn and Schopf (1969) [31] reported the discovery of eucaryotes in the Bitter Spring Formation in Central Australia, which is approximately 900 Ma in age. In the seventies, Croxford *et al* [32] and Peat *et al* [33] described similar fossils from the MacArthur Group and the Roper Group in Australia, both bearing an age of 1300 Ma. These are in accordance with the occurrence of eucaryotes with similar age from the Belt Supergroup in western North America described by Cloud in 1969 [24]. It seems that the occurrence of eucaryotes and the sparse appearance of macroalgae began as early as the beginning of Mesoproterozoic (1800 Ma), probably related to the first appearance of stable continental platforms with wide shelves. They became profuse on the carbonate platforms during the period 1450-1300 Ma, and the genus *Grypania* with exactly the same species occur both in the Gaoyuzhuang Formation in North China and in the Greyson Shales in the lower part of the Belt Supergroup in western North America [30]. This late Mesoproterozoic epoch is characterized by the first abundance of macroalgae, and may have borne some relation to the break-up of the Pangaea-1450 Ma, which was tentatively reconstructed by Li Xiang [35].

The Profusion of the Macroalgae (900-800 Ma)

Macroalgae fossils (Chuarids) were known from the Grand Canyon Group some 100 years ago and were described by Ch. Walcott. Abundant remains of the *Chuaria* group with exactly the same forms 900-800 Ma in age were discovered in North China [36] and in western North America. Profuse macroalgae with diverse forms occur in the Little Dal Group of western Canada [37] and in the Qingbaikouan "System" (1000-8000 Ma) of North China [36]. The *Tawuia-Longfengshania* assemblage 900-800 Ma in age were discovered in the seventies and eighties in China, and are now known in many parts of the world. This formed the epoch of radiation or explosion of the macroalgae (Fig. 3).

The Explosion Epoch of the Metazoan — the Ediacara Fauna (650-600 Ma)

The first appearance of Metazoa is a great concern for palaeontologists. The indubitable metazoan fossil *Rangia* was reported from South Africa at the beginning of this century. But it was not until the late forties that the famous Ediacara fauna of Australia was discovered by Sprigg [38]. This abundant naked fauna was described in the fifties by Glaessner [39]. In the following decades the Ediacara type of faunas was discovered nearly in all continents in the world, and all at similar horizons. They are known as the White Sea fauna in eastern Europe and the Wendian fauna in Siberia.

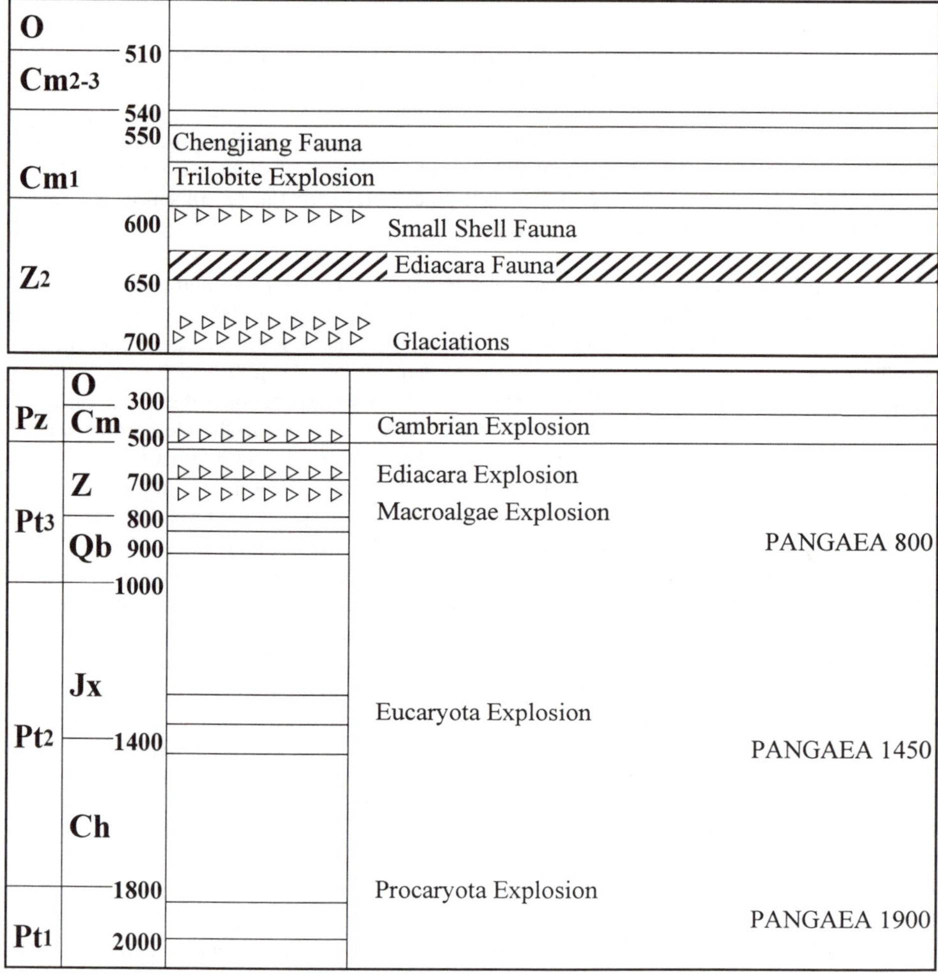

Figure 4. Proterozoic and Early Cambrian Pangaea cycles, glaciation and bioevents.

In China the study of Precambrian metazoan fossils was started in the seventies [40]. In the seventies Chen [41] reported the discovery of *Cloudina* in the Upper Sinian of the Yangtze Gorge area. The first Ediacara type fossil remains were described by Ding and Chen [42] from the same area (*Charnia*) in 1981. A rich metazoan fauna containing *Sabellidites* and *Paleolina* was described from the Jiuliqiao Formation in the Huainan region of Anhui Province [43]. The so-called Huainan fauna was once regarded as much older in age, but recent work reveals that it is probably also Late Sinian with an age of 700-650 Ma. Although the Jiuliqiao Formation contains the advanced type of fossil annelids *Huaiyuanella*, there is still the possibility that the Huainan fauna is older than the Ediacara fauna [44]. In recent years, two fossil horizons were discovered in the Neoproterozoic in Liaoning Province. The lower horizon occurs in the Changlingzi Formation and contains fossil annelids similar to the Jiuliqiao fauna, and the upper horizon belongs to the Xingmincun Formation containing *Sabellidites* and bearing an isotopic age of 650 Ma.

The discovery of the Ediacara fauna was one of the most important events in palaeontology in the middle of the century. The research and interpretation of it bears a special significance to our understanding of the procedures and mode of biological evolution. Glaessner [45] summarized and reviewed the studies on Ediacara fauna more than once. Taking the traditional view of common ancestry and continuous evolution of life, Glaessner made a classification of the Ediacara faunas on the basis of their similarity with present animal groups. On the other hand, Seilacher [46] regarded the Ediacara fauna as a failed attempt in the evolutional history of life. It became entirely extinct under hostile environments, and no direct relation exists between the Ediacara fauna and later metazoans. Buzz and Seilacher [47] even call the Ediacara fauna Vendibiota in contrast to Phanerozoic metazoa, which is called Eumetazoa. Evidently the Ediacara fauna represents an explosive evolution that was soon brought to an end. The controversy about the nature of the Ediacara fauna between Glaessner and Seilacher reflects a basic difference in the conception of evolution, a difference between the continuous and uniformitarian and the discontinuous and saltational view-points. This does not mean that Seilacher was entirely right. Biotic extinction cannot be complete and absolute. Besides the Ediacara fauna primitive lower invertebrates already appeared, and some of the Ediacara forms may have survived. But in any case the Ediacara fauna was a very special case of bioevent, quite different from those in the Phanerozoic history of biological development.

The Explosion of the Small Shell Fauna (ca 570 Ma)

In the beginning of the Early Cambrian there occurred an explosive development of the small shell fauna — fauna with hard shells. There has been a great controversy about the lower limit of the Cambrian. In China the basal Cambrian Meishucunian Stage bearing the small shell fauna was once regarded as 590-570 Ma in age. A much younger age of 544-543 Ma was obtained by Compston *et al* [48] by using the single zircon method of isotopic dating. Similar figures were obtained from the same horizon at several localities in China, Africa and Siberia. This brought difficulties to the stratigraphic interpretation of both the lower boundary of the Ordovician and the upper boundary of the Sinian. Recent studies by Yang and others [49] using Sm-Nd isotopic dating on biophosphates and sedimentary phosphates from the lower part of Meishucunian Stage yielded age values of 562 and 570 Ma. Therefore it is probably better to retain the figure 570 Ma for the Precambrian-Cambrian boundary at the base of the Meishuzunian Stage.

The small shell fauna is highly diverse, wide-spread, and bears an important significance in biological evolution. In China studies on small shell faunas began in the seventies, and over 100 papers were published up to present. According to Qian [50], the small shell fauna is well developed in the central-western parts of the Yangtze Platform, and can be well correlated with the stratigraphical sections studied in India, Pakistan and Iran. The time range of the bioevent of the small shell fauna may be designated to the early Meishucunian Age, which is mainly developed in Asia and western North America. It declined suddenly in the late Meishucunian Stage, and many forms of the fauna, according to some palaeontologists, cannot be regarded as direct ancestors of the later Phanerozoic faunas.

The Great Explosion of the Chengjiang Fauna (550-540 Ma)

The well-known Chengjiang fauna was found in the upper part of the Qiongzhusian Stage. Its discovery in 1984 was important, and it is certainly a marvel in palaeontologic resources both in regards to its content and its excellent preservation of soft parts. Trilobites and non-trilobite crustaceans were described by Zhang [51], Hou (1987) and Hou *et al* [52, 53] in 1987 and 1989. Exciting research results have been achieved through the cooperative effort of Chinese palaeontologists from the Nanjing Institute of Geology and Palaeontology and the Northwest University, and foreign scholars from Uppsala University of Sweden and the University of Cambridge, Great Britain. Most outstanding is the presence of big predators and of hemichordate and chordate animals, which were described by Chen *et al* [54], Hou *et al* [55] Shu *et al* [56]. It is noticeable that, if the suggested classification is correct, nearly all the Phanerozoic phyla of animals were represented in this treasure source.

The great biologic explosion in the Cambrian has long been a subject of discussion among palaeontologists. The renowned Burgess Shale with well-preserved soft parts has long been known. It is certainly surpassed by the older Chengjiang fauna both in taxonomic diversity and in preservation conditions. The recent report of discovery of a Middle Cambrian soft-bodied fauna in Kaili, eastern Guizhou will add to the treasure source of Cambrian fossils on the Yangtze Platform.

In the Early Cambrian, following the Meishucunian small shell explosion, came the Qiongzhusian trilobite radiation and the Chengjiang fauna explosion. In addition to these are the Middle Cambrian radiations mentioned above. If we call the Ediacara fauna an unprecedented biologic explosion and regard it as an unsuccessful attempt in evolution, then it is something different with the Cambrian bioevents. Probably the small shell fauna could be called an independent explosion because of its initiation of secretion of hardparts. The trilobite radiation seemed to be more usual rather than an explosion. The Chengjiang fauna was an explosive and innovation event, but it may be doubted whether it was succeeded by a sudden decline or extinction, especially as the diversity and abundance of the fauna is due to a large extent to the factor of preservation. In this context it may equally be questioned whether the Burgess Shale fauna should be regarded as a separate explosive event, as it bears close resemblance to the Chengjiang fauna both in its content and mode of preservation. It would seem that from Cambrian onwards, the evolution of animals has been on the whole characterized by successive and continuous innovations, and is different from the Precambrian Ediacara fauna.

REFERENCES

1. J. S. Huxley. *Evolution: the Modern Synthesis*. London, Allen Ubmin (1940).
2. G. G. Simpson. *Tempo and Mode in Evolution*. Columbia University Press, New York (1944).
3. R. Goldschmidt. *The Material Basis for Evolution.* Yale Univ. Press (1940).
4. O. H. Schindewolf. Neokatastrophism? *Deutsch. Geol. Gesell. Zeitsch.* **114**, 430-445 (1962).
5. N. Eldredge and S. J. Gould. Punctuated equilibria: an alternative to phyletic gradualism. In: *Models in Palaeobiology*. T. J. Schopf (Ed). pp. 82-115. San Francisco, Freeman, Cooper (1972).
6. S. J. Gould and N. Eldredge. Punctuated equilibria: the tempo and mode of evolution reconsidered. *Palaeobiology*, **3:1**, 115-151 (1977).
7. S. M. Stanley. Macroevolution and fossil record. *Evolution*, **36:3**, 460-472 (1982).
8. N. D. Newell. Revolution in the history of life. *The Geological Society of America, Special Paper,* **59:89**, 63-91 (1967).
9. L. W. Alvarez, W. Alvarez, F. Asard and N. V. Michel. Astronomical cause for the Cretaceous-Tertiary extinction, experimental results and theoretical interpretation. *Science*, **208**, 1095-1108 (1980).
10. W. Alvarez, E. G. Kauffman, F. Surlyk, L. W. Alvarez, P. Asars and H. V. Michel. Impact theory of mass extinction and the invertebrate fossil record. *Science*, **223**, 1135-1141 (1984).
11. D. M. Raup and J. J. Sepkoski. Mass extinction in the marine fossil record. *Science*, **211**, 1501-1503 (1982).
12. D. M. Raup and J. J. Sepkoski. Periodicities of extinctions in the geological past. *Proc. Nat. Acad. Natur. Sci., USA*, **81**, 801-805 (1984).
13. J. J. Sepkoski. Ten years in the library: new data conform palaeontological patterns. Palaeobiology, **19:1**, 43-51 (1993).
14. Wang Hongzhen and Shi Xiaoying. A scheme of the hierarchy for sequence stratigraphy. Jour. China Univ. Geosciences, **7:1**, 1-12 (1996).

15. O. H. Walliser. Global events and evolution. *Proc. 27th International Geological Congress*, **2**, 183-192 (1982).
16. O. H. Walliser. Toward a more critical approach to bioevents. In: *Global Bioevents, a Critical Approach*. O. H. Walliser (Ed.). pp. 5-11. Springer Verlag (1986).
17. J. J. Sepkoski. Mass extinctions in the Phanerozoic oceans: a review. *Geol. Soc. Amer. Spec. Pap.* **190**, 283-289 (1982).
18. B. A. Malmgren, W. A. Bergren, and G. P. Lohman. Evidence for punctuated gradualism in the late Neogene *Globorotalia timuda* lineage of planktonic foraminifera. *Palaeontology*, **9**, 377-389 (1983).
19. E. Mayr. Speciation and macroevolution . *Evolution*, **36:6**, 1119-1132 (1982).
20. W. Hennig. *Phylogenetic Systematics*. Univ. Illinois Press, Urbana (1966).
21. Chow Minchen, Zhang Meeman and Yu Xiaobo. *Translated Papers on Cladistics*. Science Press, Beijing (1983) (in Chinese).
22. Yin Hongfu. Punctuated Equilibria - some new international trends of palaeontology, part I, *Earth Science - Jour. Wuhan College of Geology*, 1983, **2**, 1-8 (1983) (in Chinese with English abstract).
23. Xu Daoyi, Zhang Qinwen and Sun Yiyin. Biomass extinction - a fundamental indicator for major natural division of geological history. Acta Geologica Sinica, **61:3**, 195-204 (1987) (in Chinese with English abstract).
24. Rong Jiayu. Meaning of microevolution and macroevolution with special reference to origin and macroevolution of early eospiriferids (Brachiopoda). In: *Selected Papers of Theoretical Palaeontology*. Rong Jiayu, Fang Zongjie and Wu Tongjia (Eds). pp. 68-90. Nanjing Univ. Press, Nanjing (1990) (in Chinese).
25. E. S. Barghoorn and J. W. Schopf. Microorganisms three billion years old from the Precambrian of South Africa. *Science*, **152**, 759-763 (1966).
26. S. M. Auramik, J. W. Schopf and M. R. Walter. Filamentous fossil bacteria from the Archaean of northern Australia. *Precambrian Research*, **10,** 157-374 (1983).
27. E. S. Barghoorn and S. S. Tyle. Microorganisms from the Gunflint chert. *Science*, **147**, 563-577 (1965)
28. Luo Qiling, Zhang Yuelin and Sun Shufen. The eucaryotes in the basal Changcheng System of Yanshan Ranges. *Acta Geologica Sinica*, **59:1**, 16-23 (1985) (in Chinese with English abstract).
29. Zhang Zhongying. Clastic facies microfossils from the Chuanlinggou Formation (1800 Ma), near Jixian, North China. *Jour. Micropalaeontology*, **5**, 9-16 (1986).
30. M. R. Walter, Du Rulin and R. J. Horodsky. Coiled carbonaceous megafossils from the Middle Proterozoic of Jixian, (Tianjin) and Montana. *Amer. Jour. Sci.*, **290**, **A**, 133-148 (1990).
31. E. S. Barghoorn and J. W. Schopf. Microorganisms from the late Precambrian of Central Australia. *Science*, **150**, 337-339 (1969).
32. C. J. W. Croxford, J. Jenneck, W. D. Muir and J. A. Plumb. Microorganism of Carpentarian (Precambrian) age from the Amelia Dolomite Group, North Territory, Australia. *Nature*, **245**, 28-30 (1973).
33. C. J. Peat, *et al.* Proterozoic microfossils from the Roper Group, North Territory, Australia. *BMR J. Austr. Geol. Geophy.* **3**, 1-17 (1978).
34. P. E. Cloud, G. R. Lichari, L. A. Wright, and B. W. Teoxel. Proterozoic eucaryotes from eastern California. *Proc. Acad. Sci.*, **62:3,** 623-630 (1969).
35. Li Xiang. Palaeocontinental reconstruction: methodology, computer, soft-ware and an example of application. Jour. China Univ. Geosciences, **4:1**, 7-13 (1993).
36. Du Rulin and Tian Lifu. *The Qingbaikouan Megaalgae in the Yanshan Region*. Science and Technology Press of Hebei Province (1986) (in Chinese).
37. H. J. Hoffman, and J. D. Aitken. Precambrian biota of the Little Dal Group, Mackenzie Mountains, Canada. *Can. J. Earth Sci.* **16:1**, 1154-166 (1979).
38. R. C. Sprigg. Early Cambrian jellyfishes from the Flinders Ranges, South Australia. *Trans. Roy. Soc. S. Austr.* **71**, 212-224 (1947).

39. M. F. Glaessner. The oldest fossil fauna of South Australia. *Geol. Rundschau*, **47:2,** 522-531 (1958).
40. Xing Yusheng and Liu Guizhi. Study on Sinian Microphyta and other fossils. In: *Sinian to Triassic Stratigraphy and Palaeontology in the Yangtze Gorge Region.* pp. 109-129. Geological Publishing House (Beijing) (1978). (in Chinese).
41. Chen Menge, Chen Xianggao and Cao Qiuyuan. Metazoan fossils in the upper part of the Sinian in Shaanxi. *Scientia Geologica*, **2**, 181-190. (1975). (in Chinese with English abstract).
42. Ding Qixiu and Chen Yiyuan. Discovery of soft-bodied metazoan from the Sinian Systems in the Yangzi Gorge, Hubei. *Earth Science — Journal of Wuhan College of Geology*, **2**, 53-57 (1981) (in Chinese with English abstract).
43. Wang Guixiang. Late Precambrian Annelida and Pogonophora from Huanan of Anhui Province. *Bull. Tianjin Inst. Geol. Min. Resources*, **6**, 56-22 (1982) (In Chinese with English abstract).
44. Sun Weiguo. Are there pre-Ediacaran metazoans? *Precambrian Research*, **31**, 409-410 (1986)
45. M. F. Glaessner. *The dawn of animal life.* Cambridge Univ. Press (1984).
46. W. E. Seilacher. Organic construction in the Proterozoic biosphere. *Lethaea*, **22**, 229-239 (1989).
47. L. W. Buzz and A. Seilacher. The Phyllum Vendibiota, sister group of Eumetazoa. *Palaeobiology*, **20:1**, 1-4 (1994).
48. W. Compston, I. S. Williams, J. L. Kirschink, Zhang Zichao and Ma Guogan. Zircon U-Pb ages for the Early Cambrian time-scale. *Jour. Geol. Soc. London*, **149**, 171-84 (1992).
49. Yang Jiedong, Sun Weiguo, Wang Zongzhe and Wang Yinxi. Sm-Nd isotopic age of Precambrian-Cambrian boundary in China. *Geological Magazine*, **133 :1**, 53-61 (1996).
50. Qian Yi. Early Cambrian small shell faunas of China with special reference to Precambrian-Cambrian boundary. *Stratigraphy and palaeontology of systematic boundaries in China: Precambrian-Cambrian boundary*, **2**. Nanjing Univ. Publ. House, Nanjing (1989).
51. Zhang Wentang. Early Cambrian Chengjiang fauna and its trilobites. *Acta Palaeontologia Sinica*, **26**, 223-235 (in Chinese with English summary) (1987).
52. Hou Xianguang. Three new large arthropods from Lower Cambrian, Chengjiang, eastern Yunnan. *Acta Palaeontologia Sinica*, **26**, 272-285 (1987). (in Chinese with English summary)
53. Hou Xianguang and Chen Junyuan. Early Cambrian arthropod-annelid intermediate animal *Luolishania* gen. nov. from Chengjiang, Yunnan. *Acta Palaeontologica Sinica*, **28**, 208-213 (1989) (in Chinese with English summary).
54. Chen Junyuan, Zhou Guiqing and L. Ramskoeld. The Cambrian lophopodian *Microdictyon sinicum. Bull. Nat. Mus. Natur. Sci.* **5**, 1-93 (1995).
55. Hou Xianguang, J. Bergstroem and P. Ahlberg. *Anomalocaris* and other large animals in the Lower Cambrian Chengjiang fauna of southwest China. *Geol. Foeren. Stockholm Foerh.* **117**, 163-183 (1995).
56. D-G. Shu, S. C. Morris and X-L. Zhang. A *Pikaia*-like chordate from the Lower Cambrian of China, *Nature* **384:14**, 157-158 (1996).

Proc. 30th Intern. Geol. Congr., Vol. 26, pp. 273-280
Wang *et al.* (Eds)

The Development of Metamorphic Geology in China

YOU ZHENDONG
Department of geology, China University of Geosciences, Wuhan 430074, P. R. China

Abstract

Since the beginning of this century metamorphic study in China has developed from the study of particular metamorphic rocks to currently playing an important role in the realm of metamorphic research over the world. There are four stages of development for metamorphic geology in China, namely; 1) before 1949, 2) 1949-1958, 3) 1958-1976 and 4) 1976-present. The development of metamorphic geology in China argues for the following points of view: 1) social demand is the most powerful motivation of scientific research; 2) the development of basic scientific research is in large measure the result of all round support from the government; 3) the exchange of ideas between scientists in the east and west is significant to the development of the geological science.

Keywords: metamorphic geology, metamorphic map, UHP metamorphism, fluids, geodynamics

INTRODUCTION

In China, the sprouting of the ideas of modern metamorphic petrology was coeval with the development of geological sciences. As a branch of geology, the development of metamorphic geology is closely connected with that of the geological sciences. The development of metamorphic geology in China can be subdivided into four major stages, namely: 1) the early stage of metamorphic study in China, before 1949; 2) study of metamorphic rocks (and rock series) in solving mineral resource problems in major perspective regions in China, 1949-1959; 3) study of metamorphic rock series in systematic regional geological surveys during the 1958-1976; 4) the growth of metamorphic geology since 1976. Some details of these four stages are given as follows.

THE EARLY STAGE OF METAMORPHIC STUDY IN CHINA (before 1949)

The introduction of modern geological science to China was as early as the establishment of the geological institute affiliated to the former Ministry of Agriculture and Commerce in the early years of this century. With the effort of V. K. Ting (Ting Wenjiang) and Doctor W. H. Wong (Weng Wenhao), over 20 geologists were graduated from this institute in 1916. In 1917 the Geological Department of the National Peking University was restored and the regular training of personnel for geological research began in China [1].

At that time geologists were few but highly energetic. Since the establishment of the Geological Society of China in 1922 the areas studied by Chinese geologists were rather wide, covering Jiangsu, Zhejiang, Fujian, Guangdong provinces and the western provinces including Sichuan,

Sikang (now western Sichuan), Qinghai and Xingjiang. Except for Xizang (Tibet), almost the entire territory of China had been covered.

In that period Weng Wenhao published two papers on the magnesian content of marbles in North China [2]. He presented a comparison of the Mg-content of Precambrian marble and that of the Cambrian limestone and concluded that the Precambrian marble is much more Mg-rich than Phanerozoic limestone. And this idea is in coincidence with those published in 1980s by western geologists.

Sun Jianchun was probably the first to study the metamorphic character of the Precambrian in China. He was the first in China to subdivide the Precambrian System into the Taishan, Wutai and Sinian (Huto) systems, each of which shows different features. During 1935-1936 Yang Jie published his study on the geology of Wutai, Shanxi Province. His paper had, to some extent, clarified the chaos in stratigraphical sequence in that area since the reconnaissance by B. Willis in 1903-1904 [3].

During the world war II (1937-1945), the Japanese invasion forced the Chinese Government and many of the educational institutions to move to SW China. Kunming and Chongqing became the centres of geological research at that time. The Southwest Associated University was then established in Kunming by the Peking, Tsinghua and Nankai universities, because of warfare geological researches were concentrated in the western territories of China. However, the geological circle of China had made splendid achievements in spite of the extremely difficult conditions.

During the 1940s a number of advanced studies on metamorphic petrology were published, such as the successive zones of regional metamorphism in Tanpa area by Cheng Yuqi in 1945 [4]. His another paper: *On the occurrence and metamorphism of late extrusive rocks of Taofu, Sikang,* together with *The prograde metamorphism near Tianwan, Sikang* by Peng Qirui (1946) and *On the rocks and metamorphism of Peyinchang volcanic series, Kaolan, Kansu* by Song Shuhe (1949) are also excellent papers in that period. These papers are characterized by careful geological observations combined with experienced microscopic studies which had helped them in discovering relic textures for the identification of protoliths.

It is noteworthy that petrofabric analysis was introduced to China by He Zuolin and Wang Jiayin at that time. In 1950 Chi Jishang published her petrofabric study on the Wissahickon schist in Pennsylvania, U.S.A. Later she studied the petrofabrics of experimentally deformed Yule marble. Her work has been cited from time to time in western literatures on metamorphic tectonites [5].

METAMORPHIC STUDY DURING THE EARLY YEARS OF THE PEOPLES' REPUBLIC(1949-1958)

After the founding of the People's Republic mineral resources were urgently needed, and geological work was facing a new upsurge. The metamorphic studies surrounding the exploration of iron, copper and their related strategic mineral resources were rapidly developed. By the end of the 1950s more than 20 research papers on metamorphic rocks were published. For instance, *Problems on the genesis of high grade ore in the Presinian (Precambrian) banded iron deposits of Anshan type in Liaoning and Shantong Provinces* written by Cheng (1957), *The metasedimentary type of phosphorites in China* by Liu Zhiyuan (1957), and *On iron ore deposits of Taye type* by a team of authors headed by Huang Yi and Pei Rongfu (1957). All of these are high quality papers coming from industrious exploration works. Among them Cheng's paper emphasized the role of regional metamorphism in the formation of high grade iron ore and considered that the accompanying hydrothermal process played only a subordinate role. This work was a summary of the exploration of banded iron ore in North China during the early years since the founding of the People's Republic.

During this period basic geological research was also flourishing. In the field of Precambriam geology, the Wutai geological team, headed by Wang Yuelun was the first to publish the paper *New findings about the Wutai Period in the Wutai mountains*, which gave a serious challenge to B. Willis' early work about the stratigraphic sequence of the Wutai system. Before long Zhao Zongfu (1954) published a paper *The precambrion stata problem.* In 1957 Ma Xingyuan published a monograph entitled *The major tectonic characteristics in the Wutaishan area* (in Chinese) which gives an overall study of the regional tectonics and was probably the first to investigate the crustal evolution in that area. Ma's results were adopted by the first national conference on the stratigraphy in China [3].

The study of metamorphism and tectonics was also active in that period. Wang Jiayin (1951) delineated the chloritoid zone in the Western Hills of Beijing. And also, his *A review on the classification of cataclastic rocks* was the first to classify dynamic metamorphic rocks based upon the nature of stress.

Granite and granitisation were a hot subject in the turn of the 40s. About the same time Mo Zhuson and Sun Nai reviewed this problem, just as pointed out by Sun Nai that it was critical to study the granites associated with metamorphic rocks in ancient massifs. It became one of the major trends in granite study. In 1958 Zhang Qiusheng published the *Granitization in the strata of Zhushan series Jiangsu Province*, emphasizing the role of alkali-metasomatism in the process of granitization.

In the 50s Soviet geologists had a great influence on the geosciences in China. In the category of metamorphic petrology, the Changchun College of Geology was the centre where Russian petrologists such as Bel'evchev gave lectures to geologists from all over China. His model on the petrogenesis of metamorphic ore deposits had an important influence in China.

METAMORPHIC ROCK STUDY IN THE SYSTEMATIC REGIONAL GEOLOGICAL SURVEY DURING 1958-1976

Due to the increasing demand of mineral resources for national construction, there was a boom of regional geological survey (mostly 1: 200 000) in the central eastern part of China since 1958. The major geological departments in universities and colleges, together with provincial geological teams, conducted extensive geological mapping.

The first regional geological survey team of Shandong Province organized and headed by Chi Jishang (C. S. Chih) did the most successful work in the western Shandong Province in 1958-1961. They were the first to clarify the lithostratigraphic units in the highly metamorphosed Taishan group: from this, they recognized the great differences in geology between the east and west parts of Shandong province across the Yi-Su megafault zone which is, in fact, part of the Tanlu in Shandong. The major tectonic framwork of the Taishan group had been revealed. At the end of 1959 *The methodology of regional geological survey in the scale of 1: 200 000* , a special issue of the Journal of Beijing College of Geology and a number of papers such as *The Presinian stratigraphy and tectonics of Zhongtiaoshan, Shanxi province* by Sun Dazhong were published and represented the major results of metamorphic study in the period.

In 1961 Cheng Yuqi (Y. C. Cheng) and his colleagues published *The major problems and methods of studying metamorphic rocks* (in Chinese). This book became an important reference book in studying metamorphic rocks. The nomenclature and classification of metamorphic rocks were especially popular with most field geologists [6].

In the same period geological surveying on a scale of 1: 200 000 (one to two hundred thousand) was widely conducted in various provinces. By the end of 1976, most of the country had been surveyed and mapped on this scale.

During this period a group of young geologists in the Changchun College of Geology rallied around Dong Shenbao, who had undertaken metamorphic study over wide regions both in North and Northeast China, and had formed a free, active, and strong academic atmosphere. Adopting new achievements from other basic sciences, a series of seminars was held and Dong gave lectures on the structures of materials. He proposed the concept of metamorphic formation and the idea of metallogeny induced by migmatization. A number of young petrologists and mineral deposit geologists have been trained and have grown mature [7]. *The petrology of metamorphic rocks* written by He Tongxing, Zhang Shuye and Lu Lianzhao was published in 1964, and became a conventional textbook for higher geological education.

THE DEVELOPMENT OF METAMORPHIC GEOLOGY SINCE 1976

The development of modern metamorphic geology is based upon: 1) the result of regional studies in metamorphic areas over mainland China during 1958-1976, which is summarized in Cheng and Zhang's paper [8]. 2) the new idea of multidiscipline studies combining the study of metamorphism with tectonism; 3) the improvement of microbeam techniques in study of metamorphic minerals (*e.g.* EPMA, HTEM, SEM techniques).

The milestone of metamorphic geology research in China was the compilation and publication of the 1:4 000 000 (one to four million) metamorphic map of China by Dong Shenbao and his group in the year 1986 [9].

As early as 1967 the subcommission for cartography of the world metamorphic belts worked out a scheme for compiling metamorphic maps. Since then many countries have compiled maps of this kind and a series of new concepts on regional metamorphism have been proposed.

Through the systematic regional geological survey of the major portion of China during 1958-1976 and some key projects in North China, Qinghai and Xizang and other important areas, abundant geological information hadaccumulated. In the early 80s, under the direction of Cheng Yuqi, Dong Shenbao, Shen Qihan, Lu Lianzhao and Sun Dazhong organized the compilation of a country-wide 1: 4 000 000 metamorphic map of China which was published in 1986. During the compilation 10 first order metamorphic units were singled out, the metamorphic series in these units were studied in great detail. The rock associations and their protoliths, metamorphic periods and sequences, metamorphic belts, metamorphic facies and facies series (facies groups), types of metamorphism, structural deformation and the granite magmatism related to the regional metamorphism were systematically summarised and theoretically discussed.

There were many new findings during the compilation of the map, such as, some new discoveries in metamorphic facies and facies series in China; four groups of regional metamorphism were subdivided and the spatial and temporal distribution of all types of metamorphism were summarized.

The compilation of the metamorphic map of China gave an impetus to the exchange of ideas between Chinese and foreign geologists. Nowadays many of the research papers of worldwide interest about the metamorphism and even tectonics in China have to refer to this map and its explanatory text [10]. The compilation of this map has stimulated the development of modern metamorphic geology in the following scopes of study.

The study of Precambrian high-grade metramorphic rocks

Contemporaneous with the compilation of the metamorphic map, a key project for exploration of iron deposits in eastern Hebei was unfolded. Joint research from different research groups stimulated the study of the Archaean highgrade rocks in China. As a result, a number of monographs and collections were published, and the study of high-grade metamorphic rocks and the evolution of the early Precambrian advanced to a new level. For instance, the monograph entitled *The early Precambrian granulites in China* published by Shen Qihan and his group has summarized the major characteristics of high grade metamorphic rocks in China, including the role of fluids in granulite facies of metamorphism.

Another book edited by Qian Xianglin of Peking University entitled *Geological evolution of the granulite terrane in the norhtern part of the North China craton* has thoroughly addressed some major problems in Precambrian tectonic evolution. Recently, Zhai Mingguo and his colleagues published a new monograph on granulites of North China. They emphasizes the link of granulites with lower continental crust of North China [11].

The early Precambrian khondalite series in North China (Lu, 1996) is a newly published monograph which studies the regional geology, the metamorphic rock association, the protolith and parent rock formation and their sedimentary environment, types of granitic rocks, patterns of metamorphic *PTt* path and the metamorphic and deformational history of the khondalite series rocks from various areas in North China. In addition, the genetic types, tectonic setting and their geodynamic processes of khondalites are also discussed. The book probes into some new categories in metamorphic geology. such as the influence of metamorphic fluids on the definition of metamorphic facies boundaries [12].

The study of metamorphism and deformation

The study of metamorphism and deformation related to orogenic belts has proceeded profoundly. In the year 1988, an international conference on *Metamophism and Crustal evolution of China* was held at Changchun University of Earth Sciences. The major idea of this conference has resulted in the publication of a special issue of *the Journal of Metamorphic Geology* on Chinese metamorphism edited by H. W. Day, J. G. Liou and Lu Liangzhao [10]. In this issue, major intracontinental mobile belts, which separate the Precambrian cratons and record the deformation and metamorphism associated with the collision and subsequent movements of the cratonic blocks as Asia was assembled, are studied. Many of these mobile belts are deformed and bounded by continental scale faults that presumably had extensive strike-slip movements. Consequently the special issue is designed to emphasize contributions from Chinese geologists by providing more detailed information on the geology, tectonics and petrology of regions of interest to metamorphic geologists.

PTt path and PT trajectories are broadly applied in metamorphic studies. Shi Yaolin is the first to study two-dimensional modeling and the *P-T-t* paths of regional metamorphism (in simple overthrust terrains). Recently he has been using this modelling in orogenic belts. Thermal modeling plays a key role in deciphering the quantitative relation between geodynamic processes and its consequent *PTt* paths [13].

In addition, a series of monographs written about the petrology and tectonics in Qinling and Dabie range has been published since the beginning of the 1990s. The relationship between metamophism and deformation related to orogenic belts has been discussed in some detail in these books, indicating a new level of research [14, 15].

The study of ultrahigh pressure (UHP) metamorphism.
Soon after the publication of the metamorphic map of China, Chinese geologists began to concentrate their mind on the high-pressure belt in Central China. The first coesite was announced to have been found by Xu Zhiqin in 1987 and the first diamond was discovered in Dabieshan by Xu Shutong in 1992. The Sulu-Dabieshan area is now the most fascinating place in the world for ultrahigh-pressure metamorphic studies. The third international eclogite field sysmposium and the second workshop of Task Group III-6 for ILP (International lithosphere program) was held in Dabieshan in 1995.

A series of new localities of UHP mineral and new UHP mineral assemblages and rock types has been found. New discoveries in the field and new ideas on the formation and exhumation of UHP metamorphic rocks can be summarized as follows: 1) All of the UHP metamorphic rocks are products of a polyphase collisional orogen between the Yangtze and the Sino-Korean blocks [16]. 2) There are different types of UHP metamorphic rocks according to their protolith: mafic, ultramafic, calcareous and quartzofeldspathic. The REE distribution patterns of the mafic and ultramafic UHPM rocks are diversified implying: E-type MORB, cumulate and island arc tholeiites indicating a pre-existing ocean between the Yangtze and North China continental blocks. 3) Isotopic dating indicates that UHP metamorphism took place during the initial collision between the Yangtze and the Sino-Korean continental blocks. Most isotopic data show Indosinian [16], however, there are also a number of data showing Late Paleozoic or even Caledonian (480 Ma) [17]. 4) The metamorphic *PTt* path of UHP rocks from Sulu to Dabieshan show 3 major stages of uplift and exhumation: in the first stage the UHP rock was uplifted from mantle to a mantle-lower crust level to be retrograted into an HP eclogite facies of metamorphism; in the second stage it was uplifted to amphibolite facies of middle crust level; in the last stage it was exhumed by erosion and eventually exposed together with the middle crust rocks at the surface. The exposure of the Dabie metamorphic complex to the surface did not take place until early Jurassic, therefore the exhumation rate has been considerably rapid since the Triassic collision [17].

The study of very-low-grade metamorphism
Very-low-grade metamorphism is widely needed in predicting the hydrocarbon resource of a region. An international symposium on very-low-grade metamorphism (IGCP project No 294) was held in Xi'an in July 1993. Some 30 manuscripts were received from 12 countries and regions. These papers deal with the burial metamorphism at low temperatures, metamorphism on the ocean floor and the orogenic belts of continental Asia [18].

Wu Hanquan published his new findings about the lawsonite in Qilian Mountains. In addition, a number of new techniques are being employed. For instance, Liu Bin of Tongji University is using immissible inclusions to calculate the themodynamic conditions of low to very low grade metamorphism. Chen Chaohsia of Cheng Kung University of Taiwan is using mica crystallinity to study very low grade metamorphism in the Central Range of Taiwan [18].

CONCLUSIONS

When we look at what is going on in Chinese metamorphic petrology today we can say: there is more precision, more experimental work and more petrogenetics especially in shedding light on the mechanism of rock formation. Due to interdisciplinary studies, metamorphism is now an important part of continental geodynamics. There are a number of conclusions we can draw from a review of the development and study of metamorphic petrology and metamorphic geology in China.

1) Social demand is the most powerful driving force in scientific research. The development of metamorphic geology was closely related to economic growth and related need of mineral resources in the past. This will continue to be true in the future, and is likely to be more pronounced than ever before.

2) The present state of metamorphic studies is in a large measure the result of open policy started in the late 1970s. The National Natural Science Foundation of China and the Ministry of Geology and Mineral Resources have organized and sponsored basic research in metamorphic geology, primarily as a result of a new awareness of the importance of these basic studies.

3) The exchange of ideas between scientists in the east and the west is a strong spur to the development of metamorphic geology in China [19]. We are looking forward to taking a more active part in international academic activities in the next century.

REFERENCES

1. Huang Jiqing. The geological study in China for the last 30 years, *Science* **28:6**, 249-263 (1945) (in Chinese).
2. W. H. Wong. L'Age du Marble de Fangchan et sa Teneur en Magnesie, *Bulletin of the Geological Society of China* **3:2**, 139-146 (1924).
3. J. Bai. *The early Precambrian geology of Wutaishan.* Tianjin Science and Technology Press, Tianjin (1996) (in Chinese with English abstract).
4. Y. C. Cheng. On successive zones of regional metamorphism in the vicinity of Tanpa, Sikang, *Science Record, Academia Sinica* **1**, 3-4 (1945).
5. F. J. Turner and L. E. Weiss. *Structural Analysis of Metamorphic Tectonites.* McGraw Hill Newyork: 244, 420, 456 (1963).
6. Chen Guoda *et al.*. *The Encyclopedia of Historic Events in Chinese Geosciences.* Shandong Science and Technology Press, Jinan (1992) (in Chinese).
7. Zhang Yixia. Under his guidance: a path to glory, in celebration of the 70 anniversary of the birth of Professor Dong Shenbao, *Journal of Changchun College of Geology*, Special Issue of metamorphic geology **1-2** (1987) (in Chinese).
8. Y. C. Cheng and S. G. Zhang. Notes on the metamorphic series and metamorphic belts of various metamorphic epochs of China and related problems, *Regional Geology in China* **2,** 1-14 (1982) (in Chinese with English abstract).
9. Lu Liangzhao. The advance in the study of regional metamorphism. In: *Review of the Study of Mineralogy, Petrology and Geochemistry in the 1980s.* The Mineralogical, Petrological and Geochemical Society of China (Ed). pp. 70-77. Seismological Press, Beijing (1991) (in Chinese).
10. H. W. Day, J. G. Liou, L. Z. Lu. Metamorphism and tectonics of China, *Jour.Metamorphic Geol.* **11,** 461-464 (1993).
11. M. G. Zhai *et al. Granulites and lower continental crust in North China Archaean Craton.* Seismological Press, Beijing (1996).
12. Lu Liangzhao, Xu Xuechun and Liu Fulan. *The early Precambrian Khondalite series in North China.* Changchun Publishing House, Changchun (1996) (in Chinese).
13. Y. Shi and C. Wang. Two-dimentional modeling of the *p-T-t* paths of regional metamorphism in simple overthrust terrains, *Geology* **15,** 1048-1051 (1987).

14. S. Suo, L. Sang, Y. Han and Z. You. *The petrology and tectonics in Dabie Precambrian metamorphic terranes, central China.* China University of Geosciences Press, Wuhan (1993) (in Chinese with extended English abstract).
15. Z. You, S. Suo, Y. Han, Z. Zhong and N. Chen. *The metamorphic processes and tectonic analyses in the core complex of an orogenic belt: an example from the eastern Qinling mountains.* China Univeristy of Geosciences Press, Wuhan (1991) (in Chinese with extended English abstract).
16. B. L. Cong, Q. C. Chen, M. G. Zhai, R. Y. Zhang, Z. Y. Zhao and K.Ye. Ultrahigh pressure metamorphic rocks in the Dabie-Su-Lu region, China: their formation and exhumation, *The Island Arc* **3**, 135-150 (1994).
17. Z. You, Y. Han, W. Yang, B. Wei and Z. Zhang. *The high-pressure and ultrahigh-pressure metamorphic belt in the East Qinling and Dabie mountains, China.* China University of Geosciences Press, Wuhan (1996).
18. H. Wu, T. Bai and Y. Liu (Eds). *Very low grade metamorphism:mechanisms and geological applications*[IGCP project 294 international symposium]. Seismological Press, Beijing (1994).
19. S. B. Dong. The development of the geological thought of metamorphism and its communication between the west and China. In: *Proceedings of XVth International Symposium of INHIGEO* 1990. H. Z. Wang, G. R. Yang and J. Y. Yang (Eds). pp. 197-213. China University of Geosciences Press, Wuhan (1991).

Proc. 30th Intern. Geol. Congr., Vol. 26, pp. 281-297
Wang *et al.* (Eds)

Theoretical Systematics and Methodology of Contemporary Geochemistry

YU CHONGWEN
Institute of Geochemistry, China University of Geosciences, Beijing, 100083, P. R. China

Abstract

The author presents a version of theoretical systematics and methodology of contemporary geochemistry on the basis of retrospect of the history of development of geochemistry and the advance of modern sciences. The definition and the four basic domains of geochemistry with their respective fundamental theories as well as the object of study and the aim of geochemistry are put forward. The scientific basis of geochemistry and fundamental disciplinary branches of geochemistry are outlined; among the latter theoretical geochemistry is especially separated. The enumerated seven important attributes endow geochemical systems with innate character of "Self-Organization of Complex Geochemical Dynamical Systems". Any geochemical system must be fitted in the reference system of geochemical processes and spatio-temporal structure. Geochemical systems once fitted in this reference system will exhibit self-organized criticality. Thereby a new proposition "Self-Organized Criticality of Complex Geochemical Dynamical Systems" is advanced which reflects the mechanisms of "Spatio-Temporal Fractal Dynamics of Self-Organized Criticality". The mechanisms and laws of generation and development of self-organized criticality should be the fundamental task of geochemical researches. The fundamental theories of the first basic domain of the system of geochemistry (structure of the Earth's material) are quantum geochemistry and cosmochemistry, and those of the other three domains (geochemical processes, geochemical courses, and geochemical fields) are the generalized geochemical dynamics and the theory of complexity and self-organization. Methodology of contemporary geochemistry involves cognitive processes of endo/exo dichotomy and subject/object dichotomy as well as a systematological approach conjugated with analytical approach in which holism is combined with reductionism and macrophysics is complemented with microphysics.

Keywords: Contemporary Geochemistry, Complex Geochemical Dynamical System, Spatio-Temporal Fractal Dynamics, Self-Organized Criticality, Quantum Geochemistry, Generalized Geochemical Dynamics, Theory of Complexity and Self-Organization

Since the birth of geochemistry in the beginning of the twentieth century, nearly one century has elapsed. During the twentieth century the world has experienced epoch-making tremendous changes. Mankind is facing with new challenges of great social reformation, rapid economic development and various global variations of nature. The development of society in turn led to great transformation of mode of thinking and intellectual structure. The present time happens to be the centennial of the birth of the discipline of geochemistry and is also confronted with the coming of the twenty first century. It seems to the author that it is advisable to present on this occasion a version of theoretical systematics and methodology of contemporary geochemistry on the basis of retrospect of the history of development of geochemistry and the advance of modern sciences.

SYSTEMATICS OF GEOCHEMISTRY

System of Geochemistry
Geochemistry may be defined as the science which studies the origin and abundance of chemical elements and their isotopes in the Earth and part of the celestial bodies, the transfer of mass, energy and momentum and the physical and chemical transformations occurring in them as well as their spatial distribution and temporal evolution. The system of geochemistry consists of four basic domains with their intrinsic qualities and fundamental theories as shown in Table 1.

<table>
<tr><th>Basic Domains</th><th>Intrinsic Quality</th><th>Fundamental Theories</th></tr>
<tr><td>Structure of the Earth's Material</td><td>Matter</td><td>Quantum Geochemistry
Cosmochemistry</td></tr>
<tr><td>Geochemical Processes</td><td>Motion</td><td rowspan="3">Generalized Geochemical Dynamics
Theory of Complexity and Self—Organization</td></tr>
<tr><td>Geochemical Courses</td><td>Time</td></tr>
<tr><td>Geochemical Fields</td><td>Space</td></tr>
</table>

Table 1. The basic domains of geochemistry and their intrinsic qualities and fundamental theories.

We take geochemical systems as the object of study and adopt a systematic approach to dispose of them. The aim of geochemistry is to explore the complexity of geochemical systems and to elucidate the laws of generation and development of geochemical self-organization.

Scientific Basis of Geochemistry and Fundamental Disciplinary Branches of Geochemistry
Among numerous disciplines of natural sciences, physics is the most fundamental science about nature, for it studies the most fundamental structure of the material world, and it has played and is still playing important role of promoting the development and advance of other natural sciences. Therefore geochemistry ought to be guided by physics, and geochemists should carry out multi-disciplinary crisscross investigation of physics with chemistry, earth sciences and biology *etc*.. The scientific basis of geochemistry combined with the earth sciences has derived various fundamental disciplinary branches of geochemistry in which quantum geochemistry, nonlinear nonequilibrium geochemical thermodynamics and generalized geochemical dynamics constitute the important field of theoretical geochemistry (Fig. 1).

Properties of Geochemical Systems
There are seven important attributes of a geochemical system which greatly influence or determine its dynamical behavior. They are: 1) Multicomponent coupling and interaction. Studies with the theories of dissipative structures and synergetics reveal that increase of number or quantity of chemical components will initiate coherence and cooperation among them and will cause the system to self-organize. Geochemical systems are "dynamic systems of coupled multicomponents". 2) Openness and dissipativeness. Open systems exchange matter and energy with their environments and thereby maintain the nonequilibrium steady state. Dissipation by the system of matter and energy input from its environments and transport of matter and energy from the system into its

environment lead to dynamic evolution of the system. 3) Nonequilibrium. In general, a geological region underwent a complicated history of sedimentation, magmatic activity, structural movement, metamophism and ore formation, consequently the terrane in a geological region as a whole deviates from the equilibrium state. Equilibrium manifests itself as stationariness and nonequilibrium is the source of variation. Nonequilibrium thermodynamics leads to dynamics, they are closely bound up. 4) Irreversibleness and the multiple coupling and interaction of processes. The majority of rock-forming and ore-forming processes are not single processes; on the contrary, they are frequently multiply-coupled processes of two or more processes. They are globally irreversible processes in the long run. 5) Nonlinear dynamic mechanisms. Geochemical systems are filled with various nonlinear factors, they belong to nonlinear systems. In nonlinear systems, a minute change of the values of control parameters in the vicinity of critical point of transformation would bring about a sudden jump. 6) Random fluctuations. As a rule, a high degree of freedom exists in a macroscopic system. It results in the emergence of fluctuations. When a system approaches the critical point, the initial minute fluctuations in it will be greatly amplified due to nonlinear dynamical mechanisms. This will cause the system to lose its stability and drive it to transform into a new state. This is the so-called "order through fluctuations". 7) Duality of continuity and descretion of matter. Matter in any geological region may be regarded as a continuous medium and constitutes a geochemical field which can be studied with macroscopic continuum physics and field theory. Nevertheless, matter consists of discontinuous molecules and atoms; in addition, the macroscopic characteristics of continuous media and fields are determined by the property and behavior of the microscopic particles in them. Consequently it is necessary to employ microscopic physics to study the discrete structure of matter.The theories of dissipative structures and synergetics demonstrate that when the nonequilibrium constraints and the nonlinear dynamical mechanisms are applied simultaneously on open, dissipative systems, they will initiate dynamical processes, leading them spontaneously to complexity and self-organization. This endows geochemical systems with the innate character of "Self-Organization of Complex Geochemical Dynamical Systems" (Fig. 2).

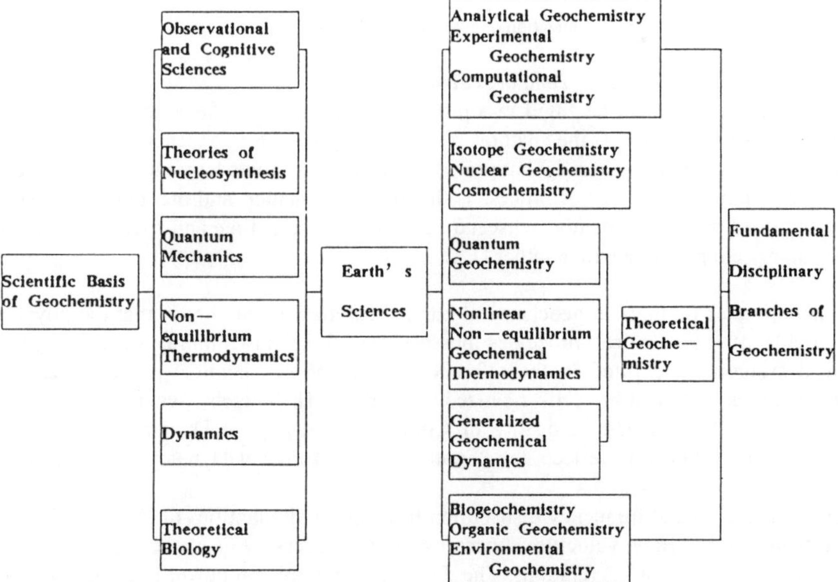

Figure 1. Scientific basis of geochemistry and fundamental disciplinary branches of geochemistry.

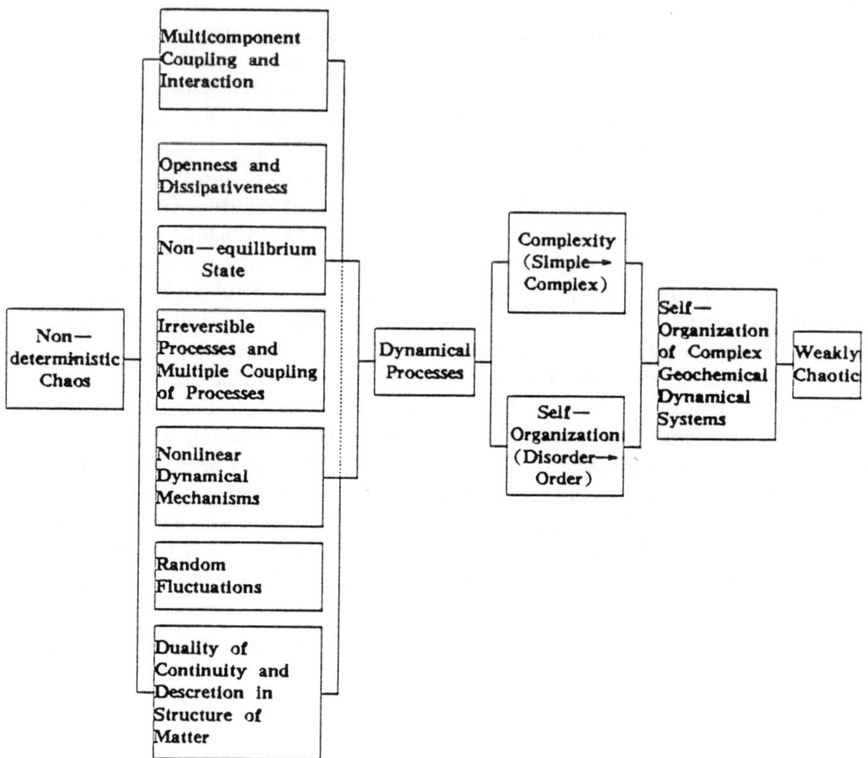

Figure 2. Self-organization of complex geochemical dynamical systems.

Geochemical Processes and Spatio-Temporal Structures

Any geochemical system must be fitted in a proper reference system. Geochemical processes and spatio-temporal structure form the natural reference system for geochemical systems. The implication of this reference system is that process is primary for the occurrence and change of any natural phenomenon. The essence of process is the motion of matter, and the motion of matter can neither be divorced from the time nor transcend the space. Process, time and space are three in one, being inseparably coupled one with another [1, 2, 3].

One of our important findings in geochemical basic theoretical researches is that the physical and chemical fields of geochemical processes possess fractal spatio-temporal structures, and that geochemical systems once fitted in the reference system of geochemical processes and spatio-temporal structures will exhibit self-organized criticality [4]. Thereby we put forward a new proposition — "Self-Organized Criticality of Complex Geochemical Dynamical Systems". The implication of this proposition reflects the "Spatio-Temporal Fractal Dynamics of Self-Organized Criticality":

1. The temporal and spatial frequency distribution of many field quantities of Earth's science fields (fields of temperature, flow velocity, concentration and stress *etc.*) possess power-law (more accurately "inverse power-law") relations. The discovery of empirical power law relations has very important significance in physics and earth sciences. It shows that the Earth's system has self-organized and is still self-organizing into critical states, and that it has the mechanism of generation

of intrinsic scale invariance [4-10].

2. The basic principle of self-organized criticality is that large, interactive, complex dynamic systems with high degree of freedom can spontaneously evolve to weakly chaotic critical state at the border of chaos by processes of self-organization without fine-tuning the external forces. This is the so-called "Self-Organized Criticality". It reveals the universal scale laws in nature, the concept of scale invariance and its intrinsic fractal dynamical mechanisms [11, 12].

3. The frequency distribution of inverse power law in space reflects the spatial fractal structure, and the inverse power-law distribution in time series reflects the temporal fractal structure. They are the manifestation of scale invariance or fractal attribute of the underlying process. The inverse power law spectra bring out the dynamic essence of random time series. Random time series with inverse power law spectra are "Fractal Dynamic Processes" [13, 14].

4. Once a system attains its critical state, the amplification and propagation of disturbances may occur at any scales of space and time (*i.e.* there are no characteristic or fundamental scales of length and time) [13]. This property of scale invariance or self-similarity is the fundamental characteristic of fractal. The power-law relation shows that the self-organized critical state possesses intrinsic scale invariance, and that the self-organized critical processes possess fractal dynamical mechanisms. Furthermore, the physical and chemical fields of these processes often manifest spatio-temporal fractal structures which are snapshots of the system at a specific instant in the slow, self-organized critical processes and at the corresponding spatial location. Physical space-time is a fractal continuum of topological dimension four. It may be simply expressed as "Fractal Space-Time" or visualized as "Zoom Space-Time" [2].

5. The shaping up of self-organized criticality is a self-organization process of nonequilibrium steady state going on at the critical state. It has the following spatio-temporal fractal dynamic mechanism. The shaping up of self-organized criticality possesses the dynamical mechanisms of "Cellular Automata" [12]. When the elemental disturbances (energy, momentum, mass) from random media interacted over a wide range of scales are sustainedly superposed upon an open system, they will give rise to coherent and cooperative dynamics of multicomponent coupled systems and the interactive dynamics of multiply coupled processes [15]. Microscopically, the highly multiple connection in three dimensions and nonlinear interaction among nearest neighbors of numerous subsystems (atoms, ions and molecules, *etc.*) induce multiplication process of the disturbances including iterated amplification and sequence of scaled pulses. Then the enhanced disturbances propagate by way of domino effects and chain reactions and spread all over the system. Afterwards the system will adjust itself to adapt the disturbances by sustained relaxation and decay, until the transport of mass, energy and momentum approaches robust nonequilibrium steady flow under the condition of global conservation. In fact this is the global attractor of dynamics [14]. We call the above mechanisms "Spatio-Temporal Fractal Dynamics of Self-Organized Criticality".

6. Scale invariance of critical states is the property of large systems with a high degree of freedom as a whole. It is independent of local properties and details of the system. The self-similar fractal structures imply that the system has the property of long-range correlation both in space and time. It is realized by way of short-range (nearest neighbors) highly multiple connection and nonlinear intensive interactions, and is not the result of direct transfer of disturbances. Therefore long-range correlation may cause disturbances at a long distance or in the remote past to trigger and influence the dynamical behavior of the system at a nearby site or the present moment [6, 16].

7. Although the spatio-temporal fractal structure of self-organized criticality arises from the empirical power law relations, but it is rooted in the "Theory of Scale Relativity", having its deep

theoretical basis. Universal scale laws exist in nature. There are various scale laws from micro-scale in quantum physics up to mega-scale in cosmology, besides physical laws themselves also depend on scale. It leads to the concept of scale invariance. But these scale laws are largely empirical summaries. There is not a fundamental theory about scales up to now which may be applied to give theoretical explanation to the scale laws. Recently Nottale [2] extended a fundamental theory about scales from the principle of relativity, and he called it "Theory of Scale Relativity". The principle of general relativity may be stated as: "The laws of physics must be such that they apply to coordinate systems whatever their state". The state of coordinate system includes origin, axes, unit and resolution. The theory of general relativity only applies the above principle to the position (origin, axes) of space-time coordinate system and state of motion, but does not involve unit and state of resolution. Consequently the theory of general relativity applies, strictly speaking, to laws of motion, it is a principle of the relativity of motion.

Based on the relativity of scales of length and time in nature, the theory of scale relativity defines the resolution of spatio-temporal measurements as the state of scale (*i.e.* state of resolution) of coordinate systems, thereby extends "Principle of Scale Relativity" from the theory of general relativity. It states: "The laws of physics must be such that they apply to coordinate systems whatever their state of scale". The theory of scale relativity expounded the existence of scale relativity in nature from the small scale in microphysics through the intermediate scale in complex system of self-organization up to the large scale of cosmology. An important result extending from the principle of scale relativity is that it endows the spatio-temporal structure with a profound revision. In the theory of general relativity, space-time is curved, while in the principle of scale relativity, space-time is fractal, or "Fractal Space-Time". That is, space-time has fractal structures.

There are two physical origins for space-time fractal structures : 1) Universal Scale Dependence in Nature, 2) Fundamental Non-differentiability of Nature. The two reach the same goal by different routes. The preceding paragraphs discussed the first aspect, now we make further analysis for the second aspect. R. P. Feynman pointed out early in 1948 and 1965 that a particle path in quantum mechanics may be described as a continuous and non-differentiable curve. The non-differentiability of trajectories of a quantum particle is an universal property of quantum mechanical objects. In physics, universal properties of physical objects may be attributed to the nature of space-time itself. It may be inferred from this that space-time is non-differentiable. But what is the essence of non-differentiability? We know that the unceasing display of new details of non-differentiable geometry as the scale becomes smaller and smaller (as represented by the Weierstrass function) implies that as resolution approaches zero as a limit, the curvature of a non-differentiable fractal space increases up to infinity. Or to state it more accurately, the curvature of a fractal surface is a fractal alternate of infinite positive and negative curvatures. In short, the non-differentiable fractal may be constructed as the limit of a family of differentiable curves. Thereby it may be concluded that: Non-differentiability implies an Explicit Dependence of Space-Time on Scale, or the Scale Invariance of Space-Time. Consequently "Fractal Space-Time is the Structural Manifestation of the Fundamental Non-differentiability of Space-Time" [2].

Geochemical systems are objects of study of geochemistry. They possess a high degree of complexity and self-organization. "Self-Organized Criticality of Complex Geochemical Dynamical Systems" centralizedly reflects the inexorable trend and fundamental law of spatio-temporal evolution of coherent and cooperative dynamics of multicomponent coupled systems and interactive dynamics of multiply-coupled processes proceeding in geochemical systems under the sustained superposition of elemental disturbances (energy, momentum, mass) from random media interacting over a wide range of scales. Therefore it is only right and proper that the mechanisms and laws of generation and development of self-organized criticality should be the fundamental task of geochemical researches (Fig. 3).

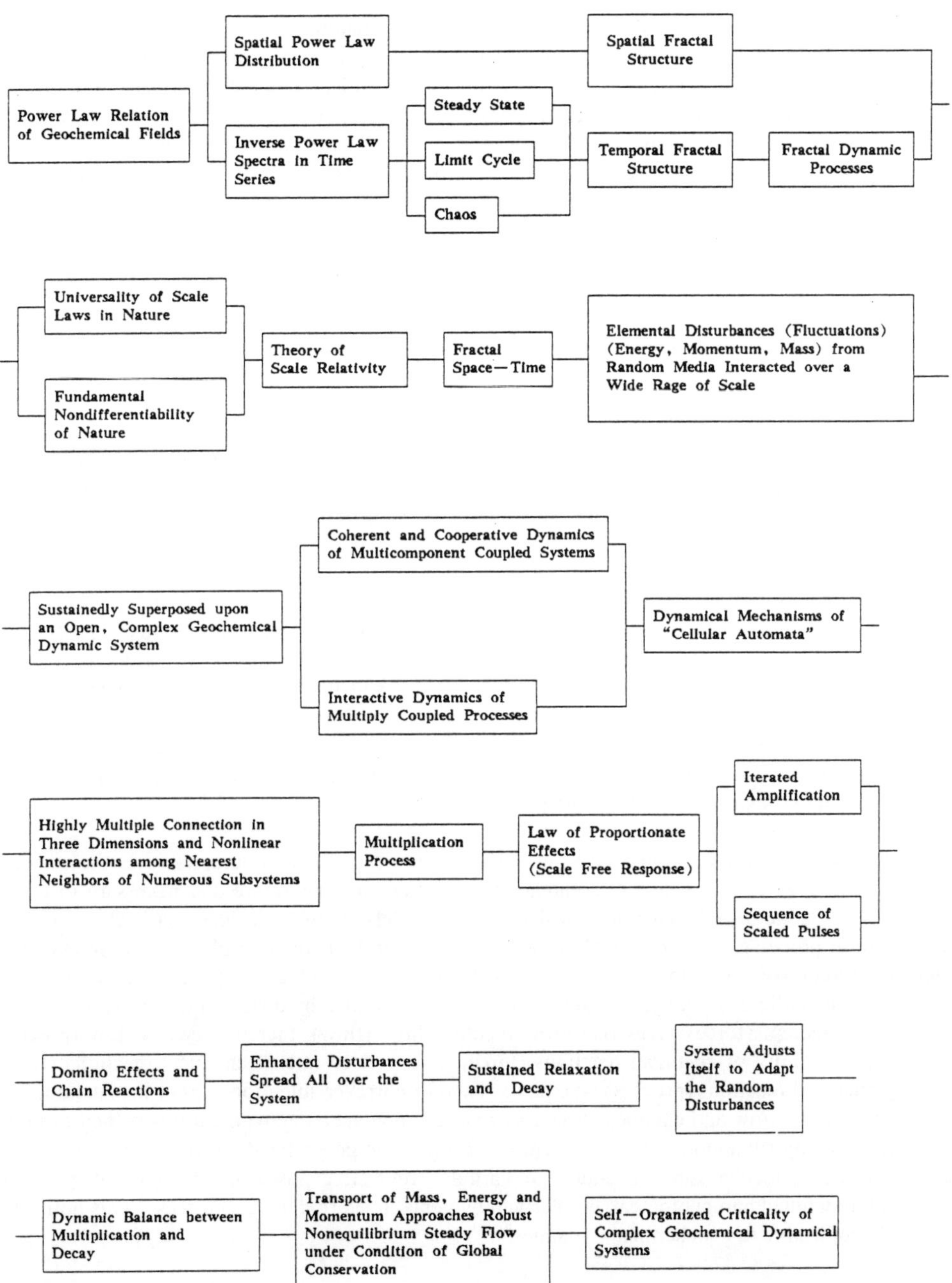

Figure 3. Spatio-temporal fractal dynamics of self-organized criticality of complex geochemical dynamical systems.

Quantum Geochemistry

Since the fifties of the twentieth century, the achievements of the analysis of crystal structure [17,

18] and the advance of the study of chemical bonds in molecular and atomic structure [19, 20] led to the rapid development of crystal chemistry [21-23] and quantitative geochemistry [24, 25]. After that, crystal field theory promoted the development of geochemistry of the transition chemical elements [26]. During the sixties and seventies, quantum chemistry and solid state quantum physics raised crystal chemistry and crystal field theory to a higher level. Eventually in the seventies quantum geochemistry was given birth to at the point of integration of mineralogy, quantum chemistry and quantum physics, constituting one of the important parts of theoretical geochemistry [27]. Simply speaking, quantum geochemistry uses quantum mechanical calculations and spectrographic analysis to study the structural physics (stereophysics) and structural chemistry (stereochemistry) of the Earth's material. The key problem of quantum geochemistry is the application of the theory of quantum mechanics and various spectrographic methods to studying chemical bonds (bond lengths, bond angles) or "electronic structure" of various geochemical species. The method, object and contents of research of quantum geochemistry are illustrated in Fig. 4.

Generalized Geochemical Dynamics

During the latest two decades a trend of shift from equilibrium thermodynamics [28-30] to nonequilibrium thermodynamics [31-34], and from thermodynamics to dynamics [35, 36] appeared in geochemical researches in the world. This manifests itself in the springing up of nonequilibrium geochemical thermodynamics after equilibrium thermodynamics and of the new disciplinary branch of geochemical dynamics [37, 38] after geochemical thermodynamics. The principal cause of the shift of focal point of research from thermodynamics to dynamics is that thermodynamics only shows the possibility, direction and limit (*i.e.* equilibrium) of the spontaneously proceeding processes, yet it does not discuss the reality and the velocity of processes. However, the purpose of dynamics consists exactly in the research on the reality and the velocity and mechanism of the process. Consequently dynamics can not only supplement and complement each other with thermodynamics in both reality and possibility for studying jointly the fundamental regularity of the motion of matter in a system, but can also make up the deficiency of thermodynamics and probe more deeply into the law of motion of matter.

So far kinetics and dynamics are divided into two disciplines in the literature; however they can not be sharply divided in geochemical researches and applications. Therefore it is necessary to give a definition of "generalized dynamics" which can be widely applied to "geochemical dynamics". Most natural processes are irreversible processes. In general irreversible phenomena, temperature gradient (force) gives rise to heat flow (flow), thereby heat conduction (process) takes place. A concentration gradient (force) gives rise to mass flow (flow), thereby diffusion (process) takes place. A velocity gradient (force) gives rise to momentum flow (flow), thereby viscous flow (process) takes place. Affinity of chemical reactions (force) gives rise to rate of chemical reactions (flow), thereby chemical reactions (process) take place. Therefore irreversible processes of heat conduction, diffusion, viscous flow and chemical reactions proceed spontaneously with various velocities under the corresponding (dynamical) forces. Thence it follows that generalized dynamics may be defined as "the velocity, mechanism and course of various irreversible processes (mechanical, physical, chemical, biological and geological, *etc.*) driven by generalized (dynamical) forces". This definition is widely applicable to geochemical dynamics.

If we regard geochemical processes as the motion of the Earth's material, then geochemical dynamics should include chemical motion, mechanical motion and magnetic motion. Chemical motion involves reactions which in turn include chemical reactions, thermonuclear reactions and interactions between living matter and environments [37, 39-45], *etc.*. Mechanical motion involves fluid mechanics, rheology and solid mechanics. Fluid mechanics in turn involves physical fluid dynamics [46-57] and physicochemical hydrodynamics [4, 15, 58-68], *etc.*. Rheology [69] involves flow which includes creeping flow of visco-elastic solids and coupled mechanics-reaction-transport [70, 71], *etc*.. Solid mechanics involves deformation which includes structure of lithosphere [69, 72], thermo-mechanical [73] evolution of continental lithosphere [72, 74, 75], structure of orogenic belts [76-78] and tectono-physical fluid dynamics [79-82]. Finally, magnetic motion involves magneto-hydrodynamics [83], *etc*.. There are various geochemical processes corresponding with all these domains of dynamics. The whole of the various domains of dynamics in geochemical dynamics constitutes a theoretical systematics of geochemical dynamics which we called "Generalized Geochemical Dynamics" (Fig. 5).

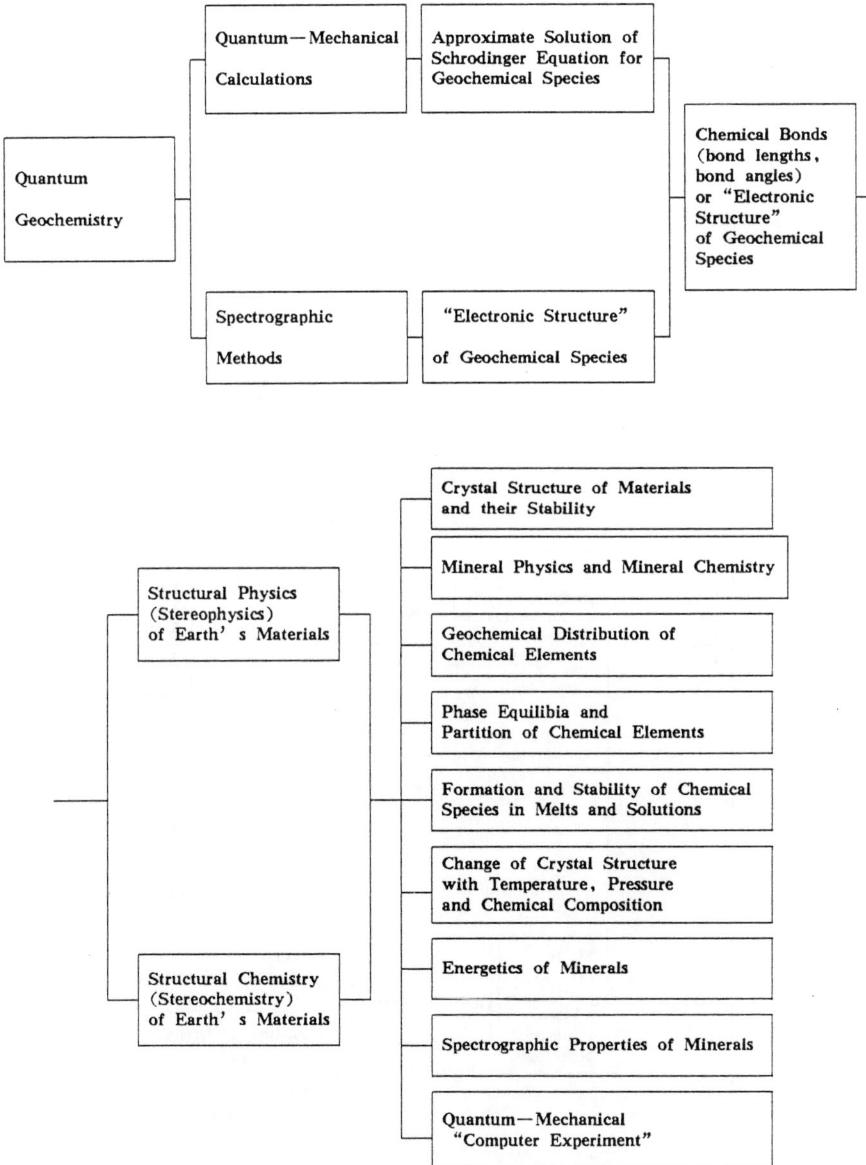

Figure 4. Method, object and contents of research of quantum geochemistry.

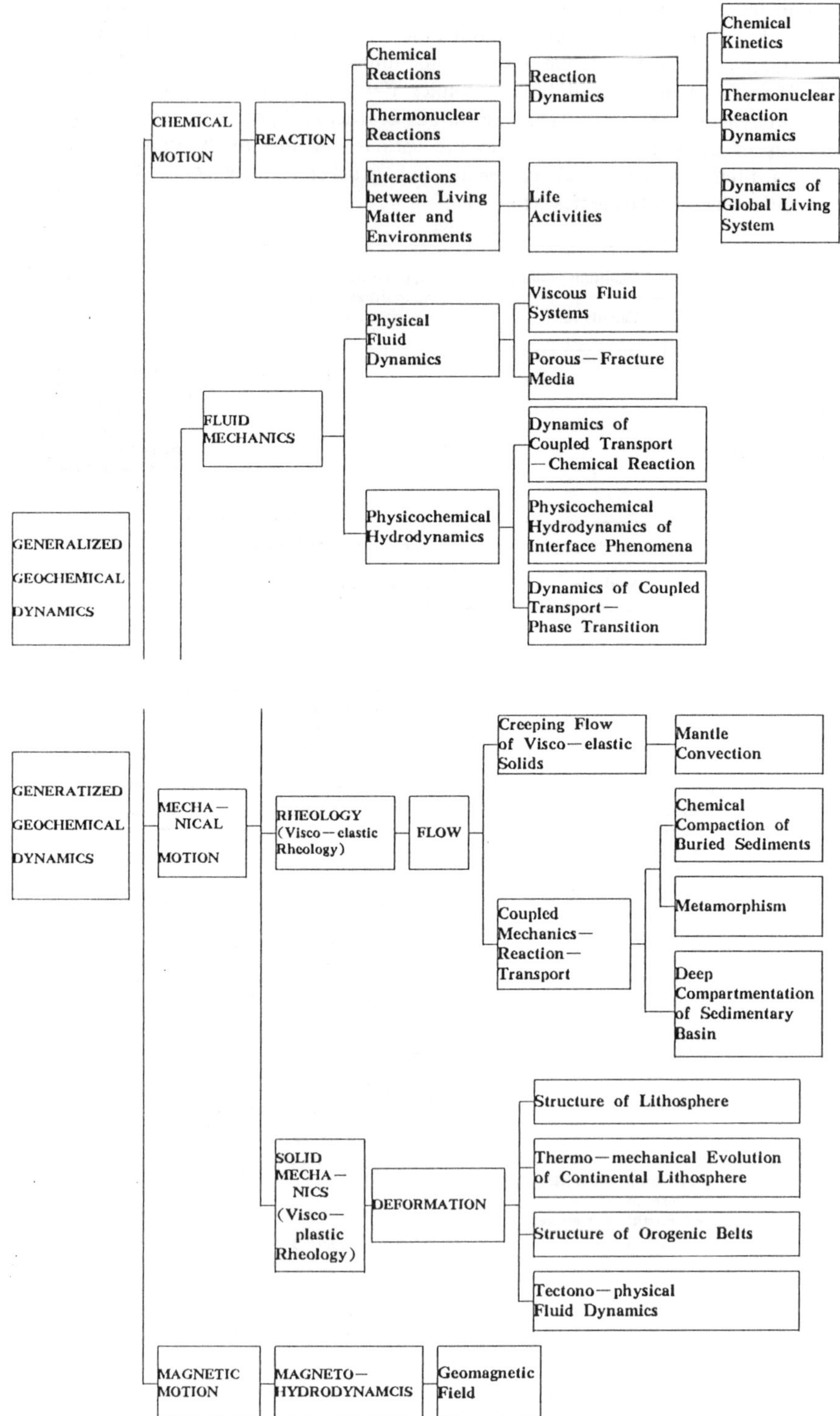
Chemical Kinetics
Chemical Reactions
Reaction Dynamics
Thermonuclear Reaction Dynamics
CHEMICAL MOTION
REACTION
Thermonuclear Reactions
Interactions between Living Matter and Environments
Life Activities
Dynamics of Global Living System
Viscous Fluid Systems
Physical Fluid Dynamics
Porous—Fracture Media
FLUID MECHANICS
Dynamics of Coupled Transport —Chemical Reaction
Physicochemical Hydrodynamics
Physicochemical Hydrodynamics of Interface Phenomena
GENERALIZED GEOCHEMICAL DYNAMICS
Dynamics of Coupled Transport— Phase Transition
Creeping Flow of Visco—elastic Solids
Mantle Convection
GENERATIZED GEOCHEMICAL DYNAMICS
MECHA— NICAL MOTION
RHEOLOGY (Visco—elastic Rheology)
FLOW
Chemical Compaction of Buried Sediments
Metamorphism
Coupled Mechanics— Reaction— Transport
Deep Compartmentation of Sedimentary Basin
Structure of Lithosphere
Thermo—mechanical Evolution of Continental Lithosphere
SOLID MECHA— NICS (Visco— plastic Rheology)
DEFORMATION
Structure of Orogenic Belts
Tectono—physical Fluid Dynamics
MAGNETIC MOTION
MAGNETO— HYDRODYNAMCIS
Geomagnetic Field

Figure 5. Theoretical systematics of generalized geochemical dynamics.

METHODOLOGY OF GEOCHEMISTRY

Methodology of contemporary geochemistry involves cognitive processes of endo/exo dichotomy and subject/object dichotomy as well as a systematological approach conjugated with analytical approach in which holism is combined with reductionism and macrophysics is complemented with microphysics.

Cognitive Processes of Endo/Exo Dichotomy and Subject/Object Dichotomy
Endo/exo dichotomy and subject/object dichotomy involve two fundamental aspects of cognitive activity, *i.e.* observation and cognition. The connotation of endo- and exo- concepts of observation and cognition consists of the collection of endo- and exo- facts, the establishment of endo- and exo-models and the provision of interrelationships between facts and models.

Endo/exo division and cognitive processes
From the practical viewpoint of geochemical research, it is reasonable to make the following endo/exo division (Fig. 6).

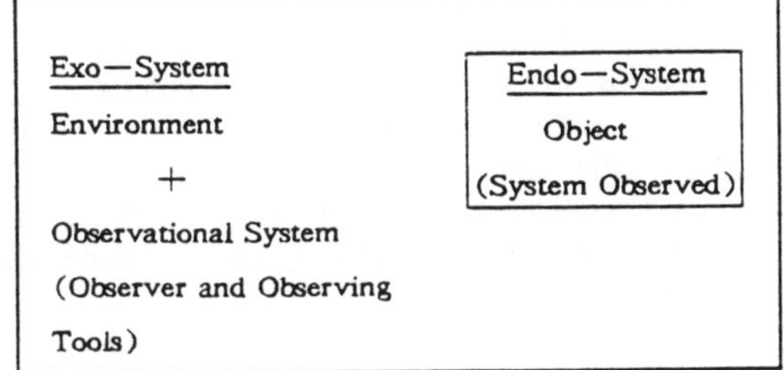

Figure 6. Endo/exo division.

The boundary between inside and outside constitutes an interface (endo/exo interface). The interface is not only a division in the sense of demarcation in material but also includes the endo/exo transition. It is of great importance that the interface not only possesses the nature of spatial structure but also the nature of dynamics in time evolution. It is a hidden window opening into dynamics. Objectification is the fundamental cognitive process of mankind. Descartes divided substance into "thinking substance" (Res Cogitans) and "extended substance" (Res Extense). The former represents abstract models and the latter represents concrete facts. The endo/exo transition corresponds to dynamical objectification, *i.e.*, objectification may be explained as an endo/exo transition. In the level of concrete facts, cognitive process transits from participation to observation, while in the level of abstract models, it transits from imagination to description. If we further combine the concepts of space and time with endo/exo dichotomy, then all observational activities return eventually to a spatial position and are displayed as structural properties of the system. Any observation concerning time eventually ends with recognition of spatial patterns. Hence we may conclude that the basis of external facts expressed by space and time consists in external and concrete operational space. Finally, dialectics of inside and outside belongs to the most essential level of human spirit. It manifests itself as the unification of separation and merging of inside and outside [84].

Holism Combined with Reductionism, Macrophysics Complemented with Microphysics Holism
Methods of study for theories of complexity and self-organization

Thermodynamic approach (Theory of Dissipative Structures)
The thermodynamic approach to the phenomena of self-organization was developed mainly by the

Brussels school of I. Prigogine *et al.* [85, 86]. They demonstrated that in a thermodynamically nonequilibrium system the energy dissipation itself determines the birth of new structures (dissipative structures). In the conditions of open and far-from-equilibrium thermodynamic systems, thermodynamic evolution can bring a system to the threshold of self-organization.

In equilibrium thermodynamics, both the thermodynamic force (the generalized force) and the thermodynamic flow (the generalized flow) vanish. Only equilibrium thermodynamic structures emerge in equilibrium systems. In linear nonequilibrium thermodynamics, the non-vanishing flows are linear functions of force. The Onsager principle (principle of the symmetry of kinetic coefficients) establishes the relations between the kinetic coefficients of cross (reciprocal) phenomena (reciprocity relations). The Prigogine's theorem of minimum entropy production states that when the external condition causes the system to deviate from equilibrium, then the nonequilibrium steady state corresponds to minimum entropy production. Thus nonequilibrium steady state emerges in this condition. The criterion of the stability of this state is that the excess entropy production be positive. Excess entropy production describes the property of a system to be stable in a state of minimum dissipation [31, 86].

In nonlinear nonequilibrium thermodynamics, flows are nonlinear function of force. Glansdorff-Prigogine's theorem shows that when the deviation from the nonequilibrium steady state gets beyond a certain critical value (instability threshold) (*i.e.* far-from-equilibrium), the excess entropy production changes sign (it becomes negative), the thermodynamic branch undergoes a bifurcation, the nonequilibrium steady state loses its stability and dissipative structures emerge in the transcritical region [85]. This is the new ordered state given birth by energy dissipation when an open system is far-from-equilibrium. But in the nonequilibrium nonlinear region up to now not a thermodynamic function has been found which could describe the properties of dissipative structures. Hence the instability threshold of the nonequilibrium steady structure marks the upper limit of the valid range of thermodynamics. After having crossed this limit, we enter the region of dynamics.

Dynamical systems approach

When an open and far-from-equilibrium dissipative system attains the instability threshold, fluctuations amplify and come up to macroscopic scale. The amplification of fluctuations causes the dynamic order to emerge. Order through fluctuation can be realized only under nonlinear conditions. Under strong nonlinear conditions, dissipative structures will eventually violate the criterion of stability — entropy production becomes negative. Just because of this situation thermodynamics can at most point out the possibility of loss of stability of a system at the critical point of transition, but it can neither reveal the condition of loss of stability, nor can it elucidate the dynamical behavior after the loss of stability. Consequently it is necessary to seek help from the theory of dynamical systems.

The theory of dynamical systems studies the temporal evolution of nonlinear dynamical systems [87-89]. It consists of linear stability theory and nonlinear bifurcation theory. The linear stability analysis of nonlinear dynamical equations may find out the concrete conditions (critical values of control parameters) of bifurcation solutions, but it can not determine the number of bifurcation solutions as well as their stability and behavior (mathematical expressions). Therefore it is necessary to study further the nonlinear dynamical equations themselves. The task for the bifurcation analysis of nonlinear dynamical equations consists firstly in finding out the existence and stability of bifurcation solutions in the vicinity of instability-critical point, and then determining their mathematical expressions (behaviors of the bifurcation solutions) by exact or approximate methods.

More than one stable branches of bifurcation solutions may simultaneously exist in the transcritical region for given values of control parameters. They may possess different spatio-temporal characteristics. What branch a system would select will be decided by the relative stability of the different branches on one hand, and also decided by the concrete forms of fluctuations (disturbances) on the other hand. When a system deviates farther from equilibrium, then the stability of bifurcation solutions will change further as the values of control parameters increase. The principle of the exchange of stability states that stability of bifurcation solutions exchange at the bifurcation point. The first bifurcation solution at the first bifurcation point (λ_1) will lose its stability at the second bifurcation point (λ_2) and transfers its stability to the second bifurcation solution and so on and so forth. Secondary bifurcation and higher order bifurcation come forth in this way. They introduce "history" and "memory" into physics and chemistry.

When higher order bifurcation occurs, complex spatio-temporal behaviors may emerge in the system due to interactions among many different unstable fluctuation components. As the branches of different orders in higher order bifurcation occur successively, more and more frequencies of oscillation will appear which correspond to the various unstable fluctuation components. The interaction of these fluctuation components may eventually lead to large fluctuations which will result in the emergence of a chaotic region in the bifurcation diagram. In chaotic regions, systems manifest deterministic randomness [12, 14, 83, 90-92].

Synergetic approach

Synergetics is an interdisciplinary field of research which studies the spontaneous, self-organized formation of structures in systems far from thermal equilibrium, as well as in non-physical systems. In short, synergetics studies the laws of self-organization. Self-organization is a process by which global external influences stimulate the start of internal mechanisms of the system which bring forth the origin of specific structures in it, whose symmetry is determined in accordance with the Curie Principle [93-96].

The synergetic approach combines the merits of thermodynamic and dynamic approaches and complements them in two important aspects: 1) it introduces randomness into the dynamical approach, and 2) it develops further the thermodynamic idea of order parameter, applying it to nonequilibrium phase transitions. So that the synergetic approach unifies nonlinearity and dissipation, randomness and coherence in open systems and expresses the symmetry change in the system through the order parameters.

The synergetic approach embodies the convergence and the eventual mergence of academic ideas of thermodynamics, nonequilibrium statistical mechanics and dynamics through the following five channels which constitute the mainstays of the synergetic approach. 1) Introduces stochastic force (fluctuations of environments and dissipative process in systems) into dynamic equations thereby transforming them into stochastic differential equations. 2) The fluctuations of the stochastic potential plays the role of the critical fluctuations. The stochastic potential determines the stationary probability distribution (ordered structure) and the critical fluctuation determines the types of structures. 3) Critical slowing down — adiabatic approximation — slaving principle. The common characteristic of all phase transitions (equilibrium and nonequilibrium) is : in the vicinity of the critical point, the response of systems to the external disturbances slows down, *i.e.*, some disturbances live longer, they relax in longer time ranges. This is the so-called "slow relaxation modes". Methods of adiabatic approximation are used to eliminate the quick relaxation modes and to express them by slow relaxation modes. The slaving principle is the principle which uses the slow relaxation modes to slave (govern) the quick relaxation modes. 4) Generalization of equilibrium and uniform (point-like) system to nonequilibrium and non-uniform (distributed) system (continuous media). Equilibrium phase transitions are generalized to nonequilibrium phase transitions and the Ginzburg-Landau equation is generalized to a stochastic evolution equation. 5) The slow modes are identified with the order parameter. Thus, a synergetic approach synergism of three disciplines and their corresponding methods is accomplished [93-95].

Reductionism

Methods of study for generalized geochemical dynamics and methods of study for quantum geochemistry are outlined in foregoing , corresponding paragraphs.

Acknowledgements

The study has been sponsored by National Natural Science Foundation No. 49633120 for the Period 1997-2000.

REFERENCES

1. Yu Chongwen, Luo Tingchuan, Bao Zhengyu, Hu Yunzhong, Liang Yuehan and Wei Xiujie. *Regional geochemistry of the Nanling District.* Geological Publishing House, Beijing (1987) (in Chinese).
2. L. Nottale. *Fractal space-time and microphysics.* World Scientific, Singapore (1993).
3. W. Schommers. *Space and time, matter and mind.* World Scientific, Singapore (1994).
4. Yu Chongwen, Bao Zhengyu and Cen Kuang. *Dynamics of the postmagmatic ore-forming processes* (in press).
5. P. Bak and C. Tang. Earthquakes as a self-organized critical phenomenon, *J. Geophys. Res.* **94,** 15635-15637 (1987).
6. D. L. Turcotte. *Fractals and chaos in geology and geophysics.* Cambridge University Press, Cambridge (1992).
7. C. G. Sammis, M. Saito and G. C. P. King (Eds). *Fractals and chaos in the Earth sciences.* Birkhauser Verlag, Basel, Boston, Berlin (1993).
8. C. C. Barton and P. R. La Pointe (Eds). *Fractals in petroleum geology and Earth processes.* Plenum Press, New York and London (1995).
9. C. C. Barton and P. R. La Pointe (Eds). *Fractals in the Earth sciences.* Plenum Press, New York and London (1995).
10. Shang-ken Ma. *Modern theory of critical phenomena.* Benjamin/Cummings Publishing Company, Inc., Reading, Massachusetts (1982).
11. P. Bak, C. Tang and K. Wiesenfeld. Self-organized criticality, *Phys. Rev.* **A38,** 364-374 (1988).
12. P. Coveney and R. Highfield. *Frontiers of complexity.* Fawcett Columbine, New York (1995).
13. P. Bak, C. Tang and K. Wiesenfeld. Scale invariant spatial and temporal fluctuations in complex systems. In: *Random fluctuations and pattern growth: Experiments and models.* H. E. Stanley and N. Ostrowsky (Eds). pp. 329-335. Kluwer Academic Publishers, Dordrecht (1988).
14. B. J. West and B. Deering. *The lure of modern science.* World Scientific, Singapore (1995).
15. Yu Chongwen. *Dynamics of ore-forming processes: systematics and methodology,* Earth Science Frontiers **1**, 54-82 (1994) (in Chinese).
16. Yu Lu and Hao Bailin. *Phase transitions and critical phenomena.* Science Publishers, Beijing (1984) (In Chinese).
17. L. Bragg and G. F. Claringbull. *Crystal structures of minerals.* G. Bell and Sons Ltd., London (1965).
18. A. F. Wells. *Structural inorganic chemistry, fifth edition.* Clarendon Press, Oxford (1984).
19. L. Pauling. *The nature of the chemical bond, third edition.* Cornell University Press (1960).
20. F. J. Berry and D. J. Vaughan (Eds). *Chemical bonding and spectroscopy in mineral chemistry.*

Chapman and Hall, London, New York (1985).
21. R. C. Evans. *An introdution to crystal chemistry, second edition.* Cambridge University Press (1976).
22. R. M. Hazen and L. W. Finger. *Comparative crystal chemistry. Temperature, pressure, composition and the variation of crystal structure.* John Wiley & Sons, New York (1982).
23. J. R. Smyth and D. L. Bish. *Crystal structures and cation sites of the rock-forming minerals.* Allen & Unwin, Boston (1988).
24. V. M. Goldschmidt. Grundlagen der quantitativen Geochemie, *Fortschr. Min. Krist.Petrog.* **17**, 112-152 (1933).
25. V. M. Goldschmidt. The principles of distribution of chemical elements in minerals and rocks, *Jour. Chem. Soc. London* 655-673 (1937).
26. R. G. Burns. *Mineralogical applications of crystal field theory, second edition.* Cambridge University Press, Cambridge (1993).
27. J. A. Tossell and D. J. Vaughan. *Theoretical geochemistry: Applications of quantum mechanics in the Earth and mineral sciences.* Oxford University Press, Oxford (1992).
28. C. J. Adkins. *Equilibrium thermodynamics, third edition.* Cambridge University Press, Cambridge (1983).
29. D. K. Nordstrom and J. L. Munoz. *Geochemical thermodynamics.* The Benjamin/Cummings Publishing Company, Inc., Reading, Massachusetts (1985).
30. G. M. Anderson and D. A. Crerar. *Thermodynamics in geochemistry.* Oxford University Press, Oxford (1993).
31. I. Prigogine. *Introduction to thermodynamics of irreversible processes, third edition.* Interscience, New York (1969).
32. H. J. Kreuzer. *Nonequlibrium thermodynamics and its statistical foundations.* Clarendon Press, Oxford (1981).
33. R. Stratonovich. *Nonlinear nonequilibrium thermodynamics. I. Linear and nonlinear fluctuation-dissipation theorems.* Springer-Verlag, Berlin (1993).
34. R. Stratonovich. *Nonlinear nonequilibrium thermodynamics. II. Advanced theory.* Springer-Verlag, Berlin (1994).
35. C. Nicolis and G. Nicolis (Eds). *Irreversible phenomena and dynamical systems analysis in geosciences.* D. Reidel Publishing Company, Dordrecht, Holland (1987).
36. J. Gorecki, A. S. Cukrowski, A. L. Kawczynski and B. Nowakowski (Eds). *Far-from-equilibrium dynamics of chemical systems.* World Scientific, Singapore (1994).
37. A. C. Lasaga and R. J. Kirkpatrick (Eds). *Kinetics of geochemical processes.* Mineralogical Society of America, Book Crafters, Inc., Chelsea, Michigan (1981).
38. V. S. Golubev. *Dynamics of geochemical processes.* The Earth's Interior, Moscow (1981) (in Russian).
39. E. T. Degens. *Perspectives on biogeochemistry.* Springer-Verlag, Berlin (1989).
40. V. Ittekkot, S. Kempe, W. Michaelis and A. Spitzy (Eds). *Facets of modern biogeochemistry.* Springer-Verlag, Berlin (1990).
41. M. H. Engel and S. A. Macko (Eds). *Organic geochemistry. Principles and applications.* Plenum Press, New York and London (1993).
42. J. A. C. Fortescue. *Environmental geochemistry.* Springer-Verlag, Berlin (1980).
43. C. R. Cowley. *An introduction to cosmochemistry.* Cambridge University Press, Cambridge (1995).
44. M. G. Edmunds and R. J. Terlevich (Eds). *Elements and the cosmos.* Cambridge University Press, Cambridge (1992).
45. P. D. Singh. *Astrochemistry of cosmic phenomena.* Kluwer Academic Publishers, Dordrecht, Boston, London (1992).
46. D. J. Tritton. *Physical fluid dynamics, second edition.* Clarendon Press, Oxford (1988).
47. H. E. Huppert. The intrusion of fluid mechanics into geology, *J. Fluid Mech.* **173,** 557-

594(1986).
48. A. S. Chehamir, A. G. Simakin and N. B. Epelbaum. *Dynamical phenomena in fluid-magmatic systems.* Science, Moscow (1991) (in Russian).
49. R. S. J. Sparks, H. E. Huppert and J. S. Turner. The fluid dynamics of evolving magma chambers, *Phil. Trans. R.. Soc.London* **A310,** 511-534 (1984).
50. V. N. Sharapov and A. N. Cherepanov. *Dynamics of magma differentiation.* Science, New Sibirsk (1986) (in Russian).
51. H. E. Huppert and D. R. Moore. Non- linear double-diffusive convection, *J. Fluid Mech.* **78,** 821-854 (1976).
52. H. E. Huppert and R. S. J. Sparks. Double-diffusive convection due to crystallization in magmas, *Ann. Rev. Earth Planet Sci.* **12,** 11-37 (1984).
53. J. S. Turner. *Buoyancy effects in fluids.* Cambridge University Press, Cambridge (1979).
54. J. S. Turner. Multicomponent convection, *Ann. Rev. Fluid Mech.* **17,** 11-44 (1985).
55. M. Y. Corapcioglu (Ed). *Advances in porous media, volume 2.* Elsevier,Amsterdam (1994).
56. M. Sahimi. *Flow and transport in porous media and fractured rock (from classical methods to modern approaches).* VCH, Weinheim (1995).
57. J. Happel and H. Brenner. *Low Reynolds Number hydrodynamics. With special applications to particulate media.* Prentice- Hall, Inc., Englewood Cliffs, N. J. (1965).
58. V. G. Levich. *Physicochemical hydrodynamics.* Prentice and Hall, New York (1962).
59. R. F. Probstein. *Physicochemical hydrodynamics. An introduction, second edition.* John Wiley & Sons, New York (1994).
60. H. E. Huppert. The fluid mechanics of solidification, *J. Fluid Mech.* **212,** 209-240 (1990).
61. S. H. Davis, H. E. Huppert, U. Muller and G. Worster (Eds). *Interactive dynamics of convection and solidification.* Kluwer Academic Publishers, Dordrecht, Boston (1992).
62. O. M. Phillips. *Flow and reactions in permeable rocks.* Cambridge University Press, Cambridge (1991).
63. P. C. Lichtner. Continuum model for simultaneous chemical reactions and mass transport in hydrothermal systems, *Geochim. Cosmochim. Acta* **49,** 779-800 (1985).
64. P. C. Lichtner. Time-space continuum description of fluid/rock interaction in permeable media, *Water Resources Research* **28:12,** 3135-3155 (1992).
65. M. P. Walsh, S. L. Bryant, R. S. Schechter and L. W. Lake. Precipitation and dissolution of solids attending flow through porous media. *AIChE J.* **30:2**, 317-328 (1984).
66. S. L. Bryant, R. S. Schechter and L. W. Lake. Mineral sequences in precipitation/dissolution waves, *AIChE J.* **33:8**, 1271-1287 (1987).
67. F. G. Helfferich. Multicomponent wave propagation: Attainment of coherence from arbitrary starting conditions, *Chem. Eng. Commun.* **44,** 275-285 (1986).
68. V. S. Golubev and V. N. Sharapov. *Dynamics of endogenic ore-forming processes.* The Earth's Interior, Moscow (1974) (in Russian).
69. G. Ranalli. *Rheology of the Earth. Deformation and flow processes in geophysics and geodynamics.* Allen & Unwin, Boston (1987).
70. T. Dewers and P. J. Ortoleva. Mechano-chemical coupling in stressed rocks, *Geochim. Cosmochim. Acta* **53**, 1243-1258 (1989).
71. T. Dewers and P. J. Ortoleva. Geochemical self-organization III : A mechano-chemical model of metamorphic differentiation, *Amer. J. Sci.* **290**, 473-521 (1990).
72. T. Engelder. *Stress regimes in the lithosphere.* Princeton University Press, Princeton, New Jersey (1993).
73. R. B. Knapp and D. Norton. Preliminary numerical analysis of processes related to magma crystallization and stress evolution in cooling pluton environments, *Amer. J. Sci.* **281**, 35-68 (1981).
74. D. M. Fountain, R. Arculus and R.W. Kay (Eds). *Continental lower crust.* Elsevier, Amsterdam (1992).

75. P. L. Hancock (Ed). *Continental deformation.* Pergamon Press, Oxford, New York (1994).
76. D. L. Turcotte and G. Schubert. *Geodynamics. Applications of continuum physics to geological problems.* John Wiley & Sons, New York (1982).
77. C. Allegre. Chemical geodynamics, *Tectonophysics* **81**, 109-132 (1982).
78. A. Zindler and S. Hart. Chemical geodynamics, *Ann. Rev. Earth Planet Sci.* **14**, 493-571 (1986).
79. W. J. Phillips. Hydraulic fracturing and mineralization, *Quarterly J. Geol. Soc. London* **128**, 337-359 (1972).
80. D. A. Spence, P. W. Sharp and D. L. Turcotte. Buoyancy-driven crack propagation: a mechanism for magma migration, *J. Fluid Mech.* **174**, 135-153 (1987).
81. J. R. Lister. Buoyancy-driven fluid fracture: the effects of material toughness and of low-viscosity precursors, *J. Fluid Mech.* **210**, 263-280 (1990).
82. H. Koide and S. Bhattacharji. Formation of fractures around magmatic intrusions and their role in ore localization, *Econ. Geol.* **70**, 781-799 (1975).
83. D. A. Yuen (Ed). *Chaotic processes in the geological sciences.* Springer-Verlag, Berlin (1992).
84. H. Atmanspacher and G.J. Dalenoort (Eds). *Inside versus outside.* Springer-Verlag, Berlin (1994).
85. P. Glansdorff and I. Prigogine. *Thermodynamic theory of structure, stability and fluctuations.* Wiley-Interscience, London (1971).
86. G. Nicolis and I. Prigogine. *Self-organization in nonequilibrium systems.* John Wiley & Sons, New York (1977).
87. S. N. Chow and J. K. Hale. *Methods of bifurcation theory.* Springer-Verlag, Berlin (1982).
88. R. Seydel. *Practical bifurcation and stability analysis (From equilibrium to chaos), second edition.* Springer-Verlag, Berlin (1994).
89. S. I. Andersson, A. E. Andersson and U. Ottoson (Eds). *Dynamical systems. Theory and applications.* World Scientific, Singapore (1993).
90. W. Guttinger and H. Eikemeier (Eds). *Structural stability in physics.* Springer-Verlag, Berlin (1979).
91. H. B. Stewart and L. M. Thompson. *Nonlinear dynamics and chaos.* Wiley, New York (1986).
92. A. Favre, H. Guitton, J. Guitton, A. Lichnerowicz and E. Wolff. *Chaos and determinism.* John Hopkins University Press, Baltimore and London (1995).
93. H. Haken. *Synergetics. An introduction, third revised and enlarged edition.* Springer-Verlag, Berlin (1983).
94. H. Haken. *Advanced synergetics. Instability hierarchies of self-organizing systems and devices.* Springer-Verlag, Berlin (1983).
95. M. Bushev. *Synergetics. Chaos, order, self-organization.* World Scientific, Singapore (1994).
96. P. J. Ortoleva. *Geochemical self-organization.* Oxford University Press, Oxford (1994).

Proc. 30th Intern. Geol. Congr., Vol. 26, pp. 299-306
Wang *et al.* (Eds)

A Brief Development History of Hydrogeology in China

ZHANG ZONGHU
Institute of Hydrogeology and Engineering Geology, Ministry of Geology and Mineral Resources, Zhengding, Hebei Province, 050803, P. R. China

Abstract

The use of groundwater in China dates back to at least 5700 years ago. Water well ruins occur in Zhejiang Province, while underground brines from depths of 100m were exploited in Sichuan Province more than 2 000 years ago. In 1931 Xie Jiarong described the geological conditions of groundwater in Nanjing City. Later the Section of Groundwater and Engineering Geology was established in the Geological Survey of China. With the assistance of U. S. S. R. hydrogeologists, hydrogeology specialties were set up in some colleges in 1952. In the mid-1950s the Institute of Hydrogeology and Engineering Geology was established and hydrogeology developed in China over the past forty years. From the fifties to the early sixties, hydrogeological work in China was mainly regional hydrogeological surveying and hydrogeological investigation mapping on the scale of 1: 500 000 and 1: 200 000. The development of urban, industrial and mine construction provided a broad demand for hydrogeological new research, particularly groundwater modelling techniques. From the sixties to the late seventies, the demand for water resources increased considerably. Hydrogeological work in China, under these conditions, was mainly focused on the exploration and research hydrogeology of field water supply and the sites of large-scale water sources. The lowering of groundwater level and land subsidence caused research for solutions to these environmental problems. In the 1970s environmental hydrogeochemistry and groundwater microorganism research developed rapidly. From the mid-1980s scientific management methods and a theory of water resources, including both groundwater and surface water, have developed and extended to the study of a united hydrosphere.

Keywords: hydrogeological mapping, groundwater modelling techniques, environmental hydrogeochemistry and groundwater microorganism research, united hydrosphere

INTRODUCTION

"Hydrogeology", a science dealing with the occurrence, movement, quality and quantity and evolution processes of groundwater as major research contents, is an important branch of geology due to the fact that groundwater is stored in rocks and is an important component part of the solid earth. With the progress of modern science and technology as well as social and economic development, the value and research content of modern hydrogeology are also developing unceasingly, thus forming many new branches with different research orientations.

The recognition and utilization of groundwater by mankind occurred much earlier than the science of hydrogeology. The earliest records about the utilization of groundwater can be traced back at least 5 700 years ago to the water well ruins in Hemudu, Yuyao of Zhejiang Province in China [1]. As early as the ancient Persian period the "kanats" with depths of 150 m were built in the vicinity of Teheran in Iran [2]. Also in the Sichuan Province of China underground brine was exploited at depths of some 100 m more than 2 240 years ago [3]. However, the scientific study of groundwater came much later.

As a special academic term, "hydrogeology" first appeared in "Hydrogeologie" written by J. B. Lamarck in 1802 [4]. Later, in 1848, Dupuit in France first studied the infiltration movement of groundwater and put forward the infiltration hypothesis [5]; soon afterwards, a French hydrological engineer, A. Darcy devised the infiltration laws of pore water through an experiment of infiltration in sand, now named "Darcy's law" [6]. In 1864, B. Lersch from Germany was the first to conduct a study of the hydrochemistry of groundwater[7]. In the United States J. W. Powell began groundwater hydrology research in 1879. Using the combination of studies of groundwater movement and geological principle as the foundation for studying the theory of groundwater hydrology and its application, he established a groundwater research section in the Geological Survey of the United States in 1907 [8]. In 1885 T. C. Chamberlin in the United States studied regional groundwater flow movement which also contributed to the understanding of the groundwater hydrology of America [8].

In 1888, И. В. Мушкетов in Russia was the first to study comprehensively groundwater action in geological processes [9]. It was A. C. Vertch, C. S. Slichtor and others from the United States who undertook early quantitative groundwater research, as in 1906 they studied the influence of external factors on groundwater table changes of artesian aquifers in New York State [8]. In 1908, А. Ф. Лебедев from Russia studied the coagulation genesis of groundwater and the migration mechanism of moisture in soil [10]. Combining with geological studies, O. E. Meinzer made groundwater investigation in America a complete scientific subject [8]. А. М. Овчеинико в from the former U. S. S. R. began the study of mineral waters in 1920, opening up a new research orientation in hydrogeology [11]. In 1933, Г. Н. Каменский (also from the former U. S. S. R.) perfected the nonstable flow movement theory of groundwater, which later became a branch of groundwater dynamics--hydrogeology [12]. In the development of groundwater dynamics, the most important contribution of the century was made by C. U. Theis who studied nonstable groundwater flow movement and developed a flow formula in 1935 [8].

After the Second World War, and particularly since 1946, there has been a more extensive content in hydrogeology and considerable multi-discipline development. In 1947, Г. Н. Каменск ий developed a new method for groundwater research — groundwater investigation and exploration, providing a working method for studying the distribution laws of regional groundwater [13]. Later, in the 1950s-60s, А. Н. Токарев *et al* (former U. S. S. R.) proposed a new research direction — isotope hydrogeology [14, 15].

From 1947 to the end of the 1980s, there was great progress in the study of groundwater models. In the United States electric simulation and mathematical simulation were developed to calculate the water resources. In 1960 the USGS set up a simulation system — simulation laboratory and in 1975 more than 100 simulation models were established. Since 1970, the simulation model has gradually been replaced by the digital computer model, which has been extensively applied in the assessment of groundwater quantity and quality, the study of the relationship between aquifers and rivers, and groundwater utilization. In 1972, S. P. Neuman and P. A. Witherspoon from the United States developed the test method for aquifers with a view to determining the hydrologic properties of aquifer systems [8]. In 1973, J. D. Bredehoeft and G. F. Pinder further developed the study of groundwater assessment by the solutant movement model. In the same year, P. C. Trescott developed groundwater flow models to simulate two-dimensional and three-dimensional groundwater flow [8]. In 1979, B. W. Hanshaw published a paper concerning the mixed zone of hydrogeology and geochemistry, pointing out that mineral solution, precipitation, dolomitization in the mixed zone are all related to the minerlization which occurs [8]. The study of prevention of groundwater from pollution has recently become a main geological focus. Thus hydrogeology has a history of over two hundred years in the world as a whole. However, as an independent subject in China, its development process is only a half century old, so hydrogeology in China is still a young subject. It developed as a sub-discipline of geology and with geology principles as its basis as part of the development of the national economy. Within a short time, it has developed as an advanced, independent subject of international significance, with abundant content and with many new branches and topics of research.

DEVELOPMENT HISTORY OF HYDROGEOLOGY IN CHINA

In 1931, the well-known geologist, Xie Jiarong, wrote the geological report entitled "Supply of Well Water in Nanjing", in which he described the geological conditions of groundwater in Nanjing City. This was the first monograph studying groundwater from the viewpoint of geology. Later the Section of Groundwater and Engineering Geology was established in the Geological Survey of China. The development of hydrogeology as an independent subject started from the early fifties. Although some work regarding groundwater investigation had been done before, yet it had not been systematic. Thus, in China, hydrogeology is a young subject with its development history much shorter than those of other geological subjects.

Under the influence of academic ideas and scientific achievements of the U. S. S. R. and with the assistance of U. S. S. R. hydrogeologists, hydrogeology sections were first set up in some colleges of geology in 1952, and groups of hydrogeological professionals were trained, thus laying the foundation for the development of hydrogeology in China. In 1950, when the economy was in its recovery period, industry and urban development required groundwater investigations for the water supply for cities, plants and mines. This was conducted under the leadership of the National Planning and Steering Commission for Geological Work of China. After the founding of the Ministry of Geology in 1952, the administrative authorities for hydrogeology and engineering geology were established, then investigation and exploration of regional groundwater was organized as the partial work of geological investigation. From 1955 to 1956, a new upsurge of attaching great importance to the science and strengthening academic study was begun. Under the Central Government leadership numerous scientists were organized to participate in the Twelve-year (1956-1965) Scientific and Technological Development Plan. Motivated by this plan, the Ministry of Geology set up the hydrogeological agency, the Institute of Hydrogeology and Engineering Geology, a nationwide institute as an important foundation for the development of hydrogeology in China over the past forty years.

From the fifties to the early sixties, hydrogeological work in China was mainly regional hydrogeological survey and hydrogeological investigation mapping on the scale of 1: 500 000 and 1: 200 000. At the same time, comprehensive scientific research and compilation of small-scale regional hydrogeological maps were also carried out.

Between 1956 and 1959, the research of regional hydrogeology was focused mainly on the Hexi Corridor of Gansu Province and the southern part of Xinjiang in northwest China and the karst areas in South China. Reports on the regional hydrogeology of arid areas were submitted and serial maps of karst development were compiled. These are the earliest comprehensive achievements concerning hydrogeology in the arid areas and karst areas, pioneering the development orientation of Chinese regional hydrogeology.

The great development of urban, industrial and mine construction provided a broad demand for hydrogeological work, many new hydrogeological issues needing to be further studied, requiring considerable new content and new research. The solution of calculation of pit water inrush and assessment of groundwater reserve facilitated the development of groundwater dynamics, thus establishing groundwater modelling techniques. In order to study the scientific basis for groundwater assessment, groundwater nonstable flow theory work was conducted in 1961.

This was also the development stage during which hydrogeology formed a more systematic, more complete scientific system, and it could be considered the first stage of hydrogeology in China.

From the sixties to the late seventies, the demand for water resources increased considerably due to the great attention being paid to agriculture, economic development and urban development and construction of large-size enterprises. Hydrogeological work in China, in the light of those conditions, was mainly focused on the exploration and research hydrogeology of field water supply and the sites of large-scale water sources for cities and enterprises. Agricultural irrigation in the eastern plain regions caused the decline of groundwater level and land subsidence. Many new aspects of hydrogeology were developed to find the mechanism and countermeasure solutions for environmental problems resulting from the exploitation of groundwater, and to resolve the contradiction between water supply and demand in large cities. In the 1970s, the study of methods and theories of groundwater resources assessment was conducted on a large scale, because of irrigation water demand of the main agricultural area in the North China Plain, and the definitive unit method was applied in the calculation of groundwater resources. At the same time, the Model DJS-6 computer was adopted to do the data processing, beginning a new development stage for hydrogeology. In 1976, the assessment of water resources in arid and semi-arid areas of northern China including seventeen provinces, autonomous regions and municipalities, was organized by the state and the monograph "Water Resources Assessment of Arid and Semi-arid Areas in China" was completed. In the meanwhile, a comprehensive hydrogeological study of various regions such as basins, intermontane great plains and loess plateaus was also carried out. The work deepened the knowledge of regional hydrogeology and summarized some new hydrogeological laws. The most important are the theory of "dual medium aquifer phreatic well flow" and the method of groundwater resources assessment in the different regions.

The chemical composition of groundwater is an important component of hydrogeology. In the mid 1950s, the study of chemical substances in groundwater was still undeveloped. About the beginning of the 1960s a hydrogeochemistry section was set up in the Institute of Hydrogeology and Engineering Geology, using hydrogeochemistry to study the aquifers. A regional hydrochemical investigation was conducted across the country and the first groundwater hydrogeochemical map was compiled. From the middle of the 1960s rapid progress was made in the study of groundwater quality, and the formation of mineral water and deep brine composition of artesian basins. Since the 1970s, a great amount of waste-water and solid-waste residual discharged by plants and domestic sewage have polluted groundwater in many areas, causing serious water environment problems. The intense exploitation of groundwater in the coastal areas induced seawater intrusion and deterioration of groundwater quality of these areas. With hydrogeochemical theory as a basis, environmental hydrogeochemistry developed rapidly and gradually formed a new branch of hydrogeology.

Since the 1970s, hydrogeochemistry has been increasingly used to seek and utilize new sources of energy, the development and utilization of medium-, and low-temperature thermal groundwater and thermal energy in North China, the southeastern coastal area and Tibet. In addition to the geochemical environment of groundwater the formation, movement and enrichment laws of chemical components in shallow groundwater, and also the temperature field, fluid field, chemical field and thermal source and water flow movement of groundwater with various depths have been studied. At the same time, isotope hydrogeology was further developed, thanks to the study of environmental isotopes and groundwater age determinations. Since then, hydrogeochemistry in China has gradually become a new branch of hydrogeology with an abundant research content and special method and theoretical basis. It can be considered as a new borderline subject with hydrogeology as a foundation, combined with geochemistry and fluid dynamics.

From the 1970s to the 1980s, another important achievement in hydrogeological development was groundwater microorganism research [18], which is the product of the combination of hydrogeology with geology and microbiology. Microorganism research is at present an advanced research subject in geoscience and has drawn extensive attention both in China and abroad. Microorganisms are widely distributed in nature due to biological activities, and are one of most active and strongest geological agents on the earth's surface, exerting a strong influence on the atmosphere, hydrosphere and lithosphere. In China although the study of microorganisms started very late, yet great progress has been made. At the beginning of the 1960s, a geological microorganism laboratory was first set up in the Institute of Hydrogeology and Engineering Geology. In the earliest stage a study on the theory and method of seeking mineral deposits through the polymetallic-deposit microorganisms [19] was carried out and the microorganism system investigated. It was noted that microorganism geochemistry plays a part in the migration and enrichment of the elements. From the 1970s to the 1980s, microorganism action in the quality evolution of groundwater in artificial reinjection wells in groundwater exploitation was studied from the viewpoint of microorganism ecology, resulting in proposing the theory of change of groundwater chemical components and the active role of microorganisms in the course of natural silting-up of reinjection wells, thus opening up the new domain of microorganism ecology [20]. In the mid 1970s, the metabolic activities of microorganisms, associated with changes of groundwater quality and the formation and migration processes of Fe, Mn trace elements were further determined and thus the idea that "groundwater quality is an ecological system taken part in by microorganisms and with the interaction of water-rock-microorganisms" was proposed [21]. In the 1980s, the research work of this kind was extended to the ore-forming action of microorganisms. At the same time, the microorganisms involved in the formation of polymetallic nodules on the ocean floor and in underground deep brines were studied and the pattern and theory of microorganism ore-formation was developed. In the 1990s, new breakthroughs in the study of modern microorganisms of deep-sea Mn nodules and palaeo (fossil) microorganism mineralization have been made.

In short, from the early 1970s to the late 1980s, there was great progress in hydrogeology research in China, and the main focal points were:

1) From study of the distribution, quantity and quality of groundwater moving to the study of its movement laws and dynamic changes;

2) From the utilization of water resources in local areas to the scientific assessment, rational exploitation and utilization of water resources on a larger scale;

3) From the study of changes of chemical compositions of inorganic substances in groundwater developing into the study of water-rock reactions with the participation of microorganisms in groundwater;

4) From a single geological theory and method moving to the combination with other subjects, such as mathematics, dynamics, thermodynamics, biology, chemistry and agriculture etc. , with the result that some new marginal subjects developed.

During the period from the late eighties to now, hydrogeology development in China entered another new stage. In this period, China, like other countries, faces the serious problem of water resources. Many big or medium-size cities and industrial districts in China are very short of water, some water sources are nearly used up, so the contradiction of water supply and demand is very obvious. The continuous development of both industry and agriculture, especially the greater development of township enterprises, rapid growth of population, sharply increased water demand. Also the pollution of groundwater by industrial wastewater and sewage results in destruction of the aquifers and decrease of groundwater quality so that the water environments deteriorated seriously. The crisis of water resources influenced the development of national economic construction and becomes a most pressing problem, it is also the new subject that hydrogeology is facing. As an engineering system, utilization and protection of water resources are inseparable. The importance of scientific water resource management should be realized. Therefore, the research content of hydrogeology in this period has mainly focused on how to establish scientific management methods and a theory of water resources (including both groundwater and surface water). The database and optimum management model of hydrogeology and water resources were established by adopting modern techniques and applying computing in order to strengthen the modernization of management .

In this period, there have also been new development in theoretical hydrogeology. First, the effect of system theory on the development of geology: concerning research on groundwater in aquifers, this developed from research on the dispersed and individual hydrogeological elements to regarding these elements as an integral whole with a certain system structure, namely, either a relatively-independent system, or a multi-level and interrelated temporal-spatial system; thus new scientific concepts such as the "groundwater system" and "aquifer system" developed. Secondly, it is realized that groundwater is a component part of the entire geological environment, whose effect and influence must be taken into account in the development and utilization of groundwater resources, thus there is a limit to resource utilization. The environmental effect should be a determination of the condition of water resources availability. Thirdly, it is realized that groundwater is a component part of the earth's hydrosphere, which interacts with the lithosphere, biosphere and atmosphere. In particular the process and mechanism of groundwater formation from atmospheric water seepage is very complicated and groundwater passing through the aeration zone (strata - soil layer above the phreatic water level) is a process in which energy exchange and substance transport continuously take place between the atmosphere and atmospheric water with groundwater. However, in the past, attention has not been paid to this natural process. In hydrogeology research concentrated on gravity water of the aeration zone because it can be directly utilized, but the important scientific significance of the aeration zone between the atmosphere and hydrosphere was neglected. Through the late eighties and the nineties, hydrogeologists noted the significance of this research realm, especially in studying the interaction of four spheres on earth research on the aeration zone is of key significance. Therefore, in recent years, great progress has been made in an important theoretical realm and research method of hydrogeology in China: the regularity of moisture migration in the aeration zone and its energy exchange and substance transport mechanism with the atmosphere [22, 23].

Since 1990 the development of hydrogeology in China faces a new stage. The research objects of hydrogeology are no longer restricted to groundwater or the water-bearing medium itself, but are extended to a united hydrosphere (surface water and groundwater together) to study the interaction and effect of this hydrosphere with other spheres [24]. The effect of human factors on the evolution of the terrestrial hydrosphere is more and more remarkable because of human activities, namely, the effect of human factors on the land hydrosphere has gradually become a important new research subject of hydrogeology. In the future research in hydrogeology will not only cover natural conditions as its main content, but will also cover the interaction between the terrestrial hydrosphere and human social and economic activities; for example, the change of the terrestrial water environment caused by sea-level fluctuation and its prediction in the coastal delta area; the trend of hydrosphere evolution in the ecological fragile arid zones; water environment evolution trends of water system watersheds of large rivers in the course of comprehensive development, systematical management and comprehensive utilization of water resources in large drainage areas, geological formation processes, environmental effect and utilization of water resources in karst areas and the establishment of terrestrial water environment monitoring systems.

In short, the development of hydrogeology in China is closely combined with the demand for national economic development. It will continue to go forward in the light of economic development and social demands in the future.

Acknowledgements

The author expresses heartfelt thanks to Ren Fuhong, Shi Dehong, Yin Zhengzhou, Yan Baorui, Zhang Xigen and Qu Huanlin for their kind help in compiling the references.

REFERENCES

1. Shen Shurong. Recognition and Utilization of Groundwater in the Ancient Times in China. In: *Reading Notes of Hydrogeological History.* pp. 11-27. Geological Publishing House, Beijing (1985) (in Chinese).
2. Liu Chifei. Source of Kanats and Its Utilization in Xinjiang. In: *Review of Geological History.* pp. 112-125. Geological Publishing House, Beijing (1989) (in Chinese).
3. Wang Dachun *et al. Basics of Hydrogeology.* Geological Publishing House, Beijing (1986) (in Chinese).
4. J. B. Lamarck. *Hydrologie*. Paris (1802).
5. A. Dupuit. *Etudes theoriques et practiques sur le movement des laux courants.* Paris, Dunod (1848).
6. A. Darcy. *Les fontaines publiques de la Ville de Dijon.* Paris, V. Dalmont (1856).
7. B. Lersch. *Hydro-Chemie*. Berlin (1864).
8. J. E. Moore and B. B. Hanshaw. Hydrological Concepts in the United States: A Historical Perspective, *Episodes* **10:4**, 315-320 (1987).
9. И В. Мушкетов."Физическая Геология" Спб (1888).
10. А. Ф. Лебедев. "Роле порообразной воды в режиме почвенных и грунтовых вод" Т р. по, С-Х. Метеорологи. Вып. **XII**, 41-165 (1913).
11. А. П. Герасимов и А. М. Овчеников. "Минеральные воды" М: Госгеолиз-даг (1947).
12. Г. Н. Каменский. "Основы Динамики Подземных вод" М. ; л. ; ОНТИ Ч. 1 (1933), Ч. 2 (1935).
13. Г. Н. Каменский "Поиски Разведка подземных вод", М. Госгеолиздат(1947).
14. А. Н. Токарев и А. В. Щербаков. "Радиогидрогеология" М: Гостопмех-издат (1956).
15. Г. Креиг, В. И. Ферронскцй, Д. цдр. Уайт. "Осноы гидрогеологии. Геологическая деятельность и история воды" Новосбрск: Наука (1982).
16. Bureau of Hydrogeology and Engineering Geology of the Ministry of Geology. *Outline of Regional Hydrogeology in China*. Geological Publishing House, Beijing (1958) (in Chinese).
17. Qu Huanlin (Ed). *Assessment of Groundwater Resources in the Arid and Semi-arid Areas of China*. Science Press, Beijing (1991) (in Chinese).
18. Yan Baorui. The Action of Microorganisms in the research of Hydrogeochemistry. In: *Theory and Method of Hydrogeochemistry.* pp. 28-29. Geological Publishing House, Beijing (1985) (in Chinese).

19. Yan Baorui. Metallic Sulphate Deposit Seeking by the Microorganism Method, *Geological Science and Technology* **2**, 102-106 (1976) (in Chinese).
20. Yan Baorui. Microorganism Ecology and Quality of Artificial Reinjection Water in Shanghai City, *Geological Review* **1**, 70-78 (1985) (in Chinese with English abstract).
21. Yan Baorui, Zhang Sheng *et al. Microorganism Activities and Its Mineralization in the Water-Rock System of the Central Pacific Ocean.* Geological Publishing House, Beijing (1994) (in Chinese with English abstract).
22. Shi Dehong. Study of Water Content in Aeration Zone, *Geology in China* **9**, 29-31 (1983) (in Chinese).
23. Shi Dehong, Jia Yongrui *et al.* Study of the Storage and Movement of Water Content in Aeration Zone of Loess Yuan in Luochuan, Shaanxi Province, *Hydrogeology and Engineering Geology* **3**, 6-13, (1983) (in Chinese).
24. Zhang Zonghu, Shi Dehong. Environmental Evolution of Groundwater Hydrosphere: An Important Component Part of Global Climatic Changes. *Proceedings of Nationwide Seminar on Climatic Changes and Environmental Issues* **27**, 1-4 (1991) (in Chinese).